SPLINES AND COMPARTMENT MODELS

An Introduction

SPLINES AND COMPARTMENT MODELS

An Introduction

Karl-Ernst Biebler
Michael Wodny

Ernst Moritz Arndt University of Greifswald, Germany

NEW JERSEY • LONDON • SINGAPORE • BEIJING • SHANGHAI • HONG KONG • TAIPEI • CHENNAI

Published by

World Scientific Publishing Co. Pte. Ltd.
5 Toh Tuck Link, Singapore 596224
USA office: 27 Warren Street, Suite 401-402, Hackensack, NJ 07601
UK office: 57 Shelton Street, Covent Garden, London WC2H 9HE

British Library Cataloguing-in-Publication Data
A catalogue record for this book is available from the British Library.

SPLINES AND COMPARTMENT MODELS
An Introduction

ISBN 978-981-4522-22-9

In-house Editor: Angeline Fong

Printed in Singapore

Preface

The application of mathematics in life sciences first requires the formulation of adequate models of biological processes that allow the quantitative evaluation of life processes by means of observations and experiments. With regard to this, the knowledge with reference to the observed biological processes, the preconditions and characteristics of the applied mathematical models as well as the conditions surrounding data collection, need to be taken into account. In this entire context it is effective to develop specific quantitative methods for the evaluation of data and to characterize attributes mathematically, thereby justifying conclusions and interpreting results comprehensively. This synopsis of problems is a characteristic of biometric work, which due to its formulation is interdisciplinary. Typical questions continue to be brought to light, for example:

- Can the conditions required by the mathematical method be seen as fulfilled in the observed examples?
- Does a solution exist for the mathematical problem associated with the general problem definition?
- Is there exactly one such solution?
- Can it be proven that this solution possess desirable characteristics?
- Are the evaluation model and possibilities for data acquisition consistent?
- Do numerical problems arise when applying processes?

Under such general points of views, two themes are discussed in this book: splines and compartment models. Their application can be seen in different areas of the sciences and technology.
Why does one deal with such different mathematical concepts in this book?

It is common for them that they both can be used for the modeling of courses. Spline functions result from a general approach. Watched courses can be reproduced very well under reference to certain optimization criteria. Splines are determined by a relatively high number of parameters. However, these are often not well interpretable according to a specific question.
Compartment models represent a description of specific processes by means of differential equations. Their parameters characterize the process. The number of model parameters is relatively small here. The data is generally not as precisely reproducible as by means of splines. The demand to judge the quality of the goodness of fit consequently arises. To do so, residence time distributions are taken to a connection with compartment models of pharmacokinetics. This way the model parameters can be calculated statistically from the measured drug concentration data. We propose the varied Minimum-χ^2 estimation. Splines are used for it. Goodness of fit can now be checked with a statistical test. This procedure is new in the field of pharmacokinetics.
Examples are given in the context of life sciences with the appropriate typical terminology. At the same time, mathematical terms are needed that cannot be explained in detail here. The reader would find it useful to be familiar with basic knowledge of analysis, algebra, statistics and probability calculus as well as the theory of ordinary differential equations. Detailed knowledge of these areas however is not assumed.
In this book we took care to include the history of the presented ideas and include references with regard to this. The historical comparison is not to be seen as just a reference to the scientists of the past. It helps to enforce the relativization of the own work, motivates students and further allow the reader to research the sources themselves. To begin, narrowing in on simple models seems to be advisable for the solid application of mathematical models in the life sciences. This simplifies the detailed clarification of their conditions of application, reduces the demand on the extent of observation and in many cases satisfies the purpose.
The literature listed in the references confines itself to the titles quoted in the text. Numerous additional publications which were evaluated but did not explicitly contribute to results are not listed. This concerns standard pharmacology or pharmacokinetics textbooks, publications concerning computer programs, applications of pharmacokinetical methods (inclusively about methods of calculation) for the characterization of certain pharmaka, literature about stochastics and their applications, contributions about numeric methods, etc.

Different spline functions are dealt with in Part I. The starting point is the task of describing an unknown functional dependency $y = f(x)$. In the life sciences this is often the change of quantity over time. The typical situation with regard to the data is the observation of a finite number of y-values that are understood as the function values of the respective independent x-values. A number of m value pairs (x_i, y_i), $i = 1, 2, \ldots, m$, are therefore given. Sometimes subject specific examples allow for the definition of a parametric function class from which $f(x)$ stems. With this it is then possible, for example with the method of least squares, to calculate the function parameters from the data. It is often the case though that no special knowledge exists about $f(x)$. This is where spline functions can be used. The natural spline functions possess a special smoothing characteristic. It can basically be assumed that the course of $f(x)$ is mostly described by the value pairs (x_i, y_i), $i = 1, 2, \ldots, m$. Unobserved oscillations are not hidden between individual points. This along with the numerically stable calculability of splines lead to their wide use. To begin, general natural splines are introduced. Following this, cubic and quadratic spline functions are described in more detail. In each case, the interpolation task is first formulated, and then their solution is developed and discussed. This makes sense and is necessary because the construction of smoothing splines is done in two steps. In the first step the still unknown values $s(x_i)$ of the spline function are calculated at x_i. In the second step, the uniquely determined interpolating spline is constructed for these. A certain optimization problem corresponds with every smoothing spline. This problem determines the main characteristics of the solution function. The individual problems are always first approached with regard to their unique solvability. If this unique solution is not given, then the problem is modified appropriately. In each case, a great emphasis is put on the numeric determination of the solution. The formal representation of each solution is given in detail. This makes the reconstruction of the problem possible with, for example, Mathematica®, Maple® or the SAS® Procedure IML.

Spline functions build a linear function space. This makes the construction of an average function that is representative of a certain group of objects or individuals possible. From the point of view of the user, this is often an advantage of the spline approach over other nonlinear models. The corresponding examples are related to the medical field. Suggestions to determine reference regions for clinical parameters are made.

Part II encompasses compartment models. The application of these models can be found in pharmacokinetics, physiology and in clinical medicine.

On hand of an individually observed course of concentration over time, so-called individual kinetics are to be described quantitatively. Mathematically speaking, this deals mainly with systems of linear ordinary differential equations with constant coefficients. Variations of this deterministic approach are shortly brought to light. For example, if a conceivable delay of effect is to be observed in the model, then differential equations with retarded argument come into question. Even for the simplest case it results that the solution of such differential equations are oscillating functions with negative values. Such models are therefore of no interest for the field of pharmacokinetics. The Two-compartment-iv model, otherwise known as the homogeneous systems of linear ordinary differential equations with constant coefficients, is the subject of additional observations. The solution to systems of linear ordinary differential equations with constant coefficients is well known. So is the existence and uniqueness of this solution. To start, the 16 possible variants of these Two-compartment-iv models and their respective solutions to the differential system of equations are given. On one hand, data regarding individual kinetics are obtained and a function from a certain function class can be fit to it. The parameters of the function that are obtained are called system parameters. On the other hand there are also the model parameters of the Two-compartment-iv model that also define the solution function. With regard to the selection of a model it is explored which model allows its parameters to be calculated from the system parameters and which model can even be identified on hand of the system parameters. With regard to this it needs to be seen which compartment can be observed. The calculation of the system parameters from available data can be done in different ways and with different results. The amazing amount of practical methods and computer programs is shown in an overview of literature.

Do the data and model even fit together? Ideally the selection of models and the calculation of parameters are handled together. A statistical process is developed for this: statistical parameter estimation and test of goodness of fit. A stochastic model is needed for this. The random variable is the residence time of a pharmacon molecule in an organism. Probability densities can be derived from solutions to differential equations. It is examined which of the Two-compartment-iv models correspond with residence time distributions For the given data, the classic PEARSON test can determine the appropriateness of the model. Since the random variable can only be observed indirectly, the asymptotically equivalent varied Minimum-χ^2 estimation occurs in place of the maximum-likelihood estimation of the param-

eters. The theory of estimation allows for the specification of characteristics of the process for calculating parameters. Time restricted observation can be taken into account by model transition to truncated distributions. An application of the two compartment iv model that is of interest to clinical medicine is finally handled: the method by DOST for the quantitative description of the multiple application of medication is detailed and extended. In Part III the reader will find Mathematica® programs for selected problems. They may support the application of methods developed in this book. These programs are written straightforward under use of the terminology of the employed formulas.
The book turns to life scientists, theoretical medical practitioners and mathematicians engaged in life sciences as well as students of these disciplines. We use this material in our lectures on biometry for students studying biomathematics, human biology and medicine.

We would like to thank our colleagues of the Medical Faculty of the Ernst-Moritz-Arndt-University in Greifswald for the helpful discussions around medical questions, students in our biometry courses for their interest in the topic, Paul Rosenthal, M.Sc.(Mathematics), Kathrin Wünsch M.Sc.(Physics) and Carolin Malsch, B.Sc.(Biomathematics) for her technical help with the manuscript and her suggestions for corrections as well as Frieda Kaiser, M.Sc. (Medical Informatics), Victoria, B.C., Canada, for her dedicated work in translating the manuscript into English.

Greifswald, Germany
2013

Karl-Ernst Biebler
Michael Wodny

Contents

PART 1

Spline models

Chapter 1

Why spline functions?

In this part, a general method for the mathematicel modeling of real processes is developed using a deterministic approach. Three aspects are important here:

- the data,
- the model and
- the criterion of goodness of fit.

The general exercise consists of the selection of an optimal model function for the respective data. We call this curve fitting or approximation.
The data consist of a set of pairs (x_i, y_i), $i = 1, 2, \dots, m$, of real numbers. The values of y_i should be understood as measured values of an unknown corresponding continuous function $g(x_i)$.
The model is represented by means of a class of continuous functions from which a function $f(x)$ should be chosen based on the available data. We assume that the function $f(x)$ of the model depend on the parameters $p_1, p_2, \dots, p_k$, so $f(x) = f_{p_1,p_2,\dots,p_k}(x)$. A relatively favorable situation exists when the function class from which $g(x)$ stems is known. This can evolve through theoretic considerations or other secured results. As a rule, this situation is advantageous because the parameters of the function equation can be interpreted well.
A criterion of goodness of fit measures the distance of the function to be chosen and the data. The parameters of the function $f(x) = f_{p_1,p_2,\dots,p_k}(x)$ are determined such that $f(x)$ is close to the data. A norm in the sense of mathematical analysis is a special distance function. Examples are the sum of the squared errors

$$H(p_1, p_2, \dots, p_k) := \sum_{i=1}^{m} (y_i - f_{p_1,\dots,p_2,\dots,p_k}(x_i))^2 \qquad \text{(SSE)}$$

and the maximum norm

$$Z(p_1, p_2, \ldots, p_k) := \max_{i \in \{1,2,\ldots,m\}} |y_i - f_{p_1,p_2,\ldots,p_k}(x_i)| \ .$$

In reality however these are norms defined on $\mathbb{R}^m$. The distances are measures between the vector of the observed values $(y_1, y_2, \ldots, y_m)$ and the vector $(f(x_1), f(x_2), \ldots, f(x_m))$.
The supremum norm

$$\|f - g\|_\infty := \max_{x \in [a,b]} |f(x) - g(x)|$$

measures the distance between functions with respect to the whole interval $[a, b]$. This is a distance in a function space.
A widely known criterion of goodness of fit is the method of least squares (MLS). A parameter set $p_1, p_2, \ldots, p_k$, is calculated so that (SSE) becomes a minimum.
If $f(x) = f_{p_1,p_2,\ldots,p_k}(x)$ is linearly dependent on the parameters $p_1, p_2, \ldots, p_k$, then the optimal parameter vector can be calculated directly from the given data. This is proven under very general conditions. A uniquely solvable linear system of equations exists due to the necessary conditions for a local minimum, $\partial H/\partial p_j = 0$ for all j from one to k.
In nonlinear cases the process becomes more complicated. A nonlinear system of equations must be solved. In general, an initial parameter set is assumed that has to be improved iteratively by individual calculation steps. However, it is not always ensured that the optimal parameters that are strived for exist and/or are uniquely defined. One cannot even assume that, for all constellations with a convergent algorithm, a local minimum of the underlying objective function H can be determined. Further, the global minimum has to be selected from the set of all local minima.
For the maximum norm

$$Z(p_1, p_2, \ldots, p_k) := \max_{i \in \{1,2,\ldots,m\}} |y_i - f_{p_1,p_2,\ldots,p_k}(x_i)| = \min!$$

as the criterion of goodness of fit, the maximum distance between the measured value and the model function value has to be a minimum for the given knots x_i. The derived concrete methods of calculation of this approach are dependent on the model functions. This will not be discussed in further detail here.
In any case, after the determination of the optimal parameter set $p_1^*, p_2^*, \ldots, p_m^*$, it is presumed that $g(x)$ is described well by $f_{p_1^*,p_2^*,\ldots,p_k^*}(x)$. In many cases, a mathematical model cannot be created for the relationship $y = g(x)$. However, the curve progression should still be described

approximately, for example, in order to be able to calculate the area under the curve, to obtain interesting function values that can not be extracted from an exact measurement, or in order to determine derivatives. Auxiliary models, that should and must replace missing theoretically founded and context related models, can be used in such cases. These are often based on deep mathematical insights. The polynomial approximation is justified by a theorem by Karl WEIERSTRASS (1815-1897). The theorem states that every continuous function $g(x)$ on a compact set can be closely approximated through a sequence of polynomials $P_n(x)$. The distance of functions is measured by the maximum norm. This type of convergence is called uniform convergence. WEIERSTRASS's theorem states

$$\lim_{n\to\infty} \|g(x) - P_n(x)\|_\infty = \lim_{n\to\infty} \left[\max_{x\in[a,b]} |g(x) - P_n(x)| \right] = 0 \ .$$

So, all polynomials up to a fixed degree n can serve as a model. The calculation of the parameters then corresponds with the determination of the polynomial coefficients. In most practical problems of curve fitting, the precondition of continuity is not a proper restriction.

A special form of curve fitting is the interpolation. Here, the strong relationship $y_i = g(x_i)$, $i = 1, 2, \ldots, m$, is assumed, and an $f(x)$ from a class of model functions is searched for such that $f(x_i) = y_i$, $i = 1, 2, \ldots, m$, holds true. One is hopeful that f can also do "a good job" of describing the unknown function g between the knots.

Well known is the possibility of interpolation by a polynomial. For every finite set of points with pairwise different x-coordinates there exists a polynomial of degree at most $(m-1)$ with $P_{m-1}(x_i) = y_i$, $i = 1, 2, \ldots, m$. This is uniquely defined and we set $f_{p_1^*, p_2^*, \ldots, p_m^*}(x) = P_{m-1}(x)$.

Despite WEIERSTRASS's theorem, problems can still arise. These are brought to light in the following. Observe for example the continuous function $g(x) = 1/(1 + x^2)$ over the interval $[-5, 5]$. The equidistant $x_i := -5 + 10(i-1)/(m-1)$, $i = 1, 2, \ldots, m$, are chosen as knots for the construction of the interpolation polynomial. The respective uniquely defined interpolation polynomials for $m = 10$ and $m = 15$ are illustrated in Figures 1.1 and 1.2. The oscillations of the polynomials between the individual knots increase with growing number of points,

$$\lim_{m\to\infty} \|g(x) - P_{m-1}(x)\|_\infty = \lim_{m\to\infty} \left[\max_{x\in[-5,5]} |g(x) - P_{m-1}(x)| \right] = \infty$$

is even true (see ISAACSON and KELLER, 1966).

With regard to the maximum norm, as the number of knots grow, the interpolating polynomials between the knots move further away from the function $g(x)$. This means that there exists at least one x_0 in $[-5, 5]$ where $\lim_{m\to\infty} |g(x_0) - P_{m-1}(x_0)| = \infty$.

The circumstance illustrated in this example can also be formulated for arbitrary intervals $[a, b]$.

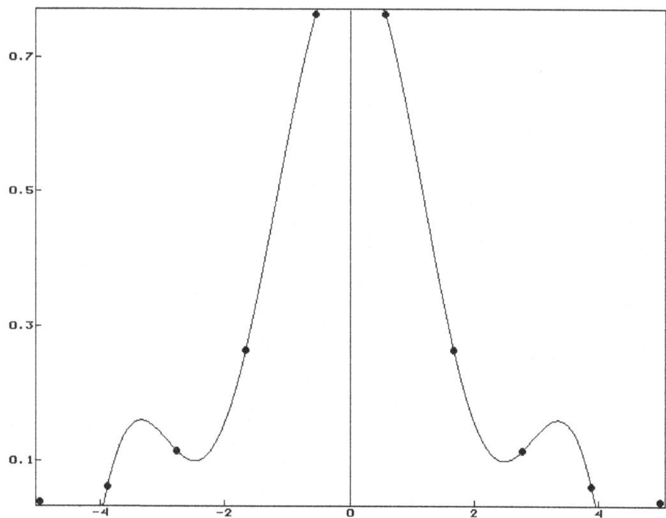

Fig. 1.1: Interpolation polynomial through 10 points of the continuous function $g(x) = 1/(1 + x^2)$.

Theorem 1.1. *Let $g(x)$ be a continuous function over a given interval $[a, b]$. A series $a = x_1 < x_2 < x_3 < \ldots < x_m = b$ of equidistant knots in $[a, b]$ is given. $P^g_{m-1}(x)$ denotes the interpolation polynomial with regard to the points $(x_i, g(x_i))$. Then a function $g_0(x)$ in the set $C([a, b])$ of all continuous functions on $[a, b]$ exists with*

$$\lim_{m\to\infty} \left\| g_0(x) - P^{g_0}_{m-1}(x) \right\|_\infty = \lim_{m\to\infty} \left[\max_{x\in[a,b]} \left| g_0(x) - P^{g_0}_{m-1}(x) \right| \right] = \infty \ .$$

Proof. With the LAGRANGIAN representation of $P^g_{m-1}(x)$ one has

$$\left\| P^g_{m-1}(x) \right\|_\infty \leq \sum_{i=1}^{m} |g(x_i)| \, \|L_i(x)\|_\infty \leq \max_i |g(x_i)| \sum_{i=1}^{m} \|L_i(x)\|_\infty \ ,$$

where

$$L_i(x) := \prod_{\substack{j=1 \\ j\neq i}}^{m} \frac{(x - x_j)}{(x_i - x_j)} \ .$$

$\lambda_m(x) := \sum_{i=1}^{m} |L_i(x)|$ is the so called LEBESGUE function. Then

$$\left\|P_{m-1}^{g}\right\|_\infty := \max_{x\in[a,b]} |P_{m-1}^{g}(x)| \le \|g\|_\infty \|\lambda_m\|_\infty$$

is true.

(1) There is a function $g_0(x)$ in $C([a,b])$ for which the equality in the last relation holds true: For fixed knots x_i, $i = 1,2,\ldots,m$, x_0 is chosen such that $\|\lambda_m\|_\infty = \lambda_m(x_0)$ is true. This is always possible, as λ_m is a positive continuous function over $[a,b]$. If we set $y = \operatorname{sign} L_i(x_0)$, $i = 1,2,\ldots,m$, and take the polygon connecting the points (x_i, y_i), $i = 1,2,\ldots,m$, for $g_0(x)$ then $\|g_0\|_\infty = 1$ and with the LAGRANGIAN representation of the interpolation polynomial

$$\begin{aligned}\|P_{m-1}^{g_0}\|_\infty &\le \|g_0\|_\infty \|\lambda_m\|_\infty = \|\lambda_m\|_\infty = \max_{x\in[a,b]} \left|\sum_{i=1}^{m} |L_i(x)|\right| \\ &= \lambda_m(x_0) = \left|\sum_{i=1}^{m} |L_i(x_0)|\right| = \left|\sum_{i=1}^{m} y_i L_i(x_0)\right| \\ &\le \max_{x\in[a,b]} \left|\sum_{i=1}^{m} y_i L_i(x)\right| = \|P_{m-1}^{g_0}\|_\infty \ .\end{aligned}$$

Therefore, the equal sign holds true everywhere in this estimation and $\|P_{m-1}^{g_0}\|_\infty = \|g_0\|_\infty \|\lambda_m\|_\infty$.

(2) The inequality $\|\lambda_m\|_\infty \ge A e^{m/2}$, $A \in \mathbb{R}$ being a constant, can be derived for equidistant knots (GOLOMB 1962, p. 209 and RIVLIN 1969, p. 99).

With (1) and (2), for all functions g_0 where $\|P_{m-1}^{g_0}\|_\infty = \|g_0\|_\infty \|\lambda_m\|_\infty$ we attain the assertion $\|P_{m-1}^{g_0}\|_\infty = \|g_0\|_\infty \|\lambda_m\|_\infty \ge \|g_0\|_\infty \alpha e^{m/2} \to \infty$ for $m \to \infty$. This evolves $\|g_0 - P_{m-1}^{g_0}\|_\infty \ge \big|\|g_0\|_\infty - \|P_{m-1}^{g_0}\|_\infty\big| \to \infty$ for $m \to \infty$. □

This behavior is not caused by the equidistance of the knots. FABER (1914) showed that for every sequence of decompositions $(\nabla^m)_{m\in\mathbb{N}}$ of the interval $[a,b]$ a continuous function $g(x)$ in $C([a,b])$ exists for which $\lim_{m\to\infty} \|g - P_{m-1}^{g}\|_\infty = \infty$ holds true.
Even if one is ready to accept the divergence at a few points, they will still be disappointed. Bernstein showed (NATANSON 1955) for the function $g(x) = |x|$ on the interval $[-1,1]$, with regard to the decomposition $x_i := -1 + 2(i-1)/(m-1)$, $i = 1,2,\ldots,m$, $1 < m \in \mathbb{N}$, that the interpolation polynomial $P_{m-1}^{g}(x)$ diverges pointwise for all $x \ne 0, \pm 1$. With polynomials as an example, one can see that the interpolating functions can display

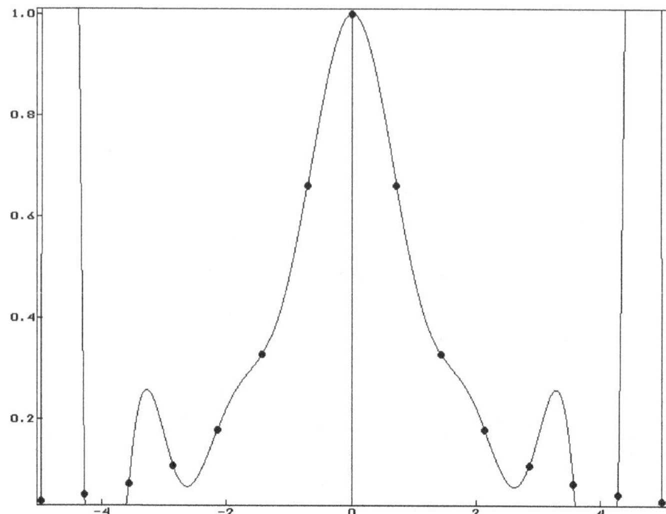

Fig. 1.2: Interpolation polynomial through 15 points of the continuous function $g(x) = 1/(1 + x^2)$.

unexpected behavior between the knots. The existing sequence of polynomials with uniform convergence to $g(x)$, in accordance with WEIERSTRASS's theorem, in general does not correspond with the sequence $P^g_{m-1}(x)$.
Another interesting class of functions for problems regarding approximation and interpolation are spline functions. SCHOENBERG(1946) first used the term spline in 1946 to describe this class. Originally spline describes a tool for constructing the hulls of ships. This spline instrument is a bendable slat that simplifies the drawing of curves. The tool is fixed at given points and takes on a role that is characterized as requiring a minimal amount of energy to bend. Characteristics of this spline are described by the spline functions. Different classes of spline functions are defined in the following chapters. Their ability to handle different problems related to interpolation and approximation is explored.
The main characteristic of interpolating natural cubic spline functions $s(x)$ is that they possess the minimal integral $\int_{x_1}^{x_m} [s''(x)]^2 dx$ for all other functions that can be continuously differentiated twice and that also fulfill the given conditions for interpolation. This measure of curvature can successfully be used as an auxiliary criterion for tasks of approximation. Adaptation to the given data should happen in such a way that the approximating function does not show any large fluctuations between the knots. Figure

1.3 illustrates that this works out well. The curves in Figures 1.2 and 1.3 can be compared directly since they were calculated from the same data. The following theorem forms the theoretical foundation for this approximation.

Theorem 1.2. *Let $g(x)$ be an $(n-1)$-times continuous differentiable function over the interval $[a,b]$. $\int_a^b [g^{(n)}(x)]^2 \mathrm{d}x \leq c^2$ is supposed to be true for the nth derivative $g^{(n)}(x)$. Furthermore, $a = x_1 < x_2 < \ldots < x_m = b$ let define a partition of $[a,b]$ and $h := \max|x_{i+1} - x_i|$, $i = 1, 2, \ldots, m-1$. $s_g(x)$ denotes the natural spline of degree $(2n-1)$ (compare Definition 2.1), which interpolates the points $(x_i, g(x_i))$, $i = 1, 2, \ldots, m$. Then $\|g^{(k)}(x) - s_g^{(k)}(x)\|_\infty := \max_{x \in [a,b]} |g^{(k)}(x) - s_g^{(k)}(x)| \leq A|h|^{n-k}c$ holds for all k from 0 to $(n-1)$, $0 < A < \infty$ a constant real number.*

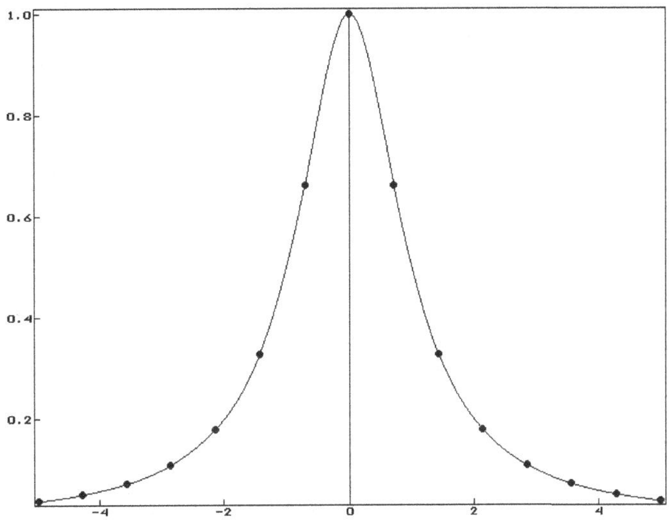

Fig. 1.3: Interpolating natural cubic spline.

The integral condition formulated in this theorem is fulfilled when the nth derivative of $g(x)$ is bounded. For the case of the approximation of continuous functions by the interpolation polynomial (see above), the preconditions of the Theorem 1.2. are fulfilled for $n = 1$ when, in addition, it is required that the first derivative $g'(x)$ over $[a,b]$ exists and is bounded by c. Then $\|g(x) - s_g(x)\|_\infty := \max_{x \in [a,b]} |g(x) - s_g(x)| \leq A|h|c$ is true for the interpolating natural spline of the first degree. This is a uniquely determined

polygon.
If continuously differentiable functions $g(x)$ with a bounded second derivative over $[a, b]$ are observed ($n = 2$ results due to Theorem 1.2), then the following inequalities for natural cubic splines, as they are described in Chapter 3, are true:

$$\|g(x) - s_g(x)\|_\infty := \max_{x\in[a,b]} |g(x) - s_g(x)| \le A|h|^2 c \ , \quad \text{and}$$

$$\|g'(x) - s_g'(x)\|_\infty := \max_{x\in[a,b]} |g'(x) - s_g'(x)| \le A|h|c \ .$$

Besides these general estimations of error for natural splines, there exist special more exact estimations, however also with stricter preconditions. The following result by SCHABACK and WERNER (1970) is used as an example.

Theorem 1.3. *Let $g(x)$ be a four times continuously differentiable function and $s_g(x)$ the interpolating cubic spline with respect to $(x_i, g(x_i))$ for which, in addition, $s_g'(a) = g'(a)$ and $s_g'(b) = g'(b)$ are true. Then*

$$\|g^{(k)}(x) - s_g^{(k)}(x)\|_\infty := \max_{x\in[a,b]} |g^{(k)}(x) - s_g^{(k)}(x)|$$

$$\le 2h^{4-k} \frac{h}{\min |x_{i+1} - x_i|} \max_{x\in[a,b]} |f^{(4)}(x)|,$$

$k = 0, 1, 2, 3.$

For approximation and interpolation problems, spline functions prove to be interesting alternatives to polynomials and other classes of functions that are not discussed here. In the first part splines are introduced, different curve fitting problems are formulated and their solutions are constructed. Strictly speaking, the set of all twice continuously differentiable functions is the model class related to the smoothing problem with regard to natural cubic splines. The objective functions however are formulated such that a spline function results as the solution. In the entire section on spline models, emphasis is put on closed analytic representation under special consideration of computational execution.
The basics of the theory of spline functions are developed in the following chapters. Real valued functions of only one real variable are considered. That spline functions with more than one variable exist, and that the term spline function was expanded into HILBERT spaces at the end of the 60's and early 70's can only be mentioned here. There is extensive literature regarding this. The majority of attention is put on cubic and quadratic splines. The construction of these functions is carried out in strict algebraic methods.

Chapter 2

Interpolating splines of degree n

The fundamental starting point of a curve fitting problem using spline functions with one variable is a finite number of points $(x_i, y_i), i = 1, 2, \ldots, m$, which describe the interesting but otherwise unknown functional dependency $y = g(x)$ at chosen knots x_i. The system of knots consists of increasingly ordered pairwise different values $x_1 < x_2 < x_3 < \ldots < x_{m-1} < x_m$.

Definition 2.1 (Spline functions of degree n). *Let $x_1, x_2, \ldots, x_{m-1}$, x_m be a strictly increasing sequence of real numbers. A **spline function** $s : \mathbb{R} \to \mathbb{R}$ **of the degree** n with the knots $x_1 < x_2 < \ldots < x_m$ is a function defined on the entire real line having the following two properties:*

a) In each interval $(-\infty, x_1]$, $[x_i, x_{i+1}]$, $i = 1, 2, \ldots, m-1$, and $[x_m, +\infty)$, $s(x)$ is given by some polynomial of degree n or less.
b) $s(x)$ and its derivatives up to the order of $(n-1)$ are continuous everywhere.

*$s(x)$ is called a **natural spline** if $n = 2k - 1$ and $s(x)$ is given in the interval $(-\infty, x_1]$ and $[x_m, +\infty)$ by some polynomial of degree $(k-1)$ or less.*

The polynomials that determine a given spline function are uniquely determined. That stems from the fact that two polynomials $P_n(x)$ and $Q_n(x)$ of degree n coincide when they are identical for $(n+1)$ points. $P_n(x) = Q_n(x)$ is the interpolation polynomial that is uniquely determined by these points. In the above definition, the polynomial degree is bounded by n, as spline functions should not oscillate too strongly between the knots. The demand for the existence of a continuous derivative to the order of $(n-1)$ ensures that the set of spline functions for a given system of knots does not only contain the appropriate polygon. Following Definition 2.1, every polynomial

of the nth degree is a spline function of degree n for any set of knots from the range of arguments. It is to be noted that all derivatives higher than to the order of $(n-1)$ are also continuous functions.
Amazing is the fact that just the demand for the existence of a continuous derivative of the order $(n-1)$ differentiates the spline functions from the polynomials. If in Definition 2.1 b) it were required that $s(x)$ possesses continuous derivatives to the order of n, then for all $x \in \mathbb{R}$, $s(x)$ would be a polynomial of degree n. This is true because, due to continuity, the nth derivative of every polynomial between the knots must be the same constant. Further examples of spline functions of degree n are illustrated by the following functions with arbitrary but fixed c,

$$(x-c)_+^n := \begin{cases} 0 & \text{for } x < c \text{ and} \\ (x-c)^n & \text{for } x \geq c \end{cases} .$$

The meaning of these so-called **elementary spline functions** is given in the following theorem.

Theorem 2.1. *Every spline function $s(x)$ of degree n for the knots $x_1 < x_2 < \ldots < x_m$ possesses a unique representation $s(x) = P_n(x) + \sum_{i=1}^{m} c_i(x - x_i)_+^n$, where $P_n(x)$ is a polynomial of degree n or less.*

Proof. Let $P_{n;i}(x)$ be a polynomial of degree n or less, which by definition represents $s(x)$ over the interval $[x_i, x_{i+1}]$ and $P_{n;i-1}(x)$ is the corresponding polynomial over $[x_{i-1}, x_i]$. Because of the continuity of $s(x)$, the difference of these polynomials must possess the root x_i. The same is true for the $(n-1)$ continuous derivatives of the spline function at this point. From this results $P_{n;i}(x) - P_{n;i-1}(x) = c_i(x-x_i)^n$. Consequently, for every k between 1 and m, $P_{n;k}(x) - P_{n;0}(x) = [P_{n;k}(x) - P_{n;k-1}(x)] + [P_{n;k-1}(x) - P_{n;k-2}(x)] + \ldots + [P_{n;1}(x) - P_{n;0}(x)]$, i.e. $P_{n;k}(x) = P_{n;0}(x) + c_1(x-x_1)^n + c_2(x-x_2)^n + \ldots + c_k(x-x_k)^n$. Therefore $s(x) = P_{n;0}(x) + \sum_{i=1}^{m} c_i(x-x_i)_+^n$ is true because the difference in the representing polynomials of two neighbouring intervals always takes the form $c_i(x-x_i)^n$.
The fact that $s(x)$ possesses $(n-1)$ continuous derivatives can be seen in the determined representation. To prove the uniqueness of this representation, first let $x < x_1$. From this follows that $s(x) = P_{n;0}(x) = P_n(x)$ is unique, as polynomials of degree n already correspond when they are identical at $(n+1)$ points. Furthermore it can be determined that $\lim_{x \to (x_j - 0)} s^{(n)}(x) = P_n^{(n)}(x_j) + n! \sum_{i=1}^{j-1} c_i$ and $\lim_{x \to (x_j + 0)} s^{(n)}(x) = P_n^{(n)}(x_j) + n! \sum_{i=1}^{j} c_i$, and therefore $c_j = \frac{1}{n!}\left(\lim_{x \to (x_j+0)} s^{(n)}(x) - \lim_{x \to (x_j-0)} s^{(n)}(x)\right)$, $j = 1, 2, \ldots, m$. □

Whether the natural splines mentioned in Definition 2.1 actually exist is initially not clear. However, examples of natural splines can be constructed based on Theorem 2.3 which is mentioned later. Given are the points (x_i, y_i), $i = 1, 2, \ldots, m$, which can be understood as points of a function $y = f(x)$. This function is otherwise unknown and will be described. Furthermore, it should always be presumed that all x_i, $i = 1, 2, \ldots, m$, are pairwise different and that $x_1 < x_2 < \ldots < x_m$ is true. The task of interpolation via spline functions of degree n can then be formulated as follows:

$$\left.\begin{array}{l} s(x) \text{ is searched for in the set of all spline functions} \\ \text{of degree } n \text{ with knots } x_1 < x_2 < \ldots < x_m \\ \text{for which } s(x_i) = y_i,\ i = 1, 2, \ldots, m, \text{ is true.} \end{array}\right\} \qquad \text{(SI)}$$

Note that the degree n of the spline function is arbitrary, however fixed. For $n = 1$, the polygon is indicated as the solution of (SI). However, as this does not possess a continuous derivative, solutions with $n > 1$ are of greater interest. Following Theorem 2.1, a unique representation $s(x) = \sum_{j=0}^{n} c_j x^j + \sum_{i=1}^{m} c_{n+i}(x - x_i)_+^n$ exists for every spline function of degree n. This is a linear combination of the $(n + m + 1)$ functions

$$F_j(x) : 1, x, x^2, \ldots, x^n, (x - x_1)_+^n, (x - x_2)_+^n, \ldots, (x - x_m)_+^n \ . \qquad (*)$$

These therefore build the basis functions for the splines of degree n.

Theorem 2.2. *The function system $(*)$ is linearly independent in the space of continuous real functions for a given system of knots $x_1 < x_2 < \ldots < x_m$.*

Proof. It must be shown that the relationships $c_i = 0$, $i = 0, 1, \ldots, m+n$, result when $s(x)$ is zero over $\mathbb{R}$.
For every $x \leq x_1$, $s(x)$ is a polynomial of degree n. This is only zero if all coefficients c_i, $i = 0, 1, 2, \ldots, n$, disappear.
If $x \in [x_1, x_2]$, $c_{n+1} = 0$ due to the definition of elementary splines. Similarly it can be determined that all other c_i, $i = n + 2, n + 3, \ldots, n + m$, must disappear. □

The linear independence allows to make the following statement:
Let $g(x)$ be an arbitrary continuous function. The distance measurement in the space of continuous functions should be a norm that is defined by a scalar product. Then there exists a uniquely determined spline function $s(x)$ of degree n that, in the sense of the norm, approximates $g(x)$ best.

For the interpolation task (SI), the y_i are understood as function values $g(x_i)$. The interpolation problem shall be solved by minimization of the sum of the squared errors SSE,

$$H(c_0, c_1, \ldots, c_{n+m}) := \sum_{i=1}^{m} (y_i - s(x_i))^2 = \text{min!} \tag{MSSE}$$

The necessary conditions $\partial H/\partial c_j = 0$ produce $(m+n+1)$ linear equations for the $(n+m+1)$ unknown parameters.
With regard to the values of y_i, the criterion MSSE formulates an approximation task in $\mathbb{R}^m$. The distance is locally measured with regard to the knots x_i. The vector $(y_1, y_2, \ldots, y_m)$ must be represented as a linear combination of the vectors $(F_j(x_1), F_j(x_2), \ldots, F_j(x_m))$, $j = 1, 2, \ldots, m+n+1$. These $(m+n+1)$ vectors cannot be linearly independent in $\mathbb{R}^m$. In general, there is no unique solution to the associated linear system of equations. Just the same, in general, $s(x_i) = y_i$ is true.
Therefore it cannot be expected that the interpolation task (SI) can always uniquely be solved by splines of degree n. Additional characteristics of splines are needed in order to guarantee a unique solution to the problem (SI).
A clue is given in Theorem 2.1. Its conclusion is also true for natural spline functions as the proof only mentioned the continuity of the first $(n-1)$ derivatives. This also holds true for natural splines.

Lemma 2.1. *Let $s(x)$ be a spline function of degree $n = 2k-1$ with, following Theorem 2.1, the unique representation*

$$s(x) = P_{k-1}(x) + \sum_{i=1}^{m} c_i (x - x_i)_+^{2k-1} .$$

$s(x)$ is a natural spline function if and only if the relationships

$$\sum_{i=1}^{m} c_i x_i^j = 0 , \quad j = 0, 1, \ldots, k-1,$$

are true for the coefficients c_i, $i = 1, 2, \ldots, m$.

Proof. For a natural spline of degree n, the uniquely determined representation is given by $s(x) = P_{k-1}(x) + \sum_{i=1}^{m} c_i (x - x_i)_+^{2k-1}$. $s(x)$ is described on the interval $(-\infty, x_1]$ by a polynomial of a degree less than or equal to $k-1$, and $n = 2k-1$.

The calculation of $(x-x_i)^{2k-1}$ with the binomial formula results in

$$(x-x_i)^{2k-1} = \sum_{j=0}^{2k-1} \binom{2k-1}{j} (-1)^j x^{2k-1-j} x_i^j \,.$$

Therefore, for $x \geq x_m$,

$$\begin{aligned} s(x) &= P_{k-1}(x) + \sum_{i=1}^{m} c_i \sum_{j=0}^{2k-1} \binom{2k-1}{j} (-1)^j x^{2k-1-j} x_i^j \\ &= P_{k-1}(x) + \sum_{j=0}^{2k-1} \binom{2k-1}{j} (-1)^j x^{2k-1-j} \sum_{i=1}^{m} c_i x_i^j \,. \end{aligned} \tag{2.1}$$

Corresponding with the definition of natural splines, $s(x)$ in $[x_m, +\infty)$ is a polynomial of degree at most $(k-1)$. For this reason, the second summand in the last equation disappears. This is exactly the case when

$$\sum_{i=1}^{m} c_i x_i^j = 0 \quad \text{is true for} \quad j = 0, 1, \ldots, k-1 \,. \tag{2.2}$$

□

To represent a natural spline function $s(x)$ exactly, $m+k$ parameters are needed. These are the c_i, $i = 1, 2, \ldots, m$, and the k coefficients of the polynomial $P_{k-1}(x)$. With (2.2), k equalities were attained that, together with the conditions for interpolation $s(x_i) = y_i$, $i = 1, 2, \ldots, m$, produce a linear system of equations to determine these parameters. It will now be shown that there is a unique solution to this system.

Theorem 2.3. *With natural spline functions of degree $n = 2k-1$, the interpolation task* (SI) *has an unique solution.*
More precisely, let $1 \leq k \leq m$ and $P_{k-1}(x) = a_0 + a_1 x + \ldots + a_{k-1} x^{k-1}$. Then the linear system of equations

$$\left.\begin{aligned} & s(x_i) = P_{k-1}(x_i) + \sum_{j=1}^{m} c_j (x_i - x_j)_+^{2k-1} = y_i \,, \quad i = 1, 2, \ldots, m \,, \\ & \textit{and} \\ & \sum_{i=1}^{m} c_i x_i^j = 0 \,, \quad j = 0, 1, 2, \ldots, k-1 \,, \end{aligned}\right\} \quad (**)$$

has exactly one solution.

Proof. It is to be shown that the system of equations $(**)$ only possesses the trivial solution $s(x) \equiv 0$ for all $y_i = 0$, $i = 1, \ldots, m$. That means $a_i = 0$,

$i = 0, 1, \ldots, k-1$, and $c_j = 0$, $j = 1, 2, \ldots, m$. First it is proven that

$$\int_a^b [s^{(k)}(x)]^2 \mathrm{d}x = (-1)^k (2k-1)! \sum_{i=1}^m c_i s(x_i)$$

is true for all $a \le x_1$ and $b \ge x_m$.
Due to the fact that $s(x)$ is already represented by a polynomial of degree $(k-1)$ on the intervals $(-\infty, x_1]$ and $[x_m, +\infty)$, one obtains

$$\begin{aligned}\int_a^b [s^{(k)}(x)]^2 \mathrm{d}x &= \int_a^{x_1} [s^{(k)}(x)]^2 \mathrm{d}x + \sum_{i=1}^{m-1} \int_{x_i}^{x_{i+1}} [s^{(k)}(x)]^2 \mathrm{d}x + \int_{x_m}^b [s^{(k)}(x)]^2 \mathrm{d}x \\ &= \sum_{i=1}^{m-1} \int_{x_i}^{x_{i+1}} [s^{(k)}(x)]^2 \mathrm{d}x \,.\end{aligned}$$

With $(k-1)$ times partial integration one obtains

$$\begin{aligned}&\int_a^b [s^{(k)}(x)]^2 \mathrm{d}x \\ &= (-1)^{k-1} \sum_{i=1}^{m-1} \int_{x_i}^{x_{i+1}} s'(x) s^{(2k-1)}(x) \mathrm{d}x \\ &= (-1)^{k-1} \sum_{i=1}^{m-1} \alpha_i [s(x_{i+1}) - s(x_i)] \\ &= (-1)^k \left[\alpha_1 s(x_1) + \sum_{i=2}^{m-1} (\alpha_i - \alpha_{i-1}) s(x_i) - \alpha_{m-1} s(x_m) \right] , \qquad (+)\end{aligned}$$

because $s^{(2k-1)}(x)$ is a constant α_i on every interval $]x_i, x_{i+1}[$, $i = 1, 2, \ldots, m-1$. This constants can be calculated from $\alpha_i = s^{(2k-1)}(x_i + 0) = s^{(2k-1)}(x_{i+1} - 0)$ (limit from the left and limit from the right, respectively). If $s(x)$ is derived $(2k-1)$ times, the representation $s^{(2k-1)}$ $(x) = (2k-1)! \sum_{j=1}^{i} c_j$ is obtained for $x \in]x_i, x_{i+1}[$. These relationships are inserted in (+). Then

$$\begin{aligned}\int_a^b [s^{(k)}(x)]^2 \mathrm{d}x &= (-1)^k (2k-1)! \left[c_1 s(x_1) + \sum_{i=2}^{m-1} c_i s(x_i) - s(x_m) \sum_{i=1}^{m-1} c_i \right] \\ &= (-1)^k (2k-1)! \sum_{i=1}^{m} c_i s(x_i)\end{aligned}$$

follows as $\sum_{i=1}^{m} c_j = 0$ for natural spline functions.
For $s(x_i) = 0$ for all i from 1 to m, the square of the kth derivation is equal to zero on the entire interval $[a, b]$. However, this is only possible when $s(x)$ is a polynomial of degree less than or equal $(k-1)$. With that, according to the fundamental theorem of tha algebra, $s(x)$ possesses at most $(k-1)$ roots in $[a, b]$.
For the system of equations $(**)$ with $y_i = 0$, $i = 1, \ldots, m$, $s(x_i) = 0$ for all i from 1 to $m > k-1$. This is only correct when $s(x)$ is identical to the zero-function and all coefficients disappear. □

At this point, several remarks regarding the condition $k \leq m$ in Theorem 2.3 should be made. It is used substantially in the proof. For $k > m$, the degree of $P_{k-1}(x)$ is greater or equal to the number m of interpolation points. In this case, the interpolation task cannot be uniquely solved anymore by a polynomial. For $k = m$ (the degree of the natural spline function is $n = 2m-1$), the solution to $(**)$ results in the interpolation polynomial through the points (x_i, y_i), $i = 1, 2, \ldots, m$. This results from the fact that the interpolating spline function of degree $(2m-1)$ and the interpolation polynomial of degree $(m-1)$ are uniquely determined by the same points. Consequently all $c_j = 0$, $j = 1, 2, \ldots, m$. Therefore, following Lemma 2.1, $s(x) = P_n(x) = P_{k-1}(x)$ is the interpolation polynomial (see Theorem 2.1). For $k = 1$ one obtains the polygon that connects the points.
Theorem 2.3 is interesting in three ways: it comprises the statement of the unique solvability of the interpolation task by natural spline functions, it thereby gives proof of existence for these functions, and it has constructive character. The linear system of equations for the determination of the unknown parameters of the interpolating natural splines is:

$$
\begin{pmatrix}
1 & x_1 & x_1^2 & \cdots & x_1^{k-1} & (x_1-x_1)^{2k-1} & 0 & \cdots & 0 \\
1 & x_2 & x_2^2 & \cdots & x_2^{k-1} & (x_2-x_1)^{2k-1} & (x_2-x_2)^{2k-1} & \cdots & 0 \\
\vdots & \vdots & \vdots & \ddots & \vdots & \vdots & \vdots & \ddots & \vdots \\
1 & x_m & x_m^2 & \cdots & x_m^{k-1} & (x_m-x_1)^{2k-1} & (x_m-x_2)^{2k-1} & \cdots & 0 \\
0 & 0 & 0 & \cdots & 0 & 1 & 1 & \cdots & 1 \\
0 & 0 & 0 & \cdots & 0 & x_1 & x_2 & \cdots & x_m \\
\vdots & \vdots & \vdots & \ddots & \vdots & \vdots & \vdots & \ddots & \vdots \\
0 & 0 & 0 & \cdots & 0 & x_1^{k-1} & x_2^{k-1} & \cdots & x_m^{k-1}
\end{pmatrix}
\circ
\begin{pmatrix}
a_0 \\ a_1 \\ \vdots \\ a_{k-1} \\ c_1 \\ c_2 \\ \vdots \\ c_m
\end{pmatrix}
$$

$$
= (y_1, y_2, \ldots, y_m, 0, 0, \ldots, 0)^T .
$$

The characteristic of natural spline functions, which is especially interesting for tasks of approximation that cannot fall back on context related models, are now analyzed.

Theorem 2.4 (DeBoor). *Let $s(x) = P_{k-1}(x) + \sum_{i=1}^{m} c_i(x - x_i)_+^{2k-1}$ be the uniquely determined natural spline function of degree $n = 2k - 1$ which solves the interpolation task $s(x_i) = y_i$, $i = 1, 2, \ldots, m$.*
Then for every interval $[a, b]$ with $a \leq x_1$ and $b \geq x_m$

$$\int_a^b \left[f^{(k)}(x)\right]^2 \mathrm{d}x \geq \int_a^b \left[s^{(k)}(x)\right]^2 \mathrm{d}x$$

holds true for all k times continuous differentiable functions $f(x)$ where $f(x_i) = y_i$, $i = 1, 2, \ldots, m$. For $k > 1$, the equality is true exactly when $f(x) \equiv s(x)$ is in $[a, b]$.

Proof. To begin, the following equality is true

$$\begin{aligned}
\int_a^b \left[f^{(k)}(x)\right]^2 \mathrm{d}x &= \int_a^b \left[s^{(k)}(x) + f^{(k)}(x) - s^{(k)}(x)\right]^2 \mathrm{d}x \\
&= \int_a^b \left[s^{(k)}(x)\right]^2 \mathrm{d}x + \int_a^b \left[f^{(k)}(x) - s^{(k)}(x)\right]^2 \mathrm{d}x \\
&\quad + 2\int_a^b s^{(k)}(x) \left[f^{(k)}(x) - s^{(k)}(x)\right] \mathrm{d}x \ .
\end{aligned}$$

Case 1 $k = 1$. Let $s'(x)$ be a constant α_i, $i = 1, \ldots, m-1$, in $]x_i, x_{i+1}[$.
Then

$$\begin{aligned}
&\int_a^b s'(x)\,[f'(x) - s'(x)]\,\mathrm{d}x \\
&= \sum_{i=1}^{m-1} \alpha_i \int_{x_i}^{x_{i+1}} [f'(x) - s'(x)]\,\mathrm{d}x \\
&= \sum_{i=1}^{m-1} \alpha_i\,[(f(x_{i+1}) - s(x_{i+1})) - (f(x_i) - s(x_i))] \\
&= 0 \ ,
\end{aligned}$$

is true due to the fact that the precondition $s(x_i) = f(x_i)$ must be true for all i from 1 to m.

Case 2 $k > 1$. After $(k-1)$ times partial integration, the same term results in

$$\int_a^b s^{(k)}(x)\left[f^{(k)}(x) - s^{(k)}(x)\right] \mathrm{d}x$$

$$= (-1)^{k-1} \sum_{i=1}^{m-1} \int_{x_i}^{x_{i+1}} s^{(2k-1)}(x)\left[f'(x) - s'(x)\right] dx \ .$$

Even here, the $(2k-1)$th derivative of $s(x)$ is respectively constant over $]x_i, x_{i+1}[$ so that, similar to the first case, this integral disappears.

The strong inequality for $f(x) \neq s(x)$ still needs to be shown.
This is fulfilled when $\left[f^{(k)}(x) - s^{(k)}(x)\right]^2 \equiv 0$ if and only if $f(x) \equiv s(x)$ is true. $f^{(k)}(x)$ and $s^{(k)}(x)$ are continuous for $k > 1$. In this case, $P_{k-1}(x) := f(x) - s(x)$ is a polynomial of degree less than or equal to $(k-1)$. However, both functions $f(x)$ and $s(x)$ fulfill the conditions of interpolation. Accordingly, $P_{k-1}(x)$ has at least $m > k-1$ roots and is therefore identically zero, i.e. $s(x) \equiv f(x)$. □

Theorem 2.3 can be generalized when the interpolation of natural splines is foregone and other auxiliary conditions are formulated instead.

Theorem 2.5. *Let (x_i, y_i), $i = 1, 2, \ldots, m$, be given points, $a \leq x_1 < x_2 < \ldots < x_m \leq b$ and $1 \leq k \leq m$. Furthermore, let I_a and I_b index sets whose elements stem from $\{0, 1, \ldots, k-1\}$. Let $f_a^{(j)}$ or $f_b^{(j)}$ be arbitrary real numbers.*
Then there exists an uniquely determined interpolating spline $s(x)$ of degree $(2k-1)$ where

$$\left.\begin{aligned} s(x_i) &= y_i, \quad i = 1, 2, \ldots, m, \\ s^{(j)}(a) &= f_a^{(j)}, j \in I_a, \\ s^{(j)}(b) &= f_b^{(j)}, j \in I_b, \\ s^{(2k-j-1)}(a) &= 0, \quad j \in \{0, 1, \ldots, k-1, \}, j \notin I_a \quad \textit{and} \\ s^{(2k-j-1)}(b) &= 0, \quad j \in \{0, 1, \ldots, k-1, \}, j \notin I_b \ . \end{aligned}\right\} \qquad (*)$$

Proof. First it is to be noted that in the case of $a = x_1$ and/or $b = x_m$ the equalities $s^{(0)}(a) = s(x_1) = f_a^{(0)}$ and/or $s^{(0)}(b) = s(x_m) = f_b^{(0)}$ cease to exist. However, $c_1(x - x_1)_+^{2k-1} = c_1(x - x_1)^{2k-1}$ so that this term can be added to the polynomial and/or $c_m(x - x_m)_+^{2k-1}$ is redundant. With that, in these marginal cases, the number of the unknowns corresponds with those of the equalities.
Similar to the proof of Theorem 2.3, the homogeneous system of equalities $(*)$ is observed and it is shown that it only possesses the trivial solution identical to zero. After $(k-1)$ times partial integration one obtains

$$\int_a^b \left[s^{(k)}(x)\right]^2 \mathrm{d}x = (-1)^k (2k-1)! \sum_{i=1}^m c_i s(x_i) = 0$$

via the necessary conditions as well as Theorem 2.1. Accordingly, $s(x)$ is a polynomial of degree $(k-1)$, which has no more than $(k-1)$ roots. However, if $s(x_i) = 0$, $i = 1, 2, \ldots, m > k - 1$, then $s(x) \equiv 0$ must be. □

Even an extremal property analogous to Theorem 2.4 is fulfilled. This will be given here without proof.

Theorem 2.6. *Let the conditions of Theorem 2.5 be fulfilled. Let $s(x)$ be the uniquely determined interpolating spline function of degree $(2k-1)$. Furthermore, let $f(x)$ be a $(k-1)$ times continuous differentiable function that fulfills $(*)$ and whose kth derivative is piecewise continuous.*
Then $\int_a^b \left[f^{(k)}(x)\right]^2 \mathrm{d}x \geq \int_a^b [s^{(k)}(x)]^2 \mathrm{d}x$ is true with regard to equality only in the case where $s(x) \equiv f(x)$.

It was already shown that, for m given points (x_i, y_i), $i = 1, 2, \ldots, m$, a natural spline of degree $(2m-1)$ just yields the interpolation polynomial. More generally, it can be formulated that for the exact representation of an arbitrary polynomial $P_r(x)$ of degree r, at least $(r+1)$ points and a natural spline function of degree $n = 2(r+1) - 1$ are necessary. $P_r(x) = s(x)$ is then true.
The system of equations to be solved can numerically cause considerable problems. In particular, $(m+k)$ grows with the number of points.
When solving large linear systems of equations one must take rounding errors into consideration; usage of system storage and processing times increase significantly.

Therefore, natural cubic splines will be addressed in greater detail in the next chapter. Especially simple constructive algorithms exist for these.

Example 2.1. Given are 13 equidistant points of the polynomial $P_4(x) = x^4 - 2x^3 + 3x^2 - 4x + 5$ in the interval from -1 to 5. Figure 2.1 shows the third derivative of the interpolating natural splines of degree 7. The third derivative of $P_4(x)$ is already well approximated (see also Part III). $P_4(x)$ is represented exactly when an interpolating natural spline of degree 9 is constructed with the corresponding points. In this case $k = 5$ holds.

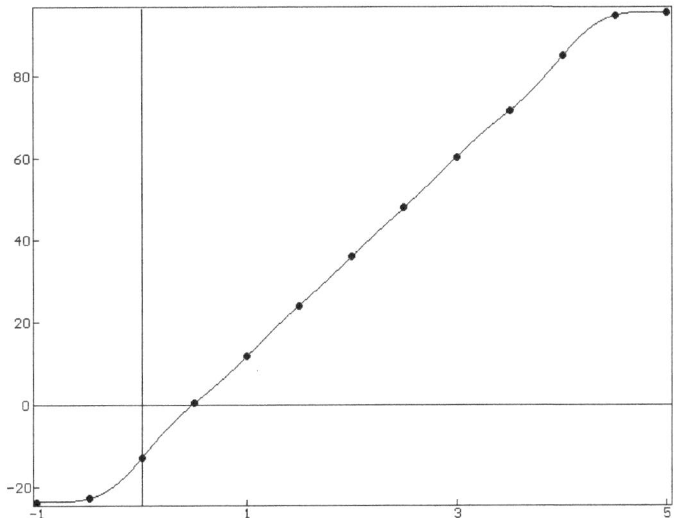

Fig. 2.1: Interpolating natural spline, for details see Example 2.1.

Chapter 3

Interpolating cubic splines

It should be noted again here that a natural cubic spline $s(x)$ on the knots $x_1 < x_2 < \ldots < x_m$ is a twice continuously differentiable function on $\mathbb{R}$ that on the intervals $(-\infty, x_1]$ and $[x_m, +\infty)$, and on $[x_i, x_{i+1}]$, $i = 1, 2, \ldots, m-1$, can be described as a straight line and a polynomial to a maximum degree of 3, respectively.
Theorem 2.3 guarantees the existence of a unique solution to the task of interpolation $s(x_i) = y_i$, $i = 1, 2, \ldots, m \geq 2$, with a natural cubic spline function. It follows from Theorem 2.4 that

$$\int_a^b [f''(x)]^2 \mathrm{d}x > \int_a^b [s''(x)]^2 \mathrm{d}x$$

is true for all twice continuously differentiable functions $f(x) \neq s(x)$ on the interval $[a, b]$ with $f(x_i) = y_i$, $i = 1, 2, \ldots, m$. Since the second derivative is a measure of the speed of variation of a function, the following definition is introduced.

Definition 3.1 (Total curvature). *Let $f(x)$ be a twice continuously differentiable function on $[a, b]$. The integral $\int_a^b [f''(x)]^2 \mathrm{d}x$ is called the total curvature of $f(x)$ on this interval.*

Using this definition, Theorem 2.4 states that for the points (x_i, y_i), $i = 1, 2, \ldots, m$, the interpolating natural cubic spline function $s(x)$ possesses minimal total curvature under all twice continuously differentiable functions $f(x)$, that likewise fulfill the interpolation conditions $f(x_i) = y_i$.
Because of the fact that $s(x)$ can be described by a polynomial of degree less than or equal to 3 between the individual knots, the following is true

for all x in $[x_i, x_{i+1}]$,

$$s(x) = A_i(x - x_i)^3 + B_i(x - x_i)^2 + C_i(x - x_i) + D_i \ .$$

Since $s(x)$ should solve the task of interpolation, it follows that

$$s(x_i) = y_i = D_i \ , \quad i = 1, 2, \ldots, m - 1 \ . \tag{3.1}$$

Furthermore

$$s''(x_i) = 2B_i \ , \quad \text{so} \quad B_i = s''(x_i)/2 \ . \tag{3.2}$$

The second derivative $s''(x)$ on $[x_i, x_{i+1}]$ is a straight line with slope $(s''(x_{i+1}) - s''(x_i))/(x_{i+1} - x_i)$ so that for A_i, $i = 1, 2, \ldots, m - 1$, the relationships

$$A_i = \frac{1}{6} \frac{s''(x_{i+1}) - s''(x_i)}{x_{i+1} - x_i} \tag{3.3}$$

result.
From (3.1), (3.2) and (3.3) one also obtains a representation of the C_i from the data and the second derivatives $s''(x_{i+1})$ and $s''(x_i)$:

$$C_i = \frac{y_{i+1} - y_i}{x_{i+1} - x_i} - \frac{1}{6}\left(x_{i+1} - x_i\right)\left(2s''(x_i) + s''(x_{i+1})\right) \ . \tag{3.4}$$

It remains to be shown that $s(x_m) = y_m$ is true.
It is true that

$$\begin{aligned} s(x_m) &= A_{m-1}(x_m - x_{m-1})^3 + B_{m-1}(x_m - x_{m-1})^2 \\ &\quad + C_{m-1}(x_m - x_{m-1}) + D_{m-1} \ . \end{aligned}$$

From this, the last condition of interpolation results from (3.1) to (3.4).
From these considerations it follows that a natural cubic spline function that fulfills the conditions of interpolation $s(x_i) = y_i$, $i = 1, 2, \ldots, m$, is uniquely determined by the data and the second derivatives $s''(x_i)$. To determine these unknown values one only needs to observe that $s'(x)$ is continuous in x_i, $i = 1, 2, \ldots, m$, and therefore $s'(x_i) = 3A_{i-1}(x_i - x_{i-1})^2 + 2B_{i-1}(x_i - x_{i-1}) + C_{i-1} = C_i$, $i = 2, 3, \ldots, m-1$, is true. When rewriting the formula and using the abbreviations $\Delta x_i := x_{i+1} - x_i$ and $\Delta y_i := y_{i+1} - y_i$, $i = 1, 2, \ldots, m - 1$, one obtains

$$\boxed{\begin{aligned} &\Delta x_{i-1} s''(x_{i-1}) + 2\left(\Delta x_{i-1} + \Delta x_i\right) s''(x_i) + \Delta x_i s''(x_{i+1}) \\ &= 6\left(\frac{\Delta y_i}{\Delta x_i} - \frac{\Delta y_{i-1}}{\Delta x_{i-1}}\right) . \end{aligned}} \tag{3.5}$$

Those are only $(m-2)$ equations $(i = 2, 3, \ldots, m-1)$ for the determination of the m unknown $s''(x_i)$. However, if one observes that $s(x)$ is a straight line over the open intervals $(-\infty, x_1]$ and $[x_m, +\infty)$ and that the second derivative is always continuous (especially in x_1 and x_m), then $s''(x_1) = 0$ and $s''(x_m) = 0$ follow.
As a matrix, the linear system of equations (3.5), with $S_i := 2(\Delta x_i + \Delta x_{i+1})$, is as follows

$$\begin{pmatrix} S_1 & \Delta x_2 & 0 & \ldots & 0 \\ \Delta x_2 & S_2 & \Delta x_3 & \ldots & 0 \\ 0 & \Delta x_3 & S_3 & \ldots & 0 \\ \vdots & \vdots & \vdots & \ddots & \vdots \\ 0 & 0 & 0 & \Delta x_{m-2} & S_{m-2} \end{pmatrix} \circ \begin{pmatrix} s''(x_2) \\ s''(x_3) \\ s''(x_4) \\ \vdots \\ s''(x_{m-1}) \end{pmatrix}$$

$$= \begin{pmatrix} 6\left[\dfrac{\Delta y_2}{\Delta x_2} - \dfrac{\Delta y_1}{\Delta x_1}\right] \\ 6\left[\dfrac{\Delta y_3}{\Delta x_3} - \dfrac{\Delta y_2}{\Delta x_2}\right] \\ 6\left[\dfrac{\Delta y_4}{\Delta x_4} - \dfrac{\Delta y_3}{\Delta x_3}\right] \\ \vdots \\ 6\left[\dfrac{\Delta y_{m-1}}{\Delta x_{m-1}} - \dfrac{\Delta y_{m-2}}{\Delta x_{m-2}}\right] \end{pmatrix} ; s''(x_1) = s''(x_m) = 0 \ . \qquad (3.6)$$

In the coefficients matrix (3.6),

$$\begin{aligned} 2(\Delta x_i + \Delta x_{i+1}) &> \Delta x_i + \Delta x_{i+1} \ , i = 2, 3, \ldots, m-3 \ , \\ 2(\Delta x_1 + \Delta x_2) &> \Delta x_2 \text{ and} \\ 2(\Delta x_{m-2} + \Delta x_{m-1}) &> \Delta x_{m-2} \end{aligned}$$

hold true.
That means the matrix is diagonally dominant. The respective linear system of equations can uniquely be solved. The interpolating natural cubic spline function, to be constructed from $s''(x_i)$, $i = 1, 2, \ldots, m$, from (3.1) to (3.4), is therefore the same as in Theorem 2.3 with $k = 2$.
At first glance the advantage over Theorem 2.3 cannot be seen. A linear system of equations with $(m+2)$ unknown values is replaced with another, with only $(m-2)$ equations. The advantages lie in the tridiagonal shape of the coefficients matrix. $(m+2)^2$ spaces are not needed anymore. Rather, only three vectors of length $(m-3)$ (or $(m-2)$) are needed.

Even the CHOLESKY decomposition and therefore the calculation of the unknown $s''(x_i)$ can successively be obtained. With this, the calculation time is now only linearly dependent on the number of points m.
The natural cubic splines can be obtained in a numerically satisfying and stable form with these results. The fact that these functions can alone be comprehensively described based on knowledge of the second derivatives at the knots has further consequences. It is also possible to provide further ancillary conditions in x_1 and x_m and to construct the appropriate spline function. However, it must then be observed that in general, no natural spline function is present and Theorem 2.4 does not hold true anymore in this form.
(3.5) provides the first possibility for an ancillary condition. If the second derivatives $y_1'' := s''(x_1)$ and $y_m'' := s''(x_m)$ are arbitrarily given, then (3.5) remains essentially unchanged. The first equation becomes

$$2(\Delta x_1 + \Delta x_2)s''(x_2) + \Delta x_2 s''(x_3) = 6\left[\frac{\Delta y_2}{\Delta x_2} - \frac{\Delta y_1}{\Delta x_1}\right] - \Delta x_1 y_1'' \qquad (3.7)$$

and the $(m-2)$th becomes

$$\begin{aligned}\Delta x_{m-2}s''(x_{m-2}) &+ 2(\Delta x_{m-2} + \Delta x_{m-1})s''(x_{m-1})\\ &= 6\left[\frac{\Delta y_{m-1}}{\Delta x_{m-1}} - \frac{\Delta y_{m-2}}{\Delta x_{m-2}}\right] - \Delta x_{m-1}y_m'' \ .\end{aligned}$$

The last equation can also be used to take into account a condition of periodicity. When the spline function to be constructed is to be periodic with the period $P := x_m - x_1$, then $s(x_1) = s(x_m)$, $s'(x_1) = s'(x_m)$ and $s''(x_1) = s''(x_m)$ hold true. From this the modified equation

$$\begin{aligned}\Delta x_{m-1}s''(x_1) + \Delta x_{m-2}s''(x_{m-2}) &+ 2(\Delta x_{m-2} + \Delta x_{m-1})s''(x_{m-1})\\ &= 6\left[\frac{\Delta y_{m-1}}{\Delta x_{m-1}} - \frac{\Delta y_{m-2}}{\Delta x_{m-2}}\right]\end{aligned}$$

results. Now if $i = 1$ is placed in (3.5) and it is taken into account that in the case of periodicity $\Delta x_0 = \Delta x_{m-1}$, $\Delta y_0 = \Delta y_{m-1}$ and $s''(x_0) = s''(x_{m-1})$, the system of equations for the calculation of the unknown values $s''(x_1), s''(x_2), \ldots, s''(x_m)$ results for the construction of a periodic interpolating spline function of degree 3 as follows:

$$\begin{pmatrix} 2(\Delta x_1 + \Delta x_{m-1}) & \Delta x_1 & 0 & \dots & \Delta x_{m-1} \\ \Delta x_1 & S_1 & \Delta x_2 & \dots & 0 \\ 0 & \Delta x_2 & S_2 & \dots & 0 \\ 0 & 0 & \Delta x_3 & \dots & 0 \\ \vdots & \vdots & \vdots & \ddots & \vdots \\ \Delta x_{m-1} & 0 & \dots & \Delta x_{m-2} & S_{m-2} \end{pmatrix} \circ \begin{pmatrix} s''(x_1) \\ s''(x_2) \\ s''(x_3) \\ s''(x_4) \\ \vdots \\ s''(x_{m-1}) \end{pmatrix}$$

$$= \begin{pmatrix} 6\left[\frac{\Delta y_1}{\Delta x_1} - \frac{\Delta y_{m-1}}{\Delta x_{m-1}}\right] \\ 6\left[\frac{\Delta y_2}{\Delta x_2} - \frac{\Delta y_1}{\Delta x_1}\right] \\ \vdots \\ 6\left[\frac{\Delta y_{m-1}}{\Delta x_{m-1}} - \frac{\Delta y_{m-2}}{\Delta x_{m-2}}\right] \end{pmatrix} ; s''(x_1) = s''(x_m) \ . \tag{3.8}$$

Again, the ability to solve the system of equations (3.8) clearly comes from the diagonal dominance of the matrix.

Since in general knowledge of the second derivative of the function to be approximated for the points (x_1, y_1) and (x_m, y_m) cannot be assumed, the following ancillary conditions offer good services, $s''(x_1) = u \cdots'' (x_2)$ and $s''(x_m) = v \cdot s''(x_{m-1})$.

The first and last equations in (3.5) become

$$[(2+u)\Delta x_1 + 2\Delta x_2]\, s''(x_2) + \Delta x_2 s''(x_3) = 6\left[\frac{\Delta y_2}{\Delta x_2} - \frac{\Delta y_1}{\Delta x_1}\right] \tag{3.9}$$

and

$$\Delta x_{m-2} s''(x_{m-2}) + [2\Delta x_{m-2} + (2+v)\Delta x_{m-1}]\, s''(x_{m-1})$$
$$= 6\left[\frac{\Delta y_{m-1}}{\Delta x_{m-1}} - \frac{\Delta y_{m-2}}{\Delta x_{m-2}}\right] , \tag{3.10}$$

respectively.

The ability to uniquely solve the new system of equations remains when $u > -(\Delta x_2/\Delta x_1 + 2)$ and $v > -(\Delta x_{m-2}/\Delta x_{m-1} + 2)$ because the diagonal dominance property is fulfilled under these conditions.

The first derivatives $y'_1 := s'(x_1)$ or $y'_m := s'(x_m)$ can also be ensured. Therefore, "initial behavior" and "ending behavior" of the spline function can be driven.

With (3.4) one obtains $s'(x_1) = C_1 = \Delta y_1/\Delta x_1 - 1/6\Delta x_1(2s''(x_1) + s''(x_2))$ and from $s'(x_m) = 3A_{m-1}(x_m - x_{m-1})^2 + 2B_{m-1}(x_m - x_{m-1}) + C_{m-1}$ follows

$\Delta x_{m-1}s''(x_{m-1}) + 2\Delta x_{m-1}s''(x_m) = 6(y'_m - \Delta y_{m-1}/\Delta x_{m-1})$ with (3.2), (3.3) and (3.4). In matrix form, the m equations for the m unknown values $s''(x_i)$, $i = 1, 2, \ldots, m$, read as

$$\begin{pmatrix} 2\Delta x_1 & \Delta x_1 & 0 & \ldots & 0 & 0 \\ \Delta x_1 & S_1 & \Delta x_2 & \ldots & 0 & 0 \\ 0 & \Delta x_2 & S_2 & \ldots & 0 & 0 \\ 0 & 0 & \Delta x_3 & \ldots & 0 & 0 \\ \vdots & \vdots & \vdots & \ddots & \vdots & \\ 0 & 0 & \ldots & \Delta x_{m-2} & S_{m-2} & \Delta x_{m-1} \\ 0 & 0 & \ldots & 0 & \Delta x_{m-1} & 2\Delta x_{m-1} \end{pmatrix} \circ \begin{pmatrix} s''(x_1) \\ s''(x_2) \\ s''(x_3) \\ s''(x_4) \\ \vdots \\ s''(x_{m-1}) \\ s''(x_m) \end{pmatrix}$$

$$= \begin{pmatrix} 6\left[\dfrac{\Delta y_1}{\Delta x_1} - y'_1\right] \\ 6\left[\dfrac{\Delta y_3}{\Delta x_3} - \dfrac{\Delta y_2}{\Delta x_2}\right] \\ 6\left[\dfrac{\Delta y_4}{\Delta x_4} - \dfrac{\Delta y_3}{\Delta x_3}\right] \\ \vdots \\ 6\left[\dfrac{\Delta y_{m-1}}{\Delta x_{m-1}} - \dfrac{\Delta y_{m-2}}{\Delta x_{m-2}}\right] \\ 6\left[y'_m - \dfrac{\Delta y_{m-1}}{\Delta x_{m-1}}\right] \end{pmatrix}. \tag{3.11}$$

The ability to solve the system (3.11) uniquely follows from the diagonal dominance property of the matrix.
All considerations made up to this point shall flow into the following theorem.

Theorem 3.1. (**Holladay**) *All cubic spline functions $s(x)$ interpolating the points (x_i, y_i), $i = 1, 2, \ldots, m$, and satisfying the following auxiliary conditions are uniquely determined:*

a) $s''(x_1) = y''_1$ *and* $s''(x_m) = y''_m$
b) $s''(x_1) = u \cdot s''(x_2)$ *and* $s''(x_m) = v \cdot s''(x_{m-1})$ *where* $u > -[\frac{\Delta x_2}{\Delta x_1} + 2]$ *and* $v > -[\frac{\Delta x_{m-2}}{\Delta x_{m-1}} + 2]$
c) $s'(x_1) = y'_1$ *and* $s'(x_m) = y'_m$
d) $s(x)$ *is periodic with the period* $P := x_m - x_1$, *i.e.* $s(x_1) = s(x_m)$, $s'(x_1) = s'(x_m)$ *and* $s''(x_1) = s''(x_m)$.

In a), *$s(x)$ has minimal total curvature on the interval $[x_1, x_m]$ in the set of all twice continuously differentiable functions $f(x)$ that also fulfill the*

conditions of interpolation $f(x_i) = y_i$, $i = 1, 2, \ldots, m$, *when* $y_1'' = y_m'' = 0$ *is true (natural cubic spline function).*
In the case of c), $s(x)$ *has minimal total curvature on* $[x_1, x_m]$ *in the set of all twice continuously differentiable functions* $f(x)$ *with* $f(x_i) = y_i$, $i = 1, 2, \ldots, m$, *and which fulfill the auxiliary conditions* $f'(x_1) = y_1'$ *and* $f'(x_m) = y_m'$.
In the case of d), $s(x)$ *has minimal total curvature on* $[x_1, x_m]$ *in the set of all twice continuously differentiable periodic functions* $f(x)$ *for which, besides the conditions of interpolation,* $f'(x_1) = f'(x_m)$ *and* $f''(x_1) = f''(x_m)$ *also hold true.*

Proof. As proof only the minimal curvature of $s(x)$ still needs to be shown in cases c) and d). Let $f(x)$ be a twice continuously differentiable function that fulfills the assumptions c) and d). The identity

$$\begin{aligned}\int_{x_1}^{x_m} [f''(x)]^2 \,\mathrm{d}x &= \int_{x_1}^{x_m} [s''(x) + f''(x) - s''(x)]^2 \,\mathrm{d}x \\ &= \int_{x_1}^{x_m} [s''(x)]^2 \,\mathrm{d}x + \int_{x_1}^{x_m} [f''(x) - s''(x)]^2 \,\mathrm{d}x \\ &\quad + 2\int_{x_1}^{x_m} s''(x)\,[f''(x) - s''(x)] \,\mathrm{d}x\end{aligned}$$

is true. Following partial integration one obtains

$$\begin{aligned}\int_{x_1}^{x_m} s''(x)\,[f''(x) - s''(x)] \,\mathrm{d}x = {}& s''(x)\,[f'(x) - s'(x)]\Big|_{x_1}^{x_m} \\ & - \int_{x_1}^{x_m} s'''(x)\,[f'(x) - s'(x)] \,\mathrm{d}x.\end{aligned}$$

Following the assumptions, the first summand is zero. Following renewed partial integration the result of the second is

$$\begin{aligned}\int_{x_1}^{x_m} s'''(x)\,[f'(x) - s'(x)] \,\mathrm{d}x = {}& \sum_{i=1}^{m-1} s'''(x)\,[f(x) - s(x)]\Bigg|_{x_i}^{x_{i+1}} \\ & - \int_{x_1}^{x_m} s^{(4)}(x)\,[f(x) - s(x)] \,\mathrm{d}x.\end{aligned}$$

Here the first summand also disappears due to the presupposed conditions of interpolation. The second part of the sum is zero because $s^{(4)}(x) \equiv 0$ over $\mathbb{R}$.
One argues for periodical functions analogously. □

Example 3.1. To illustrate the possibilities of interpolating cubic splines, the 9 points $(-3,-27)$, $(-2,-8)$, $(-1,-1)$, $(0,0)$, $(1,1)$, $(2,8)$, $(3,27)$, $(4,64)$ and $(5,125)$ are due to the polynomial $y = P_3(x) = x^3$.
In the following Figure 3.1 the uniquely determined interpolating natural cubic spline function

$$\begin{aligned} s(x) = {} & 38.40 + 21.80x - 2.89(x+3)_+^3 + 4.82(x+2)_+^3 - 1.28(x+1)_+^3 \\ & + 0.30(x-0)_+^3 + 0.060(x-1)_+^3 - 0.550(x-2)_+^3 + 2.150(x-3)_+^3 \\ & - 8.00(x-4)_+^3 + 5.340(x-5)_+^3 \end{aligned}$$

is shown.

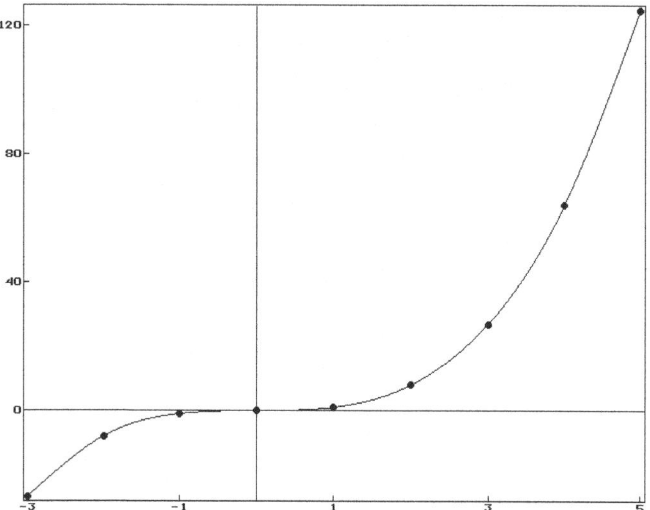

Fig. 3.1: Interpolating natural cubic spline function, see Example 3.1.

As the differences between $f(x) = x^3$ and $s(x)$ cannot be seen with this graphic resolution, the second derivative $s''(x)$ is shown in Figure 3.2. The deviations from the straight line $f''(x) = 6x$ become increasingly clear closer to the edge because for natural spline functions of degree 3, with regard to the system of knots that is used, $s''(-3) = s''(5) = 0$ must be true.

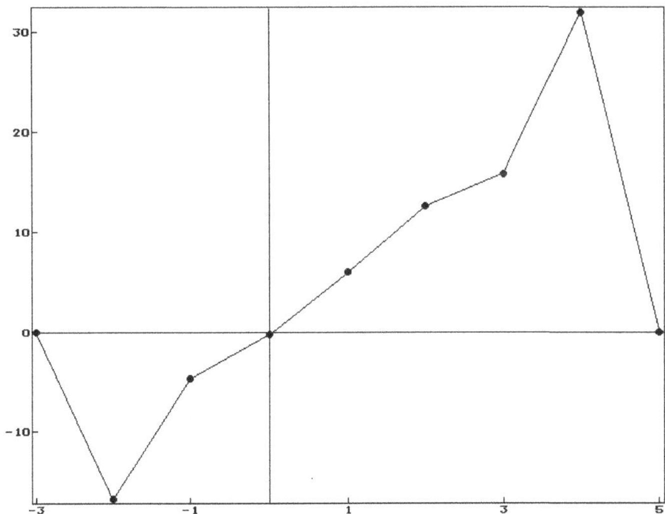

Fig. 3.2: $s''(x)$, see Example 3.1.

To improve the approximation (in this simple example it reaches the exact representation of the polynomial $f(x) = x^3$) knowledge about the function that is to be approximated can be used. $s(x) = f(x) = x^3$ can be attained by the following or other appropriate combinations of inputs:

1a)	$s''(-3)$	$=$	$-18,$	1b)	$s''(5)$	$=$	30
2a)	$s'(-3)$	$=$	$27,$	2b)	$s'(5)$	$=$	75
3a)	$s''(-3)$	$=$	$1.5 \cdot s''(-2),$	3b)	$s''(5)$	$=$	$1.25 \cdot s''(4).$

An exact representation of x^3 also results from an interpolating natural spline function of degree 7.

Example 3.2. The effects of different auxiliary conditions of the form $s'(x_1) = y'_1$ and $s'(x_m) = y'_m$, compared with the natural interpolating spline function, are demonstrated in Figure 3.3. The following eight points are used: $(2, 30)$, $(10, 25)$, $(20, 2)$, $(30, 22)$, $(40, 40)$, $(50, 68)$, $(60, 76)$ and $(90, 125)$. The related computer program one can find in Part III of this book.

The underdetermined linear system of equations (3.5) was always used as a starting point for the construction of interpolating cubic splines. Every auxiliary condition that is given (this includes the construction of the periodic spline) serves solely to obtain the two missing equations. This is easiest when the second derivatives are given for the points on the edge.

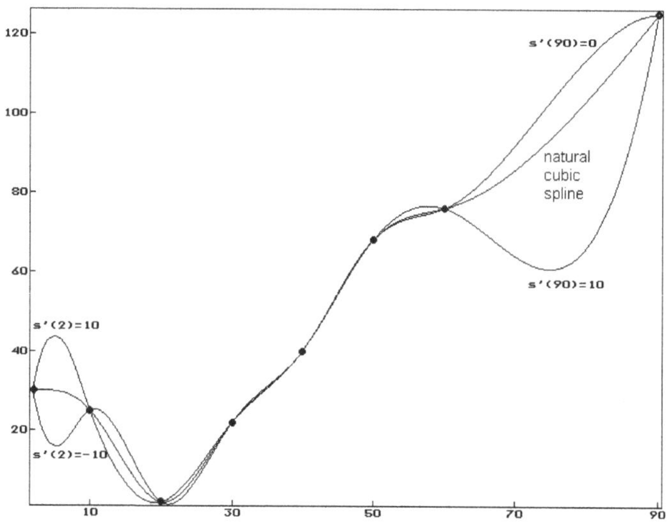

Fig. 3.3: Effects of auxiliary conditions on spline functions, see Example 3.2.

In principle, two arbitrary second derivatives $s''(x_k)$ and $s''(x_j)$, $k \neq j$ can be given and the leftover $(m-2)$ derivatives from (3.5) can therefore be determined. The described method essentially means the cancellation of columns k and j from the coefficient matrix. Consequently, diagonal dominance generally is no longer valid. A general statement about the unique solubility therefore cannot be made anymore. The calculation becomes particularly simple when the second derivatives in two neighboring points are fixed. All other unknown values $s''(x_i)$ can then successively be calculated from these. Even when two such derivatives are arbitrarily given, the unique solution to the resulting linear system of equations can be reached. The new equations $s''(x_k) = b''_k$ and $s''(x_j) = b''_j$ are inserted into (3.5). The following coefficient matrix results:

$$
\begin{pmatrix}
\Delta x_1 & 2(\Delta x_1 + \Delta x_2) & \Delta x_2 & \dots & 0 & 0 \\
0 & \Delta x_2 & 2(\Delta x_2 + \Delta x_3) & \dots & 0 & 0 \\
0 & 0 & \Delta x_3 & \dots & 0 & 0 \\
\vdots & \vdots & \vdots & \ddots & \vdots & \vdots \\
0 & 0 & \dots & \Delta x_{m-2} & 2(\Delta x_{m-2} + \Delta x_{m-1}) & \Delta x_{m-1} \\
0 & 0 & 0 & \dots 1 \dots & 0 & 0 \\
0 & \dots 1 \dots & 0 & 0 & 0 & 0
\end{pmatrix}.
$$

It is now proven that it is not singular. Suppose now $k < j$ without loss of generality. It needs to be shown that rows Z_i, $i = 1, 2, \ldots, m$, are linearly independent. For this, the equation $\sum_{i=1}^{m} \alpha_i Z_i = 0$ is observed. From the assumptions and the mainly tridiagonal form of the matrix, $\alpha_1 = 0$, $\alpha_2 = 0, \ldots, \alpha_{k-1} = 0$ and $\alpha_{m-2} = 0$, $\alpha_{m-3} = 0, \ldots, \alpha_{j-1} = 0$ immediately follow. The subsequent equations result for the remaining $(j - k + 1)$ unknown values $\alpha_k, \alpha_{k+1}, \ldots, \alpha_{j-2}, \alpha_{m-1}$ and α_m:

$$\begin{gathered}
\Delta x_k \alpha_k + \alpha_{m-1} = 0 \ , \\
2\left(\Delta x_k + \Delta x_{k+1}\right)\alpha_k + \Delta x_{k+1}\alpha_{k+1} = 0 \ , \\
\Delta x_{k+1}\alpha_k + 2\left(\Delta x_{k+1} + \Delta x_{k+2}\right)\alpha_{k+1} + \Delta x_{k+2}\alpha_{k+2} = 0 \ , \\
\vdots \\
\Delta x_{j-3}\alpha_{j-4} + 2\left(\Delta x_{j-3} + \Delta x_{j-2}\right)\alpha_{j-3} + \Delta x_{j-2}\alpha_{j-2} = 0 \ , \\
\Delta x_{j-2}\alpha_{j-3} + 2\left(\Delta x_{j-2} + \Delta x_{j-1}\right)\alpha_{j-2} = 0 \text{ and} \\
\Delta x_{j-2}\alpha_{j-2} + \alpha_m = 0 \ .
\end{gathered}$$

This results in $\alpha_{m-1} = -\alpha_k \Delta x_k$ and $\alpha_m = -\Delta x_{j-2}\alpha_{j-2}$. The coefficient matrix of the remaining $(j - k - 1)$ equations for the unknown values α_k, $\alpha_{k+1}, \ldots, \alpha_{j-3}$ and α_{j-2} also fulfills the diagonal dominance criterion. Finally, $\alpha_i = 0$ is true for all i from 1 to m.

The result of the discussed spline interpolation is extremely dependent on the specification of the second derivatives at arbitrary knots. This will be demonstrated on hand of the following calculations. The interpolating natural cubic spline function was constructed for the 20 points $(0, -0.45)$, $(1, 1.14)$, $(2, 1.02)$, $(3, 39)$, $(4, -0.55)$, $(5, -0.99)$, $(6, 0.42)$, $(7, 1.72)$, $(8, 2.47)$, $(9, 0.58)$, $(10, 0.78)$, $(11, 0.98)$, $(12, 1.08)$, $(13, 3.28)$, $(14, 2.18)$, $(15, 1.12)$, $(16, 0.58)$, $(17, 1.03)$, $(18, 2.22)$, $(20, -0.45)$ (See Figure 3.4).

For this natural cubic spline, $s''(x_1) = s''(0) = 0$ and $s''(x_2) = s''(1) = -2.56495143994279$ hold true. Using these exact values as input, the natural spline can be reconstructed. Figure 3.5 demonstrates the interpolation result for the same points and the slight change of $s''(x_2) = s''(1) = -2.564951$ in the neighboring knot to x_1.

The same results as before are obtained when two arbitrary first derivatives f'_j and f'_k are given. From (3.4), $2\Delta x_i s''(x_i) + \Delta x_i s''(x_{i+1}) = 6\left[\Delta y_i/\Delta x_i - f'_i\right]$ for all i from 1 to $(m-1)$, and $f'_m := s'(x_m) = A_{m-1}\Delta x_{m-1}^{\,2} + 2B_{m-1}\Delta x_{m-1} + C_{m-1}$ results in $\Delta x_{m-1} s''(x_{m-1}) + 2\Delta x_{m-1} s''(x_m) = 6\left[f'_m - \Delta y_{m-1}/\Delta x_{m-1}\right]$.

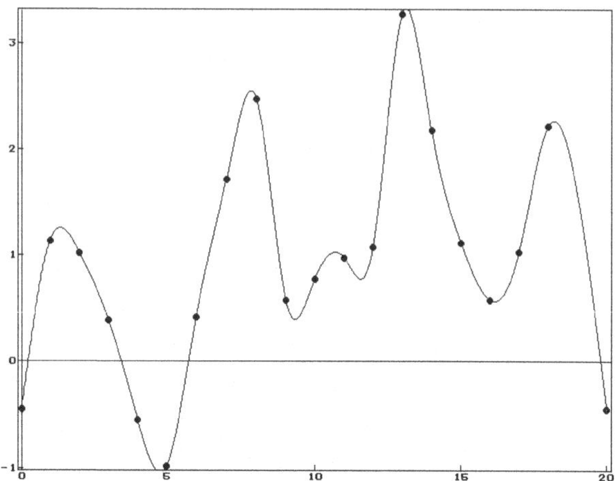

Fig. 3.4: Interpolating natural spline for 20 points.

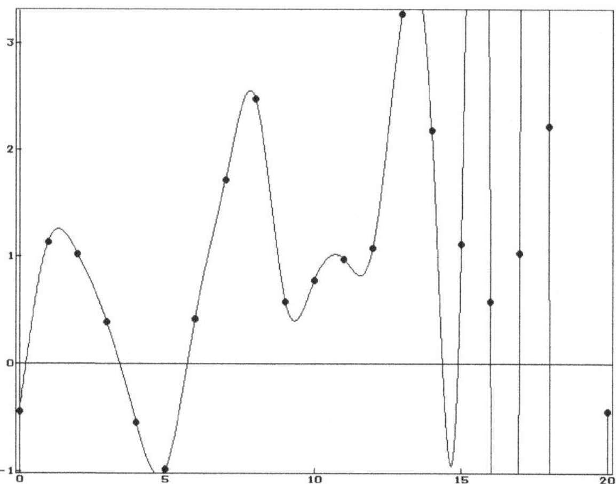

Fig. 3.5: Interpolating natural spline for the same points as in Figure 3.4, 2nd derivatives modified extremely weakly in neighboring knots; cp. Figure 3.4 and text.

Without loss of generality let be $1 \leq k < j < m$. The two additional relationships

$$2\Delta x_k s''(x_k) + \Delta x_k s''(x_{k+1}) = 6\left[\Delta y_k/\Delta x_k - f_k'\right] \text{ and} \tag{3.12}$$

$$2\Delta x_j s''(x_j) + \Delta x_j s''(x_{j+1}) = 6\left[\Delta y_j/\Delta x_j - f_j'\right] \tag{3.13}$$

are observed as the $(m-1)$th or mth equation so that the coefficient matrix for the calculation of the unknown values $s''(x_i)$, $i = 1, 2, \ldots, m$, takes the following form:

$$\begin{pmatrix} \Delta x_1 & 2(\Delta x_1 + \Delta x_2) & \Delta x_2 & \ldots & 0 & 0 \\ 0 & \Delta x_2 & 2(\Delta x_2 + \Delta x_3) & \ldots & 0 & 0 \\ 0 & 0 & \Delta x_3 & \ldots & 0 & 0 \\ \vdots & \vdots & \vdots & \ddots & \vdots & \vdots \\ 0 & 0 & \ldots & \Delta x_{m-2} & 2(\Delta x_{m-2} + \Delta x_{m-1}) & \Delta x_{m-1} \\ 0 \ldots & \ldots 2\Delta x_k & \Delta x_k & 0 & \ldots & 0 \\ 0 & 0 \ldots & \ldots 2\Delta x_j & \Delta x_j & \ldots & 0 \end{pmatrix}$$

the rows of this matrix are linearly independent and the corresponding linear system of equations can uniquely be solved when the equation $\sum_{i=1}^{m} \alpha_i Z_i = 0$ can only be fulfilled for $\alpha_i = 0$, $i = 1, 2, \ldots, m$. This approach results in m relationships for the unknown values α_i from which, due to the special form of the first $(m-2)$ rows, $\alpha_1 = 0$, $\alpha_2 = 0, \ldots, \alpha_{k-1} = 0$ and $\alpha_{m-2} = 0$, $\alpha_{m-3} = 0, \ldots, \alpha_j = 0$ result. Then $(j - k + 3)$ equations remain,

$$\begin{gathered} \Delta x_k \alpha_k + 2\alpha_k \alpha_{m-1} = 0 \ , \\ 2(\Delta x_k + \Delta x_{k+1})\,\alpha_k + \Delta x_{k+1}\alpha_{k+1} + \Delta x_k \alpha_{m-1} = 0 \ , \\ \Delta x_{k+1}\alpha_k + 2(\Delta x_{k+1} + \Delta x_{k+2})\,\alpha_{k+1} + \Delta x_{k+2}\alpha_{k+2} = 0 \ , \\ \vdots \\ \Delta x_{j-2}\alpha_{j-3} + 2(\Delta x_{j-2} + \Delta x_{j-1})\,\alpha_{j-2} + \Delta x_{j-1}\alpha_{j-1} = 0 \ , \\ \Delta x_{j-1}\alpha_{j-2} + 2(\Delta x_{j-1} + \Delta x_j)\,\alpha_{j-1} + 2\Delta x_j \alpha_m = 0 \text{ and} \\ \Delta x_{j-1}\alpha_{j-1} + \Delta x_j \alpha_m = 0 \ . \end{gathered}$$

They result in $\alpha_{m-1} = -\alpha_k/2$ and $\alpha_m = -\alpha_{j-1}$. The coefficient matrix for the calculation of the last $(j - k + 1)$ unknown values $\alpha_k, \alpha_{k+1}, \ldots, \alpha_{j-1}$ and α_j

$$\begin{pmatrix} \frac{2}{3}\Delta x_k + 2\Delta x_{k+1} & \Delta x_{k+1} & 0 & \ldots & 0 & 0 \\ 0 & \Delta x_{k+1} & 2(\Delta x_{k+1} + \Delta x_{k+2}) & \ldots & 0 & 0 \\ 0 & 0 & \Delta x_{k+2} & \ldots & 0 & 0 \\ \vdots & \vdots & \vdots & \ddots & \vdots & \vdots \\ 0 & 0 & \ldots & \Delta x_{j-1} & 2(\Delta x_{j-1} + \Delta x_j) & \Delta x_j \\ 0 & 0 & \ldots & 0 & \Delta x_j & 2\Delta x_j \end{pmatrix}$$

is diagonal dominant.

The spline function consequently is uniquely determined when two first derivatives in arbitrary knots are predefined.

With $s'(0) = 2.01749190665713$ and $s'(1) = 0.73501618668574$ given, with the same 20 points as before, the calculated spline is the natural interpolating cubic spline function as shown in Figure 3.4. The small change $s'(1) = 0.735016$ leads to similar results like in Figure 3.5.
If the auxiliary conditions concern knots lying narrowly, the interpolation is influenced very strongly then as the examples point. Giving arbitrary auxiliary conditions is possible in principle, but therefore not recommended for "neighboring" knots.
Next the following question is investigated: Can an interpolating cubic spline reproduce a polynomial of the 3rd degree, given the points $(x_i, y_i) = (x_i, P_3(x_i))$, $i = 1, 2, \ldots, m$?

Theorem 3.2. *A polynomial of degree less than or equal to 3 is unambiguously produced by a cubic spline if at least two points and two first or second derivatives of $P_3(x)$ for two arbitrary knots are given (compare Example 3.2).*
The reproduced polynomial possesses minimal total curvature exactly when $P(x) = \alpha x + \beta$ is true.

Proof. The first part of this statement is a direct consequence of Theorem 3.1 and the remarks made above, as every polynomial of degree 3 is a special spline. The statement in general is not correct for the conditions $s_3''(x_1) = u \cdot s_3''(x_2)$ and $s_3''(x_m) = v \cdot s_3''(x_{m-1})$. This is shown in the example above for $f(x) = x^3$. Obviously, $0 \neq s_3''(x_m) = v \cdot s_3''(x_{m-1}) = 0$ is not achieved for $x_{m-1} = 0$.
The second part of the statement follows from the definition. When $P_3(x) = \alpha x^3 + \beta x^2 + \gamma x + \delta$ is represented by a cubic spline with minimal total curvature with regard to the knots x_i, $i = 1, 2, \ldots, m$, $s_3(x)$ must be a *natural* spline function due to Theorem 2.4 or Theorem 3.1. From this $\alpha = 0$ and $\beta = 0$ result because $s_3''(x_1) = P_3''(x_1) = 6\alpha x_1 + 2\beta = s_3''(x_m) = P_3''(x_m) = 6\alpha x_m + 2\beta = 0$. Due to this, only straight lines possess the required characteristics. □

The observations up to this point were used for the representation of interpolating cubic splines by their second derivatives $s''(x_i)$. In a small digression it will now be shown how $s(x)$ can be represented exclusively by the first derivatives $s'(x_i)$, $i = 1, 2, \ldots m$.
Again we proceed with $s(x) = A_i(x - x_i)^3 + B_i(x - x_i)^2 + C_i(x - x_i) + D_i$, $i = 1, 2, \ldots, m - 1$.

When $s(x)$ should be continuous in x_{i+1} then

$$D_{i+1} = A_i(x_{i+1} - x_i)^3 + B_i(x_{i+1} - x_i)^2 + C_i(x_{i+1} - x_i) + D_i$$

must hold true. Analogously for the first derivative we have

$$s'(x_{i+1}) = C_{i+1} = 3A_i(x_{i+1} - x_i)^2 + 2B_i(x_{i+1} - x_i) + C_i$$

and for the second derivative we have

$$s''(x_{i+1}) = 2B_{i+1} = 6A_i(x_{i+1} - x_i) + 2B_i \ .$$

With the inserted abbreviations the following three equations result

$$A_i \Delta x_i{}^3 + B_i \Delta x_i{}^2 + C_i \Delta x_i - \Delta y_i = 0 \tag{3.14}$$

$$3A_i \Delta x_i{}^2 + 2B_i \Delta x_i + C_i - C_{i+1} = 0 \tag{3.15}$$

$$3A_i \Delta x_i + B_i - B_{i+1} = 0 \ . \tag{3.16}$$

The definitions $D_m := y_m$, $B_m := s''(x_m)/2$ and $C_m := s'(x_m)$ must be observed here. With the help of (3.14) and (3.15), A_i can be eliminated:

$$B_i = 3\frac{\Delta y_i}{\Delta x_i{}^2} - 2\frac{C_i}{\Delta x_i} - \frac{C_{i+1}}{\Delta x_i}, \qquad i = 1, 2, \ldots, m-1 \ , \tag{3.17}$$

and by the insertion of (3.17) into (3.14) one obtains

$$A_i = \frac{1}{\Delta x_i{}^2}\left(C_i + C_{i+1} - 2\frac{\Delta y_i}{\Delta x_i}\right), \qquad i = 1, 2, \ldots, m-1 \ . \tag{3.18}$$

Since $D_i = y_i$ and $s'(x_i) = C_i$ is true for all i from 1 to m, the spline function is unambiguously determined by the first derivatives in the knots. The system of equations to calculate these unknown values is obtained when (3.17) and (3.18) are inserted into (3.16). Then

$$\begin{aligned} &\frac{1}{\Delta x_i}C_i + 2\left(\frac{1}{\Delta x_i} + \frac{1}{\Delta x_{i+1}}\right)C_{i+1} + \frac{1}{\Delta x_{i+1}}C_{i+2} \\ &= 3\left(\frac{\Delta y_{i+1}}{\Delta x_{i+1}^2} + \frac{\Delta y_i}{\Delta x_i^2}\right), i = 1, 2, \ldots, m-2, \end{aligned} \tag{3.19}$$

are true. With the specification of the first derivatives for the edge points $y_1' = s'(x_1) = C_1$ and $y_m' = s'(x_m) = C_m$ and after multiplication of (3.20) with $\Delta x_i \Delta x_{i+1}$, the following uniquely solvable linear system of equations results,

$$\begin{pmatrix} S_1 & \Delta x_1 & 0 & 0 & \dots & 0 \\ \Delta x_3 & S_2 & \Delta x_2 & 0 & \dots & 0 \\ \vdots & \vdots & \vdots & \vdots & \ddots & \vdots \\ 0 & 0 & \dots & \Delta x_{m-2} & S_{m-3} & \Delta x_{m-3} \\ 0 & 0 & \dots & 0 & \Delta x_{m-1} & S_{m-2} \end{pmatrix} \circ \begin{pmatrix} C_2 \\ C_3 \\ \vdots \\ C_{m-2} \\ C_{m-1} \end{pmatrix}$$

$$= \begin{pmatrix} 3\left(\dfrac{\Delta x_1^2 \Delta y_2 + \Delta x_2^2 \Delta y_1}{\Delta x_1 \Delta x_2}\right) - y_1' \Delta x_2 \\ 3\left(\dfrac{\Delta x_2^2 \Delta y_3 + \Delta x_3^2 \Delta y_2}{\Delta x_2 \Delta x_3}\right) \\ \vdots \\ 3\left(\dfrac{\Delta x_{m-3}^2 \Delta y_{m-2} + \Delta x_{m-2}^2 \Delta y_{m-3}}{\Delta x_{m-3} \Delta x_{m-2}}\right) \\ 3\left(\dfrac{\Delta x_{m-2}^2 \Delta y_{m-1} + \Delta x_{m-1}^2 \Delta y_{m-2}}{\Delta x_{m-2} \Delta x_{m-1}}\right) - y_m' \Delta x_{m-2} \end{pmatrix}, \quad (3.20)$$

$$C_1 = y_1', C_m = y_m' \quad (S_i := 2(\Delta x_i + \Delta x_{i+1})) \ .$$

From Theorem 3.1 is known that the spline function determined by the C_i, $i = 1, 2, \dots, m$, is the function with minimal total curvature in the set of all twice continuously differentiable functions $f(x)$ that fulfill the conditions $f(x_i) = y_i$, $i = 1, 2, \dots, m$, $f'(x_1) = y_1'$ and $f'(x_m) = y_m'$.

Specifying the second derivatives $y_1'' = s''(x_1) = 2B_1$ and $y_m'' = s''(x_m) = 2B_m$ for the edge points is also possible. From (3.17) follows

$$2C_1 + C_2 = 3\frac{\Delta y_1}{\Delta x_1} - \frac{\Delta x_1}{2} y_1''$$

and from (3.17) and (3.18) with $s''(x_m) = 6A_{m-1}\Delta x_{m-1} + 2B_{m-1}$ one has

$$C_{m-1} + 2C_m = 3\frac{\Delta y_{m-1}}{\Delta x_{m-1}} + \frac{\Delta x_{m-1}}{2} y_m'' \ .$$

These are two additional equations form together with (3.20) a uniquely solvable linear system of m equations. It allows the determination of the C_i, $i = 1, 2, \dots, m$. This is how, for example, an interpolating natural spline function is constructed when setting $y_1'' = y_m'' = 0$.

Finally it should be mentioned that auxiliary conditions of the form $C_1 =$

$u \cdot C_2$ and $C_m = v \cdot C_{m-1}$ can be incorporated into (3.20). The first and the $(m-2)$th equations then are slightly modified,

$$(2\Delta x_1 + 2\Delta x_2 + u\Delta x_2)C_2 + \Delta x_1 C_3 = 3\left(\frac{\Delta x_1^2 \Delta y_2 + \Delta x_2^2 \Delta y_1}{\Delta x_1 \Delta x_2}\right) \quad \text{and}$$

$$\Delta x_{m-1}C_{m-2} + (2\Delta x_{m-1} + 2\Delta x_{m-2} + v\Delta x_{m-2})C_{m-1} = 3\left(\frac{\Delta x_{m-2}^2 \Delta y_{m-1} + \Delta x_{m-1}^2 \Delta y_{m-2}}{\Delta x_{m-2}\Delta x_{m-1}}\right).$$

3.1 Interpolating cubic splines with other extreme characteristics

The interpolating natural cubic spline $s_3(x)$ possesses minimum total curvature in the set of all other twice continuously differentiable functions that also solve the interpolation task. This spline is characterized by the conditions $s_3''(x_1) = 0$ and $s_3''(x_m) = 0$. On the other hand, every interpolating cubic spline is uniquely determined by the specifications $s_3''(x_1) = 2B_1$ and $s_3''(x_m) = 2B_m$.
As an example of another condition of construction, the interpolating cubic spline that minimizes

$$AQ := \int_{x_1}^{x_m} [s_3'(x)]^2 \mathrm{d}x$$

should be constructed for the given points (x_i, y_i), $i = 1, 2, \ldots, m$.
AQ measures the curvature of $s_3(x)$ on the interval $[x_1, x_m]$. When $s_3(x)$ is represented as a polynomial of the form

$$A_i(x - x_i)^3 + B_i(x - x_i)^2 + C_i(x - x_i) + D_i$$

over the intervals $[x_i, x_{i+1}]$ then $s_3(x_i) = y_i = D_i$, $A_i = 1/3(B_{i+1} - B_i)/\Delta x_i$ and $C_i = \Delta y_i/\Delta x_i - 1/3\Delta x_i(2B_i + B_{i+1})$ are true.
Since $s_3(x)$ is uniquely and completely determined by the data and B_i, an expression for $\int_{x_1}^{x_m} [s_3'(x)]^2 \mathrm{d}x$ in the B_i is now derived.

$$\int_{x_1}^{x_m} [s_3'(x)]^2 \mathrm{d}x = \sum_{i=1}^{m-1} \int_{x_i}^{x_{i+1}} [s_3'(x)]^2 \mathrm{d}x = \sum_{i=1}^{m-1} \int_{x_i}^{x_{i+1}} \left(3A_i(x - x_i)^2 + 2B_i(x - x_i) + C_i\right)^2 \mathrm{d}x$$

$$= \sum_{i=1}^{m-1} \int_{x_i}^{x_{i+1}} \left[9A_i^2(x-x_i)^4 + 12A_iB_i(x-x_i)^3\right.$$

$$\left. + (4B_i^2 + 6A_iC_i)(x-x_i)^2 + 4B_iC_i(x-x_i) + C_i^2\right] \mathrm{d}x$$

$$= \sum_{i=1}^{m-1} \left[\frac{9}{5}A_i^2\Delta x_i^5 + 3A_iB_i\Delta x_i^4 + (\frac{4}{3}B_i^2 + 2A_iC_i)\Delta x_i^3\right.$$

$$\left. +2B_iC_i\Delta x_i^2 + C_i^2\Delta x_i\right] .$$

Here, if the above expressions for A_i and C_i are inserted, after some formula manipulations one gets

$$\int_{x_1}^{x_m} [s_3'(x)]^2 \mathrm{d}x = \sum_{i=1}^{m-1} (4B_i^2\Delta x_i^4 + 7B_iB_{i+1}\Delta x_i^4 + 4B_{i+1}^2\Delta x_i^4 + 45\Delta y_i^2)/(45\Delta x_i)$$

$$\int_{x_1}^{x_m} [s_3'(x)]^2 \mathrm{d}x = \frac{1}{45}\sum_{i=1}^{m-1} \Delta x_i^3(4B_i^2 + 7B_iB_{i+1} + 4B_{i+1}^2) + \sum_{i=1}^{m-1} \frac{\Delta y_i^2}{\Delta x_i} .$$

Since the second sum in the last formula is not dependent on B_i anymore,

$$G(B_1, \dots, B_m) := \sum_{i=1}^{m-1} \Delta x_i^3(4B_i^2 + 7B_iB_{i+1} + 4B_{i+1}^2)$$

must be minimized with regard to the B_i. By means of the matrix

$$H := \begin{pmatrix} 4\Delta x_1^3 & \frac{7}{2}\Delta x_1^3 & 0 & 0 & \dots & 0 & 0 \\ \frac{7}{2}\Delta x_1^3 & \gamma_1 & \frac{7}{2}\Delta x_2^3 & 0 & \dots & 0 & 0 \\ 0 & \frac{7}{2}\Delta x_2^3 & \gamma_2 & \frac{7}{2}\Delta x_3^3 & \dots & 0 & 0 \\ \vdots & \vdots & \vdots & \vdots & \ddots & \vdots & \vdots \\ 0 & 0 & 0 & 0 & \dots & \gamma_{m-2} & \frac{7}{2}\Delta x_{m-1}^3 \\ 0 & 0 & 0 & 0 & \dots & \frac{7}{2}\Delta x_{m-1}^3 & 4\Delta x_{m-1}^3 \end{pmatrix}$$

one obtains the representation $G(B_1, \dots, B_m) = B^T \circ H \circ B$ where $B = (B_1, B_2, \dots, B_m)^T$ and $\gamma_i = 4(\Delta x_i^3 + \Delta x_{i+1}^3)$.
$H \circ B = 0$ is necessary for a local minimum of G. Since H is a diagonal dominant matrix, H^{-1} exists and $B_i = 0$, $i = 1, 2, \dots, m$, is the uniquely determined solution. With this, $C_i = \Delta y_i/\Delta x_i$ and all $A_i = 0$.
In this constellation, the polygon is described by the given points. However, this is not twice continuously differentiable and therefore not a cubic spline. To obtain a cubic spline as a solution the following equations need to be

observed,

$$\Delta x_{k-1}B_{k-1} + 2(\Delta x_{k-1} + \Delta x_k)B_k + \Delta x_k B_{k+1} = 3\left(\frac{\Delta y_k}{\Delta x_k} - \frac{\Delta y_{k-1}}{\Delta x_{k-1}}\right), \quad k = 2, 3, \ldots, m-1. \tag{3.21}$$

With two additional auxiliary conditions each, as formulated in the theorem by HOLLADAY, $s_3(x)$ is already uniquely determined. In this case, the task of minimizing AQ does not make any sense: this spline naturally also minimizes $\int_{x_1}^{x_m}[s_3'(x)]^2 dx$ because the set of permissible functions is a singleton.

Since every interpolating cubic spline is uniquely determined when B_1 and B_m are given, these two parameters should be inserted as variables: $s := B_1 = s''(x_1)/2$ and $t := B_m = s''(x_m)/2$. With the invertible matrix

$$\overline{Z} := \begin{pmatrix} 1 & 0 & 0 & \ldots & 0 & 0 & 0 & 0 \\ \Delta x_1 & S_1 & \Delta x_2 & \ldots & 0 & 0 & 0 & 0 \\ 0 & \Delta x_2 & S_2 & \ldots & 0 & 0 & 0 & 0 \\ 0 & 0 & \Delta x_3 & \ldots & 0 & 0 & 0 & 0 \\ \vdots & \vdots & \vdots & \ddots & \vdots & \vdots & \vdots & \vdots \\ 0 & 0 & 0 & \ldots & \Delta x_{m-3} & S_{m-3} & \Delta x_{m-2} & 0 \\ 0 & 0 & 0 & \ldots & 0 & \Delta x_{m-2} & S_{m-2} & \Delta x_{m-1} \\ 0 & 0 & 0 & \ldots & 0 & 0 & 0 & 1 \end{pmatrix}$$

(3.21) becomes $\overline{Z} \circ B = (s, \beta_2, \beta_3, \ldots, \beta_{m-1}, t)^T$, from which $B = \overline{Z}^{-1} \circ (s, \beta_2, \beta_3, \ldots, \beta_{m-1}, t)^T$ results.

The β_i are abbreviations for the right side of (3.21). The objective function G can be represented as a function of the both variables s and t; $G(B_1, \ldots, B_m) = G(s,t) := (s, \beta_2, \beta_3, \ldots, \beta_{m-1}, t) \circ (\overline{Z}^{-1})^T \circ H \circ \overline{Z}^{-1} \circ (s, \beta_2, \beta_3, \ldots, \beta_{m-1}, t)^T$.

With this it is to be noted that $E := (e_{ij})_{i,j=1,\ldots,m} := (\overline{Z}^{-1})^T \circ H \circ \overline{Z}^{-1}$ is a symmetric matrix.

$$E \circ (s, \beta_2, \beta_3, \ldots, \beta_{m-1}, t)^T = \begin{pmatrix} e_{11}s + \sum_{k=2}^{m-1} e_{1k}\beta_k + e_{1m}t \\ e_{21}s + \sum_{k=2}^{m-1} e_{2k}\beta_k + e_{2m}t \\ \vdots \\ e_{m1}s + \sum_{k=2}^{m-1} e_{mk}\beta_k + e_{mm}t \end{pmatrix}$$

is obtained and therefore

$$
\begin{aligned}
(s,\beta_2,\beta_3,\dots,\beta_{m-1},t)\circ E\circ(s,\beta_2,\beta_3,\dots,\beta_{m-1},t)^T \\
&= e_{11}s^2 + s\sum_{k=2}^{m-1} e_{1k}\beta_k + e_{1m}st \\
&+ \beta_2 e_{21}s + \beta_2\sum_{k=2}^{m-1} e_{2k}\beta_k + \beta_2 e_{2m}t \\
&+ \beta_3 e_{31}s + \beta_3\sum_{k=2}^{m-1} e_{3k}\beta_k + \beta_2 e_{3m}t \\
&+ \cdots + e_{m1}ts + t\sum_{k=2}^{m-1} e_{mk}\beta_k + e_{mm}t^2 .
\end{aligned}
$$

The two conditions $\partial G(s,t)/\partial s = 0$ and $\partial G(s,t)/\partial t = 0$ are necessary for a local minimum of G. They result in the linear system of equations

$$
\begin{aligned}
2e_{11}s + (e_{1m}+e_{m1})t &= -\sum_{k=2}^{m-1} e_{1k}\beta_k - \sum_{k=2}^{m-1} e_{k1}\beta_k \\
(e_{1m}+e_{m1})s + 2e_{mm}t &= -\sum_{k=2}^{m-1} e_{mk}\beta_k - \sum_{k=2}^{m-1} e_{km}\beta_k \ .
\end{aligned}
$$

Due to the symmetry of E, this means

$$
\left.
\begin{aligned}
e_{11}s + e_{1m}t &= -\sum_{k=2}^{m-1} e_{1k}\beta_k \\
e_{1m}s + e_{mm}t &= -\sum_{k=2}^{m-1} e_{mk}\beta_k \ , \\
\text{with } \beta_k &= 3\left(\frac{\Delta y_k}{\Delta x_k} - \frac{\Delta y_{k-1}}{\Delta x_{k-1}}\right) \ , k = 2,3,\dots,m-1 \ .
\end{aligned}
\right\} \tag{3.22}
$$

It remains to be determined when this linear system of equations has a unique solution.

Example 3.3. For the 6 points $(-1.0,-1.0)$, $(0.0,3.5)$, $(1.0,1.0)$, $(2.0,0.26)$, $(4.0,-1.0)$ and $(5.0,1.5)$, the respective uniquely determined interpolating cubic spline with minimal total curvature (natural spline) and minimal $AQ := \int_{x_1}^{x_m} [s'_{AQ}(x)]^2 \mathrm{d}x$ was constructed (see Part III).

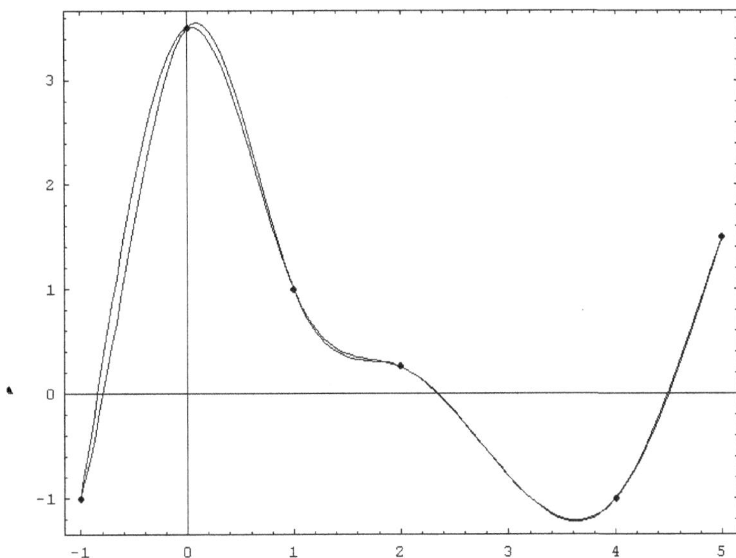

Fig. 3.6: The interpolating natural cubic spline $s_n(x)$ and the cubic spline $s_{AQ}(x)$ with minimal AQ.

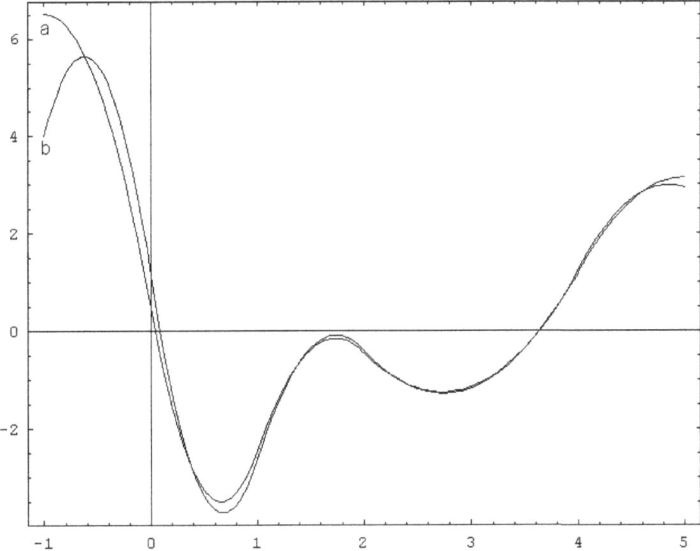

Fig. 3.7: The first derivatives of $s_n(x)$ (a) and $s_{AQ}(x)$ (b).

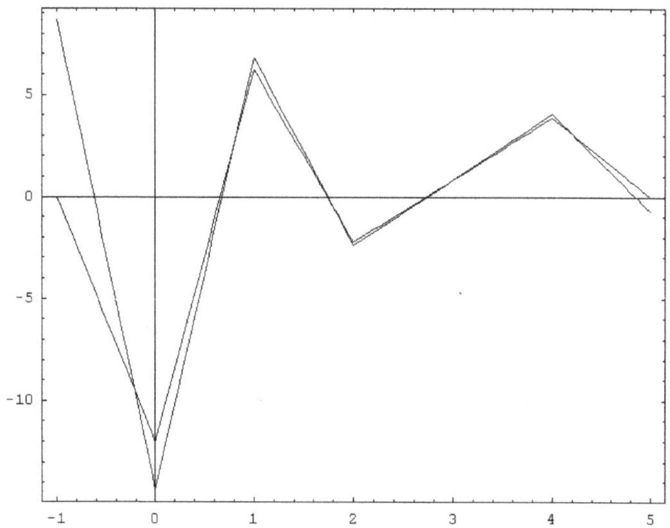

Fig. 3.8: The second derivatives of $s_{AQ}(x)$ and $s_n(x)$ $(s_n(-1) = s_n(5) = 0)$.

The presented concept allows for the construction of interpolating cubic splines which should fulfill the different objective functions. The condition is such that this optimum criterium is given over the vector B by a matrix H. For example, if H is replaced by

$$H_{GK} := \begin{pmatrix} 2\Delta x_1 & \Delta x_1 & 0 & 0 & \dots & 0 & 0 \\ \Delta x_1 & S_1 & \Delta x_2 & 0 & \dots & 0 & 0 \\ 0 & \Delta x_2 & S_2 & \Delta x_3 & \dots & 0 & 0 \\ \vdots & \vdots & \vdots & \vdots & \ddots & \vdots & \vdots \\ 0 & 0 & 0 & 0 & \dots & S_{m-2} & \Delta x_{m-1} \\ 0 & 0 & 0 & 0 & \dots & \Delta x_{m-1} & 2\Delta x_{m-1} \end{pmatrix},$$

then this matrix gives the total curvature $GK := \int_{x_1}^{x_m} \left[s_3''(x)\right]^2 \mathrm{d}x$, because with (3.3) one obtains

$$\begin{aligned}
\int_{x_1}^{x_m} \left[s_3''(x)\right]^2 \mathrm{d}x &= \sum_{i=1}^{m-1} \int_{x_i}^{x_{i+1}} \left(6A_i(x-x_i)+2B_i\right)^2 \mathrm{d}x \\
&= \sum_{i=1}^{m-1} \int_{x_i}^{x_{i+1}} \left[36A_i^2(x-x_i)^2 + 24A_iB_i(x-x_i) + 4B_i^2\right] \mathrm{d}x \\
&= \sum_{i=1}^{m-1} \left[12A_i^2\Delta x_i^3 + 12A_iB_i\Delta x_i^2 + 4B_i^2\Delta x_i\right] \\
&= \frac{4}{3}\sum_{i=1}^{m-1} \Delta x_i \left(B_i^2 + B_iB_{i+1} + B_{i+1}^2\right) \quad .
\end{aligned}$$

Then the system of equations (3.22) has a unique solution and $s = t = 0$. Since the equation $B_i = s'(x_i)/2$ is true for all i, $i = 1, \ldots, m$, the missing derivatives can be obtained from (3.6). The constructed spline is the interpolating natural cubic spline. To ensure that the solutions $s = 0$ and $t = 0$ result, the right sides of (3.22) must disappear.

It must explicitly be noted that in general, by means of this concept, only the respective minimum task within the set of all **interpolating cubic splines** is solved for the given knots $x_1 < x_2 < \ldots < x_m$.

Other solutions can result if the "global" minimum in the set $C^2([x_1, x_m])$ of all in $[x_1, x_m]$ twice continuously differentiable functions is sought. **Splines under tension** are observed as an example. As it is generally known, the interpolating natural spline minimizes $\int_{x_1}^{x_m} [f''(x)]^2 \mathrm{d}x$. This corresponds with physically minimal bending energy. If tension is added to the ends then the functional

$$\int_{x_1}^{x_m} [f''(x)]^2 \mathrm{d}x + \mu^2 \int_{x_1}^{x_m} [f'(x)]^2 \mathrm{d}x \tag{T}$$

results. μ^2 is the parameter representing tension. One also speaks of a spline under tension. A twice continuously differentiable function $f(x)$ is sought that minimizes (T) and fulfills the conditions of interpolation $f(x_i) = y_i$, $i = 1, 2, \ldots, m$. If this is minimized with the above described process, the solution is an interpolating cubic spline. The LAGRANGE principle from the calculus of variations can be applied to determine the global solution in $C^2([x_1, x_m])$. This approach results in the necessary condition for a local minimum

$$\left.\begin{aligned}&\int_{x_1}^{x_m} f''(x)h''(x)\mathrm{d}x + \mu^2 \int_{x_1}^{x_m} f'(x)h'(x)\mathrm{d}x = 0\\ &\text{for all } h(x) \in C^2([x_1,x_m]) \text{ with } h(x_1) = h(x_m) = 0\ .\end{aligned}\right\}$$

We start with

$$\int_{x_1}^{x_m} f''(x)h''(x)\mathrm{d}x = [f''(x_m)h'(x_m) - f''(x_1)h'(x_1)] - \int_{x_1}^{x_m} f'''(x)h'(x)\mathrm{d}x\ .$$

From partial integration one obtains

$$\begin{aligned}-\int_{x_1}^{x_m} f'''(x)h'(x)\mathrm{d}x &= -\sum_{i=1}^{m-1}\int_{x_i}^{x_{i+1}} f'''(x)h'(x)\mathrm{d}x\\ &= -\sum_{i=1}^{m-1} f'''(x)h(x)\Big|_{x_i}^{x_{i+1}} + \sum_{i=1}^{m-1}\int_{x_i}^{x_{i+1}} f^{(4)}(x)h(x)\mathrm{d}x\ .\end{aligned}$$

In addition,

$$\int_{x_1}^{x_m} f'(x)h'(x)\mathrm{d}x = [f'(x_m)h(x_m) - f'(x_1)h(x_1)] - \int_{x_1}^{x_m} f''(x)h(x)\mathrm{d}x\ .$$

With this, the necessary condition becomes

$$\begin{aligned}&[f''(x_m)h'(x_m) - f''(x_1)h'(x_1)] - \sum_{i=1}^{m-1} f'''(x)h(x)\Big|_{x_i}^{x_{i+1}}\\ &\quad + \sum_{i=1}^{m-1}\int_{x_i}^{x_{i+1}} f^{(4)}(x)h(x)\mathrm{d}x - \mu^2\int_{x_1}^{x_m} f''(x)h(x)\mathrm{d}x\\ &\quad + \mu^2[f'(x_m)h(x_m) - f'(x_1)h(x_1)] = 0.\end{aligned}$$

This equation can only be fulfilled for *all* twice continuously differentiable functions $h(x)$ when

a) $f^{(4)} = \mu^2 f''(x)$ is true over $[x_i, x_{i+1}]$ and

b) $f''(x_1) = f''(x_m) = 0$ is true.

From a) follows that the solution $f(x)$ on each of the intervals $[x_i, x_{i+1}]$ is $f(x) = A_i\mathrm{e}^{-\mu(x-x_i)} + B_i\mathrm{e}^{\mu(x-x_i)} + C_i(x-x_i) + D_i$. That is why the solution to (T) is called an exponential spline. When setting $M_i := f''(x_i) = \mu^2(A_i + B_i)$, $A_i + B_i + D_i = y_i$ follows due to $D_i = y_i - M_i/\mu^2$.

Due to the continuity of $f(x)$, $A_i e^{-\mu\Delta x_i} + B_i e^{\mu\Delta x_i} + C_i\Delta x_i + D_i = y_{i+1}$ is true. $M_{i+1}/\mu^2 + C_i\Delta x_i + D_i = y_{i+1}$ results due to $\mu^2 A_i e^{-\mu\Delta x_i} + \mu^2 B_i e^{\mu\Delta x_i} = M_{i+1}$ ($f''(x)$ is continuous) and then $C_i = \Delta y_i/\Delta x_i + (M_i - M_{i+1})/(\mu^2\Delta x_i)$. From $(A_i + B_i) = M_i/\mu^2$ and $A_i e^{-\mu\Delta x_i} + B_i e^{\mu\Delta x_i} = M_{i+1}/\mu^2$ one obtains

$$B_i = -\frac{1}{\mu^2(e^{2\mu\Delta x_i} - 1)} M_i + \frac{e^{\mu\Delta x_i}}{\mu^2(e^{2\mu\Delta x_i} - 1)} M_{i+1} \text{ and}$$

$$A_i = \frac{e^{2\mu\Delta x_i}}{\mu^2(e^{2\mu\Delta x_i} - 1)} M_i - \frac{e^{\mu\Delta x_i}}{\mu^2(e^{2\mu\Delta x_i} - 1)} M_{i+1} .$$

With this, also the spline under tension is uniquely and comprehensively defined because of the knowledge of the second derivatives in the knots. From the continuity of the first derivative in the x_i results

$$-\mu A_{i-1} e^{-\mu\Delta x_{i-1}} + \mu B_{i-1} e^{\mu\Delta x_{i-1}} + C_{i-1} = -\mu A_i + \mu B_i + C_i.$$

One obtains with the insertion of the expressions for the spline coefficients

$$\left[\frac{1}{\mu^2\Delta x_{i-1}} - 2\frac{e^{\mu\Delta x_{i-1}}}{\mu(e^{2\mu\Delta x_{i-1}} - 1)}\right] M_{i-1} + \left[-\frac{1}{\mu^2\Delta x_{i-1}} - \frac{1}{\mu^2\Delta x_i}\right.$$
$$\left. + \frac{1}{\mu(e^{2\mu\Delta x_{i-1}} - 1)} + \frac{1}{\mu(e^{2\mu\Delta x_i} - 1)} + \frac{e^{\mu\Delta x_{i-1}}}{\mu(e^{2\mu\Delta x_{i-1}} - 1)} + \frac{e^{\mu\Delta x_i}}{\mu(e^{2\mu\Delta x_i} - 1)}\right] M_i$$
$$+ \left[\frac{1}{\mu^2\Delta x_i} - 2\frac{e^{\mu\Delta x_i}}{\mu(e^{2\mu\Delta x_i} - 1)}\right] M_{i+1} = \frac{\Delta y_i}{\Delta x_i} - \frac{\Delta y_{i-1}}{\Delta x_{i-1}} \quad , i = 2, 3, \ldots, m-1.$$

With b) it is true that $M_1 = M_m = 0$. So we have obtained a linear system of equations with $(m-2)$ equations for the $(m-2)$ unknown values M_2 to M_{m-1}. This system has a unique solution as the matrix of coefficients is diagonal dominant. The notation is simplified with the definitions $\alpha_i := e^{\mu\Delta x_i}$ and $\beta_i := 1/(\mu(\alpha_i^2 - 1))$,

$$\left[\frac{1}{\mu^2\Delta x_{i-1}} - 2\beta_{i-1}\alpha_{i-1}\right] M_{i-1} + \left[-\frac{1}{\mu^2\Delta x_{i-1}} - \frac{1}{\mu^2\Delta x_i} + \beta_{i-1}\right.$$
$$\left. + \beta_{i-1}\alpha_{i-1}^2 + \beta_i + \beta_i\alpha_i^2\right] M_i + \left[\frac{1}{\mu^2\Delta x_i} - 2\beta_i\alpha_i\right] M_{i+1} = \frac{\Delta y_i}{\Delta x_i} - \frac{\Delta y_{i-1}}{\Delta x_{i-1}} .$$

The natural interpolating cubic spline results for $\mu = 0$. The polygon is obtained for μ going to infinity.

In the numerical handling of the above system of equations there is a certain additional effort required for stabilizing the exponential function for small or large μ. This however is not described in any further detail here.

Example 3.4. The (T)-minimizing interpolating cubic spline $s_k(x)$ was constructed with $\mu = 10$ for the 10 points $(9.30, 15.90)$, $(10.20, 20.70)$,

(11.40, 1.86), (11.80, 20.40), (13.10, 19.00), (13.40, 18.00), (14.00, 22.50), (14.20, 21.50), (15.30, 23.90) and (17.20, 26.20).
The exponential spline $s_e(x)$ was determined for the same parameter of tension $\mu = 10$ (see Figure 3.9 and Part III).

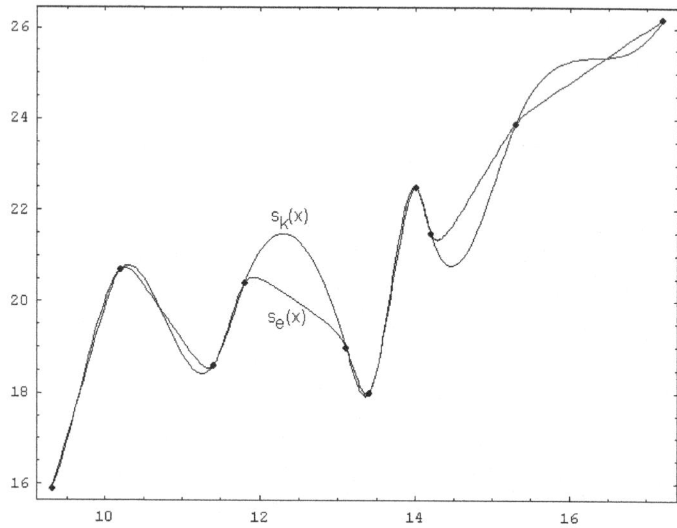

Fig. 3.9: The splines $s_k(x)$ and $s_e(x)$, see Example 3.4.

Chapter 4

Smoothing natural cubic splines and the choice of the smoothing parameter

Up to this point tasks of interpolation for spline functions were discussed. Given are the points (x_i, y_i), $i = 1, 2, \ldots, m$, that can be understood as points of a function $y = g(x)$ that is to be described, but is otherwise unknown.
This attempt is too restrictive, especially in the area of life sciences. For the determination of functional dependencies, both biological variations and measurement errors should be taken into account. The data are seen as inaccurate. Seeking interpolation of the measurements therefore does not make sense.
An "optimal" curve that describes the points closely is sought. This problem can be dealt with using stochastic models. Specific assumptions are made with regard to this. In particular, the data must come from random samples. Calculations of parameters for functions of models, when speaking of stochastic parameter estimations, occur with regard to the maximum likelihood principle. This principle, under certain conditions of linearity, leans towards the stochastic model for the calculation of the unknown model parameters by the method of least squares. A substantial statement regarding this is made in the GAUSS-MARKOV theorem.
On the other hand, a norm in $\mathbb{R}^m$ is defined by the sum of the squared errors. The method of least squares enables a purely algebraic-geometric approach to the problem of curve fitting.
Subsequently, curve fitting is handled with respect to cubic splines. Different problems of optimization are formulated. For these problems, a solution in the set of twice continuously differentiable functions is sought after. The individual objective functions and/or auxiliary conditions always include the total curvature. With this it is ensured that a natural cubic spline function $s(x)$ is obtained as the solution to the problem of optimiza-

tion. Each of these splines is, with regard to a fixed system of knots x_i, $i = 1, 2, \ldots, m$, uniquely and completely determined by the vector of the $D_i = s(x_i)$, $i = 1, 2, \ldots, m$. This makes it possible to carry over the original problem of optimization regarding the space $C^2([x_1, x_m])$ of the twice continuously differentiable functions over the interval $[x_1, x_m]$ into equivalent problems regarding $\mathbb{R}^m$. These are then solved in two steps. In the first step the unknown $D_i = s(x_i)$ are calculated. For these (x_i, D_i), the uniquely determined interpolating natural cubic spline must be constructed in a second step. This is why dealing with the task of interpolation could not be avoided.

Each of the subsequently considered problems of optimization has its certain advantages and disadvantages. To begin, a problem is observed for which the smoothing is steered by a smoothing parameter μ. The bigger this parameter is, the closer the solution lies to the regression line. The smoothing effect is also dependent on the data. Relief can be found through a problem where a maximum sum of the squared errors is given. Something that may be problematic here though is that the distance of the solution from the individual measured values could be very different. For this case, a problem is formulated where it can be specified how far away the function that is to be constructed can be from the measured values.

The uniquely determined interpolating natural cubic spline possesses minimal total curvature in the subset of functions from $C^2([x_1, x_m])$ that solve the same task of interpolation. In addition, the curve fitting should regard the sum of the squared errors. That is why the following optimization problem is formulated:

> Given are the points (x_i, y_i), $k_i > 0$, $i = 1, 2, \ldots, m$, and a fixed $0 \leq \mu$. A twice continuously differentiable function $f(x)$ in $C^2([x_1, x_m])$ on the interval $[x_1, x_m]$ is sought which minimizes the objective function (OP3)
>
> $$Z(f) := \mu \int_{x_1}^{x_m} [f''(x)]^2 \, dx + \sum_{i=1}^{m} \left(\frac{f(x_i) - y_i}{k_i} \right)^2 .$$

The function $f(x)$ should be constructed such that the weighted sum of total curvature and the sum of the squared errors becomes a minimum. $\mu \in \mathbb{R}^+$ is an arbitrary but fixed smoothing parameter. It determines the ratio of total curvature and weighted sum of the squared errors to be entered into the objective function $Z(f)$. The additive combination of these opposing components of the objective function should effect the expected "smoothing".

The original optimization problem (OP3) is now only observed with regard to the set of all natural cubic splines $s(x)$ for the given and fixed knots x_i, $i = 1, 2, \dots, m$: Minimize

$$\mu \int_{x_1}^{x_m} [s''(x)]^2 \, dx + \sum_{i=1}^{m} \left(\frac{s(x_i) - y_i}{k_i} \right)^2 . \tag{4.1}$$

According to the following lemma, it is sufficient to observe the task (4.1) in place of the problem (OP3). Both optimization tasks are equivalent such that

(1) every solution to (OP3) is a natural cubic spline function,
(2) every solution to (4.1) is a solution to (OP3) and
(3) (OP3) does not possess a solution if and only if (4.1) does not have a solution.

Lemma 4.1. *In the preceding sense the problems* (OP3) *and* (4.1) *are equivalent.*

Proof. Suppose (OP3) can be solved. There then exists a function $g(x) \in C^2([x_1, x_m])$ with $Z(g) \le Z(f)$ for all $f(x)$ in $C^2([x_1, x_m])$. If $g(x)$ is a natural cubic spline function, then nothing needs to be shown. In the other case one constructs the uniquely determined interpolating cubic spline function $s_g(x)$ for the points $(x_i, g(x_i))$, $i = 1, 2, \dots, m$. For these, $s_g(x_i) = g(x_i)$, $i = 1, 2, \dots, m$, is true. The one with the k_i weighted sum of the squared errors of $s_g(x)$ and $g(x)$ with regard to y_i is alike. However, with Theorem 3.1, $\int_{x_1}^{x_m} [g''(x)]^2 \mathrm{d}x > \int_{x_1}^{x_m} [s_g''(x)]^2 \mathrm{d}x$. Consequently, $g(x)$ cannot be the solution to (OP3). Rather, $s_g(x)$ solves (OP3) and (4.1).

On the other hand, if (4.1) can be solved through $s(x)$, then with the same thinking, there cannot exist another function $f(x)$ from $C^2([x_1, x_m])$ with $Z(f) < Z(s)$. That is, $s(x)$ not only solves (4.1) but also (OP3). It remains to be shown that it is not possible for (4.1) to have a solution but (OP3) not. Let $s(x)$ be a solution to (4.1) and $Z(s) = c > 0$ ($c = 0$ means that $s(x)$ also solves (OP3) because Z(*) can't become less than zero). If (OP3) could not be solved, then there would exist a sequence $f_n(x) \in C^2([x_1, x_m])$ and an

$c_1 \geq 0$ with $\lim_{n\to\infty} Z(f_n) = c_1 < c$, but no $f(x)$ where $Z(f) = c_1$. There is a n_o with $Z(f_{n_o}) < c$. If the uniquely determined interpolating natural cubic spline function $s_{n_o}(x)$ is reconstructed for the points $(x_i, f_{n_o}(x_i))$, $i = 1, 2, \ldots, m$, then $Z(s_{n_o}) < c$ is true for this function. However, this contradicts the assumption that $s(x)$ solves problem (4.1). □

For the interpolating natural cubic splines it was shown:

a) $s(x) = A_i(x-x_i)^3 + B_i(x-x_i)^2 + C_i(x-x_i) + D_i$ for x from $[x_i, x_{i+1}]$,

b) $s(x)$ is uniquely determined by the $B_i = s''(x_i)/2$ and

c) the B_i fulfill the linear system of equations

$$\begin{aligned}\Delta x_{i-1}B_{i-1} + 2(\Delta x_{i-1} + \Delta x_i)B_i + \Delta x_i B_{i+1} \\ = 3\left(\frac{D_{i+1}-D_i}{\Delta x_i} - \frac{D_i - D_{i-1}}{\Delta x_{i-1}}\right)\end{aligned}, \qquad (4.2)$$

$i = 2, 3, \ldots, m-1$, and $B_1 = B_m = 0$.

To solve the optimization problem (4.1), besides B_i, $i = 2, 3, \ldots, m-1$, the unknown values $D_i = s(x_i)$, $i = 1, 2, \ldots, m$, need to be determined.
It is observed that $s''(x)$ is a straight line of the form $s''(x) = 6A_i(x-x_i) + 2B_i$ between the knots x_i and x_{i+1}. Then from (3.2) and (3.3) follows

$$\begin{aligned}\int_{x_1}^{x_m}[s''(x)]^2\mathrm{d}x &= \sum_{i=1}^{m-1}\int_{x_i}^{x_{i+1}}[s''(x)]^2\mathrm{d}x = \sum_{i=1}^{m-1}\int_{x_i}^{x_{i+1}}\left(6A_i(x-x_i)+2B_i\right)^2\mathrm{d}x \\ &= \frac{4}{3}\sum_{i=1}^{m-1}\Delta x_i\left(B_i^2 + B_iB_{i+1} + B_{i+1}^2\right) \\ &= \frac{2}{3}\sum_{i=1}^{m-1}\Delta x_i\left(B_i^2 + (B_i + B_{i+1})^2 + B_{i+1}^2\right) \quad . \end{aligned} \qquad (4.3)$$

This way, (4.1) can be reformulated by (4.3). Due to Lemma 4.1, the following problem results:

Minimize $\frac{2}{3}\mu\sum_{i=1}^{m-1}\Delta x_i(B_i^2 + (B_i+B_{i+1})^2 + B_{i+1}^2) + \sum_{i=1}^{m}\left(\frac{D_i - y_i}{k_i}\right)^2$.
With this, the minimum should be realized through the appropriate choice of the B_i, $i = 2, 3, \ldots, m-1$, $(B_1 = B_m = 0)$ and D_i, $i = 1, \ldots, m$.

Since $s(x)$ is uniquely determined by (4.2), these equations can be incorporated into the objective function as constraints. Finally, by inserting the LAGRANGE multipliers $V_2, V_3, \ldots, V_{m-1}$, the problem reads now:
Minimize

$$\begin{aligned}
&L\left(B_2, \ldots, B_{m-1}, D_1, \ldots D_m, V_2, \ldots, V_{m-1}\right) \\
&\quad := \frac{2}{3}\mu \sum_{i=1}^{m-1} \Delta x_i \left[B_i^2 + (B_i + B_{i+1})^2 + B_{i+1}^2\right] + \sum_{i=1}^{m} \left(\frac{D_i - y_i}{k_i}\right)^2 \\
&\qquad + \sum_{i=2}^{m-1} V_i \left[\Delta x_{i-1} B_{i-1} + 2(\Delta x_{i-1} + \Delta x_i) B_i + \Delta x_i B_{i+1}\right. \\
&\qquad \left. - 3\left(\frac{(D_{i+1} - D_i)}{\Delta x_i} - \frac{(D_i - D_{i-1})}{\Delta x_{i-1}}\right)\right] .
\end{aligned}$$

Necessary conditions for a local minimum of L are the relationships

$$\begin{aligned}
\frac{\partial L}{\partial B_i} &= \frac{2}{3}\mu \Delta x_{i-1}(4B_i + 2B_{i-1}) + \frac{2}{3}\mu \Delta x_i (4B_i + 2B_{i+1}) \\
&\quad + V_{i-1}\Delta x_{i-1} + 2V_i(\Delta x_{i-1} + \Delta x_i) + V_{i+1}\Delta x_i = 0
\end{aligned}$$

or

$$\begin{aligned}
\frac{\partial L}{\partial B_i} &= \frac{4}{3}\mu[\Delta x_{i-1}B_{i-1} + 2(\Delta x_{i-1} + \Delta x_i)B_i + \Delta x_i B_{i+1}] + V_{i-1}\Delta x_{i-1} \\
&\quad + 2V_i(\Delta x_{i-1} + \Delta x_i) + V_{i+1}\Delta x_i = 0 \text{ for } i = 2, 3, \ldots, m-1 \ , \qquad (4.4) \\
\frac{\partial L}{\partial D_i} &= \frac{2}{k_i^2}(D_i - y_i) - 3\frac{V_{i-1}}{\Delta x_{i-1}} + 3V_i\left[\frac{1}{\Delta x_{i-1}} + \frac{1}{\Delta x_i}\right] - 3\frac{V_{i+1}}{\Delta x_i} = 0 \ , \qquad (4.5)
\end{aligned}$$

$i = 1, 2, \ldots, m$, and for $i = 2, 3, \ldots, m-1$

$$\begin{aligned}
\frac{\partial L}{\partial V_i} &= \Delta x_{i-1}B_{i-1} + 2(\Delta x_{i-1} + \Delta x_i)B_i + \Delta x_i B_{i+1} \\
&\quad - 3\left[\frac{(D_{i+1} - D_i)}{\Delta x_i} - \frac{(D_i - D_{i-1})}{\Delta x_{i-1}}\right] = 0 \ . \qquad (4.6)
\end{aligned}$$

Note that $B_1 = B_m = 0$, the constants $V_0 = V_1 = V_m = V_{m+1} = 0$ and $\Delta x_0 = \Delta x_m \neq 0$ were inserted for the purpose of easier notation in (4.5). The relationships (4.4) and (4.5) together with (4.6) form a linear system of equations for the determination of the B_i, V_i and D_i. An attempt could be made to solve this equation system directly. This is not recommended though because it can be separated into two different systems of equations for the D_i and B_i. To simplify the notation the vectors $B := (B_2, B_3, \dots, B_{m-1})^T$, $V := (V_2, V_3, \dots, V_{m-1})^T$, $D := (D_1, D_2, \dots, D_m)^T$, $Y := (y_1, y_2, \dots, y_m)^T$ and the matrices

$$K := \begin{pmatrix} k_1 & 0 & 0 & \dots & 0 \\ 0 & k_2 & 0 & \dots & 0 \\ \vdots & \vdots & \vdots & \ddots & \vdots \\ 0 & 0 & 0 & \dots & k_m \end{pmatrix},$$

$$Z := \frac{1}{3} \begin{pmatrix} S_1 & \Delta x_2 & 0 & \dots & 0 \\ \Delta x_2 & S_2 & \Delta x_3 & \dots & 0 \\ 0 & \Delta x_3 & S_3 & \dots & 0 \\ \vdots & \vdots & \vdots & \ddots & \vdots \\ 0 & \dots & \Delta x_{m-3} & S_{m-3} & \Delta x_{m-2} \\ 0 & \dots & 0 & \Delta x_{m-2} & S_{m-2} \end{pmatrix}$$

$(S_i := 2(\Delta x_i + \Delta x_{i+1}))$ and

$$Q := \begin{pmatrix} \frac{1}{\Delta x_1} & -q_1 & \frac{1}{\Delta x_2} & 0 & \dots & 0 \\ 0 & \frac{1}{\Delta x_2} & -q_2 & \frac{1}{\Delta x_3} & \dots & 0 \\ \vdots & \vdots & \vdots & \vdots & \ddots & \vdots \\ 0 & 0 & \dots & \frac{1}{\Delta x_{m-2}} & -q_{m-2} & \frac{1}{\Delta x_{m-1}} \end{pmatrix}$$

$(q_i := \frac{1}{\Delta x_i} + \frac{1}{\Delta x_{i+1}})$ are introduced.
The system of equations (4.6) or (4.2) then has the form $Z \circ B = Q \circ D$ from which B results in:

$$B = Z^{-1} \circ Q \circ D \ . \tag{4.7}$$

The sytem (4.4) becomes $\frac{4}{3}\mu 3Z \circ B = -3Z \circ V$. Then

$$V = -\frac{4}{3}\mu B \tag{4.8}$$

is true. From (4.5) results the relationship

$$2K^{-2} \circ [D - Y] = 3Q^T \circ V. \tag{4.9}$$

Equation (4.8) followed by (4.7) are inserted in (4.9). So $2K^{-2} \circ [D - Y] = -3Q^T(\frac{4}{3})\mu Z^{-1} \circ Q \circ D$ is obtained. From this

$$[I + 2\mu K^2 \circ Q^T \circ Z^{-1} \circ Q] \circ D = Y \tag{4.10}$$

is developed.

Lemma 4.2. *Let Z and Q be the matrices defined above and $D \in \mathbb{R}^m$. Further to this, let $s(x)$ be the interpolating natural spline function where $s(x_i) = D_i$, $i = 1, 2, \ldots, m$. Then*

$$\int_{x_1}^{x_m} [s''(x)]^2 dx = 2B^T \circ Z \circ B = 2D^T \circ Q^T \circ Z^{-1} \circ Q \circ D$$

is true. Z is positive definite and $E := Q^T \circ Z^{-1} \circ Q$ is symmetrical and positive semi-definite.

Proof. E is a symmetrical matrix because Z and therefore Z^{-1} are symmetric and $(Q^T \circ Z^{-1} \circ Q)^T = Q^T \circ Z^{-1} \circ Q$ is true. A short calculation (observe $B_1 = B_m = 0$) results in $B^T \circ Z \circ B = \frac{2}{3}\sum_{i=1}^{m-1} \Delta x_i (B_i^2 + B_i B_{i+1} + B_{i+1}^2) = \frac{1}{2}\int_{x_1}^{x_m} [s''(x)]^2 \mathrm{d}x \geq 0$. The statement of the lemma then results from (4.7). □

Given $0 \leq \mu$, equation (4.10) is a linear system of equations for the determination of the unknown values $D_i = s(x_i)$, $i = 1, 2, \ldots, m$. It has a unique solution because the matrix of coefficients $I + 2\mu K^2 \circ Q^T \circ Z^{-1} \circ Q$ for $0 \leq \mu$ is positive definite and therefore invertible (compare Lemma 4.2). With the help of (4.2), B_i, $i = 2, 3, \ldots, m-1$, can be calculated from the D_i so that the spline function $s(x)$ that is searched for can be constructed by means of (3.2) to (3.4).
The considerations that have been made so far for the construction of the uniquely determined cubic spline function that solves (4.1) can now be summarized.

Theorem 4.1. *The problem* (OP3) *has a unique solution for every set of points (x_i, y_i), $i = 1, 2, \ldots, m > 2$, and for all $0 \leq \mu$. The solution is a natural cubic spline function that can be calculated via* (4.10) *and* (4.2).

Proof. The task (4.1) and the problem (OP3) are equivalent according to Lemma 4.1. The construction of the unique solution to (4.1) for $0 \leq \mu$ was developed above. (OP3) cannot be uniquely solved for $\mu = 0$. The interpolating natural cubic spline can be chosen as a solution. □

The smoothing matrix H, $H := [I + 2\mu K^2 \circ Q^T \circ Z^{-1} \circ Q]^{-1}$, is exclusively determined by the knots, weights k_i and the smoothing parameter μ. If these values are fixed (and so $D = H \circ Y$), the optimization problem

$$\mu \int_{x_1}^{x_m} [f''(x)]^2 \,\mathrm{d}x + \sum_{i=1}^{m} \left(\frac{f(x_i) - y_i}{k_i} \right)^2 = \min!$$

is uniquely solvable for all vectors Y. H then does not need to be calculated again. This is also specially true for $y_i = ax_i + b$, $i = 1, 2, \ldots, m$. From this it follows that $H \circ Y = aH \circ X + bH = aX + b$. When all y_i lie on a straight line, the total curvature is zero and the sum of the squared errors also disappears. This relationship can only be true for arbitrary real values a and b when $H \circ X = X$ and $H \circ \mathbf{1} = \mathbf{1}$. Here, $\mathbf{1} \in \mathbb{R}^m$ is the vector that contains only ones. In other words: H is a matrix for which all of the sums of the rows equal one. The same is true for H^{-1}. These considerations result in

$$\begin{aligned}\sum_{i=1}^{m}(y_i - s(x_i)) &= \sum_{i=1}^{m} y_i - \sum_{i=1}^{m}\sum_{k=1}^{m} h_{ik} y_k = \sum_{k=1}^{m} y_k \left(1 - \sum_{i=1}^{m} h_{ik}\right) \\ &= \sum_{k=1}^{m} y_k \left(\sum_{i=1}^{m} h_{ki} - \sum_{i=1}^{m} h_{ik}\right).\end{aligned}$$

The sum of the residuals disappears when H is symmetrical (for instance in the case $K = I$). In addition, the following is true:

Lemma 4.3. *The objective function $F(\mu) := \mu GK(\mu) + SSE(\mu)$ is strictly monotone increasing for $\mu \geq 0$. $GK(\mu)$ represents the total curvature and $SSE(\mu)$ the weighted sum of the squared errors of the solution to* (4.1) *as a function of μ.*

Proof. Let $0 \leq \mu_1 < \mu_2$. From Theorem 4.1 it follows that there exist uniquely determined natural cubic splines $s_{\mu_1}(x)$ and $s_{\mu_2}(x)$, respectively, that minimize the objective function $F(D_{\mu_j})$. The spline functions are represented by the vectors $D_{\mu_j} = (s_{\mu_j}(x_1), \ldots, s_{\mu_j}(x_m))$. Consequently, $F(D_{\mu_j}) = F(\mu_j) = \mu_j GK(\mu_j) + SSE(\mu_j)$, $j = 1, 2$, is true.
To prove this assertion a uniquely determined interpolating natural cubic spline $s_{int}(x)$ is constructed for D_{μ_2}. It possesses the same weighted sum of the squared errors SSE_{int}, with regard to the y_i, as $s_{\mu_2}(x)$; $SSE(\mu_2) = SSE_{int}$. The total curvature GK_{int} is less than or at most equal to $GK(\mu_2)$ due to the minimal total curvature construction.

From this $\mu_2 GK(\mu_2) + SSE(\mu_2) > \mu_1 GK_{int} + SSE_{int} \geq \mu_1 GK(\mu_1) + SSE(\mu_1)$ follows, as $s_{\mu_1}(x)$ was constructed such that $\mu_1 GK(\mu_1)+SSE(\mu_1)$ is minimal. □

Note 4.1. The unique solution to (OP3) one can get otherwise.
From $2K^{-2} \circ [D - Y] = 3Q^T \circ V$ and (4.8) follows $D = Y - 2\mu K^2 \circ Q^T \circ B$. Then (4.7) is seen as $B = Z^{-1} \circ Q \circ [Y - 2\mu K^2 \circ Q^T \circ B]$ or equivalently $[Z + 2\mu Q \circ K^2 \circ Q^T] \circ B = Q \circ Y$.
This linear equations system has a unique solution because the matrix of coefficients is positive definite for $\mu \geq 0$. This equation has the advantage that one does not need inverted matrices. In addition, $[Z + 2\mu Q \circ K^2 \circ Q^T]$ is symmetric and has band form with five non-zero diagonals.

Example 4.1.
To the points used in Example 3.2, the uniquely determined natural cubic spline $s(x)$ was determined for which

$$19 \int_{x_1}^{x_m} [s''(x)]^2 \mathrm{d}x + \sum_{i=1}^{m} (y_i - s(x_i))^2$$

becomes minimal (see Part III). So, $k_i = 1$ for all i from 1 to m. The following is true for the solution:

$$\begin{aligned} &\text{Sum of squared errors } SSE &&= 22.75328 \\ &\text{Total curvature } GK &&= 2.63391 \\ &\text{Objective function } \mu \cdot GK + SSE &&= 72.79747. \end{aligned}$$

Two crucial disadvantages are tied to the solution to the problem (OP3). For one, the selection of the parameter μ is arbitrary. That is why there is no reference to errors of the measured values y_i, $i = 1, 2, \ldots, m$. On the other hand, the effect of μ on the smoothing is not directly foreseeable. This is due to the unknown value of the total curvature of $s(x)$ over the interval $[x_1, x_m]$. If the interpolating cubic spline function possesses a large total curvature for the given points (x_i, y_i), $i = 1, 2, \ldots, m$, then visible smoothing already results for small μ. For the reverse case, visible smoothing is only attained for large μ. In general it can be said that the interpolating spline function can be obtained for $\mu = 0$ and the regression line for $\mu \to \infty$.
The described disadvantages can be partially eliminated by the following problem formulation (Reinsch 1967).

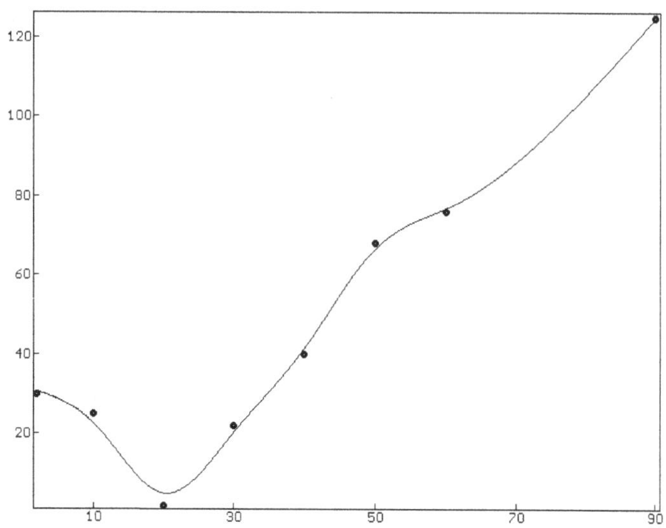

Fig. 4.1: Smoothing spline for Example 4.1, $\mu = 19$ and $k_i = 1$.

> A function $f(x)$ is sought for which $\int_{x_1}^{x_m} [f''(x)]^2 \, \mathrm{d}x$ becomes minimal in the set $C^2([x_1, x_m])$ of all twice continuously differentiable functions $g(x)$ with the additional property
> $$\sum_{i=1}^{m} \left(\frac{g(x_i) - y_i}{k_i} \right)^2 \leq S \; . \tag{R}$$

The real numbers k_i, $i = 1, 2, \ldots, m$, and $S \geq 0$ are arbitrary but fixed. Subsequently, a constructive procedure is developed that calculates the uniquely determined solution to the problem (R) in the set of natural cubic spline functions. This set of all natural cubic spline functions with knots $x_1 < \ldots < x_m$ forms a linear subspace of $C^2([x_1, x_m])$.
The problem (R) can be reformulated for this subspace with the help of Lemma 4.2 into:

$$\left.\begin{array}{l} \text{A vector } D \text{ of } \mathbb{R}^m \text{ is searched for that fulfills the constraint} \\ ||K^{-1} \circ (D - Y)||^2 \leq S \text{ and minimizes } 2D^T \circ E \circ D. \end{array}\right\} \quad \text{(RS)}$$

Here K is the earlier described $(m \times m)$ diagonal matrix in which the k_i, $i = 1, 2, \ldots, m$, can be found on the main diagonal. The norm is the

EUKLIDEAN norm of the $\mathbb{R}^m$. According to Lemma 4.2, E is positive semi-definite. Accordingly, the objective function $2D^T \circ E \circ D$ of (RS) is convex on $\mathbb{R}^m$ and takes on a minimum on the convex set of admissible vectors. If an auxiliary variable h is used, the constraint in (R) can be transferred into the equation $\sum_{i=1}^{m} \left((f(x_i) - y_i)/k_i\right)^2 + h^2 - S = 0$ and can be built into the objective function with the LAGRANGE multiplier p. With (4.3) the following task results:

$$\frac{2}{3}\sum_{i=1}^{m-1} \Delta x_i (B_i^2 + (B_i + B_{i+1})^2 + B_{i+1}^2) + p\left[\sum_{i=1}^{m}\left(\frac{D_i - y_i}{k_i}\right)^2 + h^2 - S\right] = \min!$$

Minimization results due to the appropriate selection of the real numbers B_i, $i = 2, 3, \ldots, m-1$ ($B_1 = B_m = 0$), D_i, $i = 1, 2, \ldots, m$, p and h.
If one proceeds in a similar way as in problem (4.1), one arrives at

$$\begin{aligned} &L^*(B_2, \ldots, B_{m-1}, D_1, \ldots, D_m, V_2, \ldots, V_{m-1}, p, h) \\ &:= \frac{2}{3}\sum_{i=1}^{m-1} \Delta x_i \left[B_i^2 + (B_i + B_{i+1})^2 + B_{i+1}^2\right] + p\left[\sum_{i=1}^{m}\left(\frac{D_i - y_i}{k_i}\right)^2 + h^2 - S\right] \\ &\quad + \sum_{i=2}^{m-1} V_i \left[\Delta x_{i-1} B_{i-1} + 2(\Delta x_{i-1} + \Delta x_i) B_i + \Delta x_i B_{i+1}\right. \\ &\quad \left. - 3\left(\frac{(D_{i+1} - D_i)}{\Delta x_i} - \frac{(D_i - D_{i-1})}{\Delta x_{i-1}}\right)\right] = \min! \end{aligned}$$

and the respective necessary conditions $\partial L^*/\partial B_i = 0$, $\partial L^*/\partial V_i = 0$, $i = 2, 3, \ldots, m-1$, and $\partial L^*/\partial D_i = 0$, $i = 1, 2, \ldots, m$.
The relationships $B = Z^{-1} \circ Q \circ D$, $V = -\frac{4}{3}B$ and

$$2pK^{-2} \circ [D - Y] = 3Q^T \circ V \tag{4.11}$$

and the linear system of equations

$$\left[I + \frac{2}{p}K^2 \circ Q^T \circ Z^{-1} \circ Q\right] \circ D = Y \tag{4.12}$$

result. If the LAGRANGE multiplier p were known, the vector D_i, $i = 1, 2, \ldots, m$, for $p > 0$ in (4.12) and with this the interpolating natural spline function could be constructed. This is because for $p > 0$, the matrix of coefficients for (4.12) is positive definite (Lemma 4.2).
$2pK^{-2} \circ [D - Y] = 3(-4/3)Q^T \circ B$ follows from (4.11), so

$$D = Y - \frac{2}{p}K^2 \circ Q^T \circ B\ . \tag{4.13}$$

This is substituted into $B = Z^{-1} \circ Q \circ D$ such that $B = Z^{-1} \circ Q \circ [Y - (2/p)K^2 \circ Q^T \circ B]$ or

$$\left[\frac{p}{2}Z + Q \circ K^2 \circ Q^T\right] \circ B = \frac{p}{2}Q \circ Y \ . \tag{4.14}$$

Because $rank(Q) = m-2$, the matrix $Q \circ Q^T$ is positive definite. Since Z is also positive definite, this property is also true for the matrix of coefficients $[(p/2)Z + Q \circ K^2 \circ Q^T]$ in (4.14). This is why $[(p/2)Z + Q \circ K^2 \circ Q^T]$ is invertible.

With this, when $p \geq 0$ is known, a second possibility for the construction of a solution to (R) in the set of the natural cubic spline functions for the given system of knots has been found. First B in (4.14) is determined, followed by the unknown values $D_i = s(x_i)$, $i = 1, 2, \ldots, m$, in (4.13). The relationships (4.13) and (4.14) lead to $D = Y - K^2 \circ Q^T \circ [(p/2)Z + Q \circ K^2 \circ Q^T]^{-1} \circ Q \circ Y$. This equation has the advantage that it is also defined for $p = 0$.

Because of (4.12), for $p > 0$ the identity $[I + (2/p)K^2 \circ Q^T \circ Z^{-1} \circ Q]^{-1} = I - K^2 \circ Q^T \circ [(p/2)Z + Q \circ K^2 \circ Q^T]^{-1} \circ Q$ is true. The goal is therefore the calculation of p.

With the two equations (4.13) and (4.14) one obtains

$$\begin{aligned} K^{-1} \circ [D - Y] &= K^{-1} \circ [Y - (2/p)K^2 \circ Q^T \circ B - Y] \\ &= -K \circ Q^T \circ [(p/2)Z + Q \circ K^2 \circ Q^T]^{-1} \circ Q \circ Y \ . \end{aligned}$$

The following is specified:

$$\begin{aligned} F(p)^2 &:= ||K^{-1} \circ [D - Y]||^2 \\ &:= ||K \circ Q^T \circ [(p/2)Z + Q \circ K^2 \circ Q^T]^{-1} \circ Q \circ Y||^2 \\ F(p)^2 &:= Y^T \circ Q^T \circ [(p/2)Z + Q \circ K^2 \circ Q^T]^{-1} \circ Q \circ K^2 \circ Q^T \\ &\quad \circ [(p/2)Z + Q \circ K^2 \circ Q^T]^{-1} \circ Q \circ Y = S - h^2 \ . \end{aligned}$$

Two possible situations arise from the necessary condition $\partial L^*/\partial h = 2ph = 0$:

Case 1 $p = 0$. $F(0)^2 \leq S$ results, i.e. the regression line with regard to the weighted sum of the squared errors fulfills the constraint $S_0 := \sum_{i=1}^m \left(\frac{y_i - g(x_i)}{k_i}\right)^2 \leq S$, as (4.14) turns into $Q \circ K^2 \circ Q^T \circ B = 0$ so that $B_i = 0$ for all $i = 1, 2, \ldots, m$.

From $D = Y - K^2 \circ Q^T \circ [(p/2)Z + Q \circ K^2 \circ Q^T]^{-1} \circ Q \circ Y$ the following is obtained for $p = 0$: $D = Y - K^2 \circ Q^T \circ [Q \circ K^2 \circ Q^T]^{-1} \circ Q \circ Y$.

Case 2 $p \neq 0$. Because of the necessary condition $\partial L^*/\partial h = 2hp = 0$, it results that $h = 0$. That is why p could be determined from the equation $F(p)^2 = S$.

To verify which of the solutions to p defines the solution spline, a result of JEROMIN (1972) is used:

Lemma 4.4. *Let* $0 \neq Y \in \mathbb{R}^m$ *and let* M *and* N *be positive definite* $(m-2) \times (m-2)$ *-matrices. Then* $G(p) := Y^T \circ Q^T \circ [pN+M]^{-1} \circ M \circ [pN+M]^{-1} \circ Q \circ Y$ *has the following characteristics:*

(1) G *is continuous over* $[0,\infty)$, $G(0) = Y^T \circ Q^T \circ M^{-1} \circ Q \circ Y$, $\lim\limits_{p\to\infty} G(p) = 0$;

(2) G' *is continuous over* $[0,\infty)$, $G'(0) = -2Y^T \circ Q^T \circ M^{-1} \circ N \circ M^{-1} \circ Q \circ Y < 0$, $\lim\limits_{p\to\infty} G'(p) = 0$;

(3) G'' *is continuous over* $[0,\infty)$, $G''(0) = 6Y^T \circ Q^T \circ M^{-1} \circ N \circ M^{-1} \circ N \circ M^{-1} \circ Q \circ Y > 0$, $\lim\limits_{p\to\infty} G''(p) = 0$.

If $G = F(p)^2$, $M = Q \circ K^2 \circ Q^T$ and $N = Z$, a uniquely determined positive solution p to the equation $F(p)^2 = S$ exists for the case $0 \leq S \leq S_0$. According to a result of KUHN and TUCKER (1951) it is also necessary and sufficient for the solution to (RS) that a $p > 0$ exists so that $p\left[\sum\limits_{i=1}^{m}\left(\frac{f(x_i)-y_i}{k_i}\right)^2 - S\right] = 0$ is true.

In summary it can be said, (R) possesses a uniquely determined solution for the given system of knots in the set of all natural cubic splines. The equation $F(p)^2 = S$ or $F(p) - \sqrt{S} = 0$ must be solved for the positive p for further calculation of the solution. With regard to this,

$$U := (2/p)B = [(p/2)Z + Q \circ K^2 \circ Q^T]^{-1} \circ Q \circ Y \ .$$

Then

$$F(p)^2 = ||K \circ Q^T \circ U||^2 = (K \circ Q^T \circ U)^T \circ K \circ Q^T \circ U = U^T \circ Q \circ K^2 \circ Q^T \circ U$$

is true.

On the other hand

$$
\begin{aligned}
F(p)F'(p) &= \frac{1}{2}(F^2(p))' = \frac{1}{2}\frac{\mathrm{d}(U^T \circ Q \circ K^2 \circ Q^T \circ U)}{\mathrm{d}p} \\
&= \frac{1}{2}\left(\frac{\mathrm{d}U^T}{\mathrm{d}p} \circ Q \circ K^2 \circ Q^T \circ U + U^T \circ Q \circ K^2 \circ Q^T \circ \frac{\mathrm{d}U}{\mathrm{d}p}\right) \\
&= U^T \circ Q \circ K^2 \circ Q^T \circ \frac{\mathrm{d}U}{\mathrm{d}p} \\
&= \frac{p}{2}U^T \circ Z \circ \left(\frac{p}{2}Z + Q \circ K^2 \circ Q^T\right)^{-1} \circ Z \circ U - U^T \circ Z \circ U.
\end{aligned}
$$

It is possible to iteratively determine the unknown value p with these two relationships from

$$
p_{k+1} := p_k - \frac{F(p_k) - \sqrt{S}}{F(p_k)'} = p_k - \frac{F(p_k)^2 - \sqrt{F(p_k)^2 S}}{F(p_k)F'(p_k)}
$$

by application of the NEWTON method, and therefore one also calculates the unique solution to (RS).

Reinsch algorithm

Begin with $p_0 = 0$.

(1) Calculation of $e := F(p_k)^2$. First U is determined from the linear system of equations $[(p_k/2)Z + Q \circ K^2 \circ Q^T] \circ U = Q \circ Y$. For this one starts with the CHOLESKY decomposition $R^T \circ R$ of the above matrix of coefficients. The vector $V := K \circ Q^T \circ U$ is calculated with U so that $e := V^T \circ V = F(p_k)^2$ results. If $e > S$, then proceed to (2), otherwise the iterative process ends here, and $D = Y - K \circ V$ and $B = (p/2)U$ are true.

(2) Determination of $F(p_k)F'(p_k)$.
Since $[(p_k/2)Z + Q \circ K^2 \circ Q^T]^{-1} = R^{-1} \circ (R^T)^{-1}$, the linear system of equations $R^T \circ W = Z \circ U$ is solved first. Then $W = (R^T)^{-1} \circ Z \circ U$ is true and due to $Z = Z^T$ it follows that $F(p_k)F'(p_k) = p_k W^T \circ W - U^T \circ Z \circ U$. With this, p_{k+1} can be calculated as

$$
p_{k+1} := p_k - \frac{e - \sqrt{eS}}{p_k W^T \circ W - U^T \circ Z \circ U}
$$

Continue with (1).

The preceding considerations with regard to problem (R) resulted in a constructive process for the calculation of the uniquely determined solution. This solution refers to the subset of the functions from $C^2([x_1, x_m])$ which fulfill the constraint mentioned in (R). The gained knowledge is summarized in a theorem.

Theorem 4.2. *Let $g(x)$ be the uniquely determined regression line with regard to the weighted sum of the squared errors and $S_0 := \sum_{i=1}^{m} \left(\frac{y_i - g(x_i)}{k_i}\right)^2$. Then the problem (R) can be uniquely solved for all $0 \leq S \leq S_0$ in the set of the twice continuously differentiable functions. The solution is a natural cubic spline function $s(x)$ for which $\sum_{i=1}^{m} \left(\frac{y_i - s(x_i)}{k_i}\right)^2 = S$ is true.*

Proof. The statement made in the theorem is already proven for the set of all natural cubic spline functions. It remains to be shown that no other twice continuously differentiable function $f(x)$ exists that also fulfills the constraint $\sum_{i=1}^{m} \left(\frac{y_i - f(x_i)}{k_i}\right)^2 \leq S$ but possesses a smaller total curvature. If such an $f(x)$ exists, then the vector $D_f := (f(x_1), f(x_2), \ldots, f(x_m))$ fulfills the constraint. If the uniquely determined interpolating natural cubic spline function $s_{D_f}(x)$ is calculated for D_f, then the statement follows from Theorem 2.4 (and Theorem 3.1). This is because the total curvature of $s_{D_f}(x)$ is smaller than that of $f(x)$ and then this function cannot be a solution to (R). Parallel to Lemma 4.1, it is shown that the problems (R) and (RS) are equivalent. If $S > S_0$, then besides $g(x)$, other straight lines exist that also fulfill the constraint and whose total curvature is zero. □

Example 4.2. The structure of a crystalline substance should be clarified when it is excited by light of a defined wavelength.
The light emission of the substance was measured for 81 wavelengths. The obtained points represent a spectrum of energy. No theoretical conclusions exist with regard to the functional relationship of wavelength and energy. In the relatively strongly disturbed measured values, the structure of this relation is made more visible through smoothing. An appropriate compensatory spline function is constructed using REINSCH's approach. The degree of smoothing can be steered with the parameter S. It is known that the mean error for all measured values is equal and is estimated on 2.5 units. $S = 81(2.5)^2 = 506.25$ can then be used.
Through this process, the uniquely determined natural cubic spline function with the sum of the squared errors 506.25 is calculated when all $k_i = 1$ are

set. Through the iterative determination of the LAGRANGE parameter p, the sequence of p_k became:

$$\begin{aligned} p_0 &= 0.00000 \quad , \quad p_1 = 2.19179 \quad , \quad p_2 = 0.53027, \\ p_3 &= 0.19920 \quad , \quad p_4 = 0.10580 \quad , \quad p_5 = 0.07862, \\ p_6 &= 0.07403 \quad , \quad p_7 = 0.07386 \quad , \quad p_8 = 0.07386. \end{aligned}$$

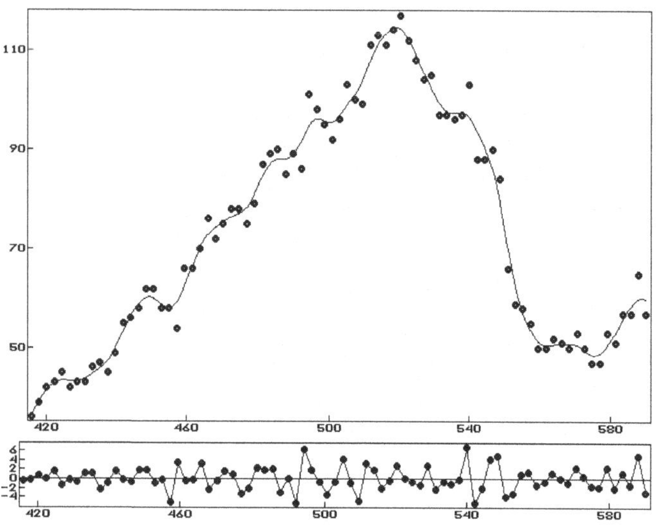

Fig. 4.2: Smoothing splines and residuals (below) according to the problem (R) (Example 4.2).

In Figure 4.2, besides the spline function the residuals being shown, the residuals $r_i := y_i - s(x_i)$, $i = 1, 2, \ldots, 81$, are tied together via a polygon for better illustration. The deviation at concrete knots can be smaller but also greatly larger than 2.5 units. However, the sum of the squared errors corresponds exactly with the given 506.25 (or $\sum_{i=1}^{81} \dfrac{(y_i - s(x_i))^2}{2.5^2} = 81$).

Note 4.2. In 1967, REINSCH indicated the solution to (R) and published a program for its numeric calculation. He used the EULER-LAGRANGE differential equation for the derivation. Unfortunately he made an error. This will be explained subsequently in detail.

In order to properly make use of the principle of variation, it is not sufficient to observe functions from $C^2([x_1, x_m])$. In addition, it is required that

a) The function $f'''(x)$ has discontinuities of the first kind, if at all, only in the knots and

b) $f^{(4)}(x)$ is continuous on the intervals $[x_i, x_{i+1})$.

When $f(x)$ is a solution to (R), then this function also minimizes

$$G(f) := \int_{x_1}^{x_m} [f''(x)]^2 \, dx + p \left[\sum_{i=1}^{m} \left(\frac{f(x_i) - y_i}{k_i} \right)^2 + Z^2 - S \right].$$

The slack variable Z allows for the inclusion of the constraint. One obtains from the derivative

$$\frac{\mathrm{d}G(f + \varepsilon h)}{\mathrm{d}\varepsilon} = \int_{x_1}^{x_m} 2(f''(x) + \varepsilon h''(x))h''(x)\mathrm{d}x + 2p \left[\sum_{i=1}^{m} \left(\frac{(f(x_i) + \varepsilon h(x_i) - y_i)h(x_i)}{{k_i}^2} \right) \right]$$

the necessary conditions, after dividing by two and for ε towards zero, as

$$\left. \begin{array}{l} \int_{x_1}^{x_m} f''(x)h''(x)dx + \sum_{i=1}^{m} p \left(\frac{(f(x_i) - y_i)h(x_i)}{{k_i}^2} \right) = 0 \\ \text{for all } h(x) \in C^2([x_1, x_m]) \text{ where } h(x_1) = h(x_m) = 0. \end{array} \right\} \qquad (1')$$

Next it is true

$$\int_{x_1}^{x_m} f''(x)h''(x)\mathrm{d}x = [f''(x_m)h'(x_m) - f''(x_1)h'(x_1)] - \int_{x_1}^{x_m} f'''(x)h'(x)\mathrm{d}x. \qquad (2')$$

Partial integration can now take place because of a) and b) :

$$-\int_{x_1}^{x_m} f'''(x)h'(x)\mathrm{d}x = -\sum_{i=1}^{m-1} \int_{x_i}^{x_{i+1}} f'''(x)h'(x)\mathrm{d}x = -\sum_{i=1}^{m-1} f'''(x)h(x)\Big|_{x_i}^{x_{i+1}} + \sum_{i=1}^{m-1} \int_{x_i}^{x_{i+1}} f^{(4)}(x)h(x)\mathrm{d}x.$$

If $f(x)$ were three times continuously differentiable, then

$$-\sum_{i=1}^{m-1} f'''(x)h(x)\Big|_{x_i}^{x_{i+1}} = 0$$

because $h(x_1) = h(x_m) = 0$ is true. But due to a),

$$-\sum_{i=1}^{m-1} f'''(x)h(x)\Big|_{x_i}^{x_{i+1}}$$

$$= f'''(x_1)_+ h(x_1) - \sum_{i=2}^{m-1} \left[f'''(x_i)_- - f'''(x_i)_+\right] h(x_i) - f'''(x_m)_- h(x_m)$$

where $f'''(x_i)_\pm := \lim_{w\downarrow 0} f'''(x_i \pm w)$. In addition,

$$-\int_{x_1}^{x_m} f'''(x)h'(x)\mathrm{d}x = \sum_{i=1}^{m} -[f'''(x_i)_- - f'''(x_i)_+]h(x_i)$$

$$+\sum_{i=1}^{m-1}\int_{x_i}^{x_{i+1}} f^{(4)}(x)h(x)\mathrm{d}x. \qquad (3')$$

The following is obtained when (3') is inserted into (2') and then (2') into (1') :

$$[f''(x_m)h'(x_m) - f''(x_1)h'(x_1)] + \sum_{i=1}^{m-1}\int_{x_i}^{x_{i+1}} f^{(4)}(x)h(x)\mathrm{d}x$$

$$+\sum_{i=1}^{m}\left\{-[f'''(x_i)_- - f'''(x_i)_+] + p\left(\frac{f(x_i) - y_i}{k_i{}^2}\right)\right\} h(x_i) = 0.$$

If this expression is to disappear for all $h(x)$ in $C^2([x_1, x_m])$, then the following must be true:

(1) $f^{(4)}(x) \equiv 0$ over $[x_i, x_{i+1})$, so $f(x)$ is a polynomial of at most 3rd degree over $[x_i, x_{i+1})$

(2) $f''(x_1) = f''(x_m) = 0$

(3) $f'''(x_i)_- - f'''(x_i)_+ = p\left(\frac{f(x_i)-y_i}{k_i{}^2}\right)$, $i = 1, 2, \ldots, m$.

f is twice differentiable. It follows that $f(x)$ is a natural cubic spline with regard to the knots x_1 to x_m. With the approach $s(x) = A_i(x - x_i)^3 + B_i(x - x_i)^2 + C_i(x - x_i) + D_i$ over $[x_i, x_{i+1}]$ due to (3') and the technical assertion $A_0 = A_m = B_0 = B_{m+1} = 0$ it follows that

$$6A_{i-1} - 6A_i = p\left(\frac{f(x_i) - y_i}{k_i{}^2}\right) = 6\left[\frac{B_i - B_{i-1}}{3\Delta x_{i-1}} - \frac{B_{i+1} - B_i}{3\Delta x_i}\right]$$

$$= -2\left[\frac{1}{\Delta x_{i-1}}B_{i-1} - \left(\frac{1}{\Delta x_{i-1}} + \frac{1}{\Delta x_i}\right)B_i + \frac{1}{\Delta x_i}B_{i+1}\right] \quad (4')$$

for $i = 1, 2, \ldots, m$.
The known matrix definitions and the conditions $B_1 = B_m = 0$ turn (4') into $-2 \circ Q^T \circ B = pK^{-2} \circ [D - Y]$. With $B = Z^{-1} \circ Q \circ D$ one obtains $-2 \circ Q^T \circ Z^{-1} \circ Q \circ D = pK^{-2} \circ [D - Y]$ from which finally the equation $[I + (2/p)K^2 \circ Q^T \circ Z^{-1} \circ Q] \circ D = Y$ results.
In the original paper from REINSCH (1967), only the conclusions from the EULER-LAGRANGE equation were illustrated. There, relationship (3) has the form

$$f'''(x_i)_- - f'''(x_i)_+ = 2p\left(\frac{f(x_i) - y_i}{{k_i}^2}\right). \tag{5'}$$

The equation $[I+(1/p)K^2 \circ Q^T \circ Z^{-1} \circ Q] \circ D = Y$ results. It is different from the relation given four lines before! This is not a problem for the numerical solution to (R) because the LAGRANGE parameter has to be determined such that the weighted sum of the squared errors is equal to S.
However, if the program published by REINSCH (1967) is used to solve other smoothing problems, the error would possibly have to be taken into account.
(OP3) is such an example. The equation $[I + \mu K^2 \circ Q^T \circ Z^{-1} \circ Q] \circ D = Y$ results when using the faulty relationship (5').
The correct linear system of equations for the determination of the unique $D_i = s(x_i)$, $i = 1, 2, \ldots, m$, reads $[I + 2\mu K^2 \circ Q^T \circ Z^{-1} \circ Q] \circ D = Y$.
Since in the program published by REINSCH (1967) $F(1/p_R) - \sqrt{S} = 0$ is calculated to achieve better convergence, the comparison of (4.12) and (4.10) makes the use of this program to numerically solve (OP3) possible. When the process of iteration for the determination of the LAGRANGE multiplier p_R is turned off and the remaining procedure is run using $p_R = 2\mu$ then the spline coefficients of the unique solution to

$$\mu \int_{x_1}^{x_m} [s''(x)]^2 \, dx + \sum_{i=1}^{m} \left(\frac{y_i - s(x_i)}{k_i}\right)^2 = \min!$$

are obtained.
If μ is used without the given correction, (OP3) is not solved in $C^2([x_1, x_m])$, rather

$$\frac{\mu}{2} \int_{x_1}^{x_m} [f''(x)]^2 \, dx + \sum_{i=1}^{m} \left(\frac{f(x_i) - y_i}{k_i}\right)^2 = \min!.$$

This error did not seem to be known up to this point. In renown programs such as SAS®, for given μ, up to now the problem (OP3) has been solved in reality with $\mu/2$.

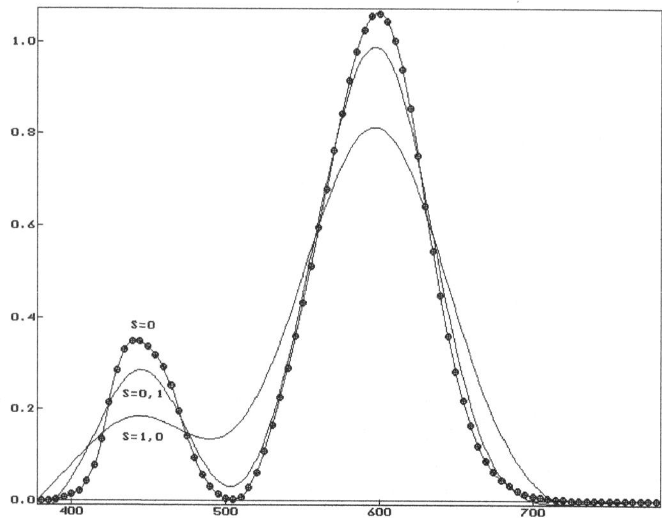

Fig. 4.3: Smoothing splines according to the problem (R), effect of the constant S on smoothing.

The influence of the constant S on the smoothing is illustrated in Figure 4.3. All $k_i = 1$ were set such that the value of S corresponds with the value of the sum of the squared errors between the spline function and the measured values y_i, $i = 1, 2, \ldots, 81$. It can be clearly seen that with growing S, smoothing continues to progress. The interpolating spline function results for $S = 0$. The regression line not shown in the Figure 4.3 corresponds with the maximum possible $S = S_0 = 8.7272$.
As can be seen in the Figure 4.3, smoothing mainly occurs in the areas of the modeled course of the function with large local curvature. If S is selected inadequately, it is easily possible that information is "smoothed away". In particular this is the case for strongly oscillating functions. The same problem occurs when only few points are available and these only poorly describe the course of the function (Figure 4.4).
In principle it is possible to try to eliminate this disadvantage by two procedures. First, the number of points in these areas should be increased. Second, the k_i are decreased so that $[(y_i - s(x_i))/k_i]^2$ is weighted more strongly at these points and the smoothing is moved to other areas. The effect cannot always be predicted though and the arbitrariness of this process is not satisfying. If the described problem needs to be avoided, a new optimization problem can be used. Here the need for smaller total curvature

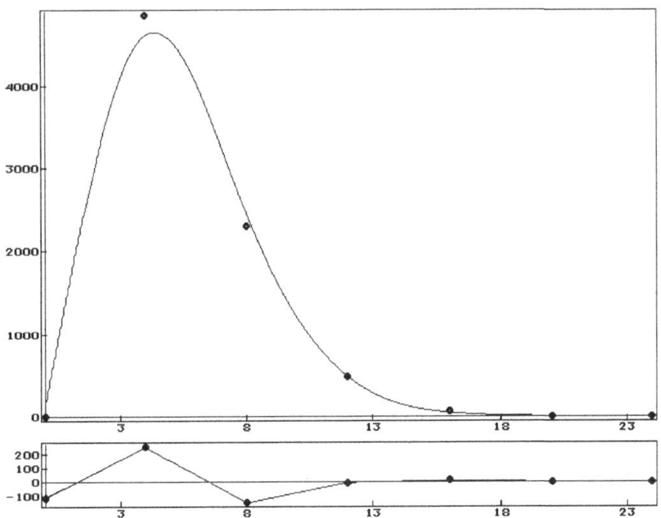

Fig. 4.4: Smoothing spline and residuals (below) according to the problem (R).

and "uniform" smoothing are combined.

Let the points (x_i, y_i), i=1,2,...,m, be given. A twice continuously differentiable function $f(x)$ with the constraints $|f(x_i) - y_i| \leq k_i$, $i = 1, 2, \ldots, m$, is searched for so that

$$\int_{x_1}^{x_m} [f''(x)]^2 dx \leq \int_{x_1}^{x_m} [g''(x)]^2 \mathrm{d}x \qquad \text{(KO)}$$

for all permissible $g(x) \in C^2([x_1, x_m])$.

In more detail, (KO) means: a function from $C^2([x_1, x_m])$ is searched for that proceeds with minimal total curvature within fixed given boundaries k_i around y_i.
To begin, the existence and unambiguity of such a solution to (KO) is to be examined.

Theorem 4.3. *Let $x_1 < x_2 < \ldots < x_m$. The problem* (KO) *has a solution for all points (x_i, y_i) and all $k_i \geq 0$, $i = 1, 2, \ldots, m$. Every solution is a natural cubic spline function $s(x)$ with regard to the knots x_i.*

Proof. To begin, (KO) is only observed in the set of all cubic spline functions with regard to the given knots x_i, $i = 1, 2, \ldots, m$.
Each of these functions is uniquely defined by the points (x_i, D_i), $i = 1, 2, \ldots, m$, $D_i = s(x_i)$.
With Lemma 4.1, (KO) can be reformulated:
A vector $D := (D_1, D_2, \ldots, D_m)^T \in \mathbb{R}^m$ is searched for that fulfills the conditions $|D_i - y_i| \leq k_i$, $i = 1, 2, \ldots, m$, and becomes minimal for $2D^T \circ E \circ D$. The matrix E is a positive semi-definite matrix that is dependent on the knots x_i. As a result, the real valued and continuous function $2D^T \circ E \circ D$ is convex and possesses at least one minimum in the compact convex set of permissible solutions. Each of these vectors $D_{min} \in \mathbb{R}^m$ determines an interpolating natural cubic spline function. These natural cubic spline functions solve the problem (KO) via the known conclusion from Lemma 4.1. □

The uniqueness of a minimum is not guaranteed for convex functions. This can be seen for this problem on hand of a simple constellation. When the points (x_i, y_i) and k_i, $i = 1, 2, \ldots, m$, are chosen such that various lines can be layed within the corridor, then all of these straight lines are solutions to (KO): they fulfill the constraints and their total curvature is zero (see Figure 4.5).

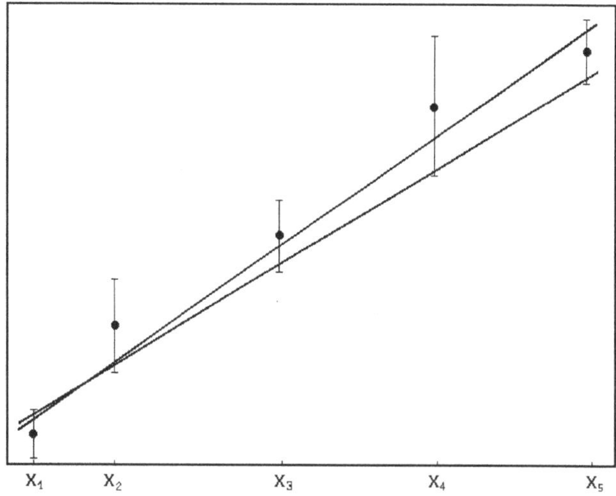

Fig. 4.5: The problem (KO) in general cannot be uniquely solved.

The convex nature of the objective function is not sufficient to secure the uniqueness of the solution to the problem (KO). To ensure uniqueness, a strong convex objective function is needed. For the described constellation, this could occur by choosing one of the permissible straight lines, more exactly, the one with minimal sum of the squared errors with regard to y_i. The objective function of (KO) should therefore be modified:

A function $f(x) \in C^2([x_1, x_m])$ for the given points (x_i, y_i) and fulfilling the constraints $\lvert f(x_i) - y_i \rvert \leq k_i$ for k_i, $i = 1, 2, \ldots, m$, is searched for so that $$\int_{x_1}^{x_m} [f''(x)]^2 \mathrm{d}x + \sum_{i=1}^{m} (f(x_i) - y_i)^2 \leq \int_{x_1}^{x_m} [g''(x)]^2 \mathrm{d}x + \sum_{i=1}^{m} (g(x_i) - y_i)^2$$ is true for all $g(x)$ in $C^2([x_1, x_m])$, which also fulfill the conditions $\lvert g(x_i) - y_i \rvert \leq k_i$, $i = 1, 2, \ldots, m$.	(MKO)

Similar to the past argument, the problem is equivalent to the following due to Lemma 4.2:

$$\left.\begin{array}{l} \text{A vector } D := (D_1, D_2, \ldots, D_m)^T \in \mathbb{R}^m, \text{ is searched for} \\ \text{that fulfills the conditions } |D_i - y_i| \leq k_i, i = 1, 2, \ldots, m, \\ \text{and for which the objective function} \\ \qquad F(D) := 2D^T \circ E \circ D + \sum\limits_{i=1}^{m} (D_i - y_i)^2 \\ \text{approaches a minimum.} \end{array}\right\} \tag{4.15}$$

The unique solution to problem (MKO) is indeed obtained. $\sum_{i=1}^{m}(D_i - y_i)^2$ is a strong convex and continuous function on $\mathbb{R}^m$. That is why $F(D)$ has the same properties. From this it follows that $F(D)$ possesses an unique minimum in the compact and convex set of permissible solutions.
Under the mentioned conditions, $|D_i - y_i| \leq k_i$ is true if and only if $D_i - y_i + k_i \geq 0$ and $-D_i + y_i + k_i \geq 0$, $i = 1, 2, \ldots, m$. In matrix form, this corresponds with the inequality system

$$\begin{pmatrix} 1 & 0 & \cdots & 0 & 0 \\ -1 & 0 & \cdots & 0 & 0 \\ 0 & 1 & \cdots & 0 & 0 \\ 0 & -1 & \cdots & 0 & 0 \\ \vdots & \vdots & \ddots & \vdots & \vdots \\ 0 & 0 & \cdots & 0 & 1 \\ 0 & 0 & \cdots & 0 & -1 \end{pmatrix} \circ D + \begin{pmatrix} -y_1 & + k_1 \\ y_1 & + k_1 \\ -y_2 & + k_2 \\ y_2 & + k_2 \\ & \vdots \\ -y_m & + k_m \\ y_m & + k_m \end{pmatrix} \geq \begin{pmatrix} 0 \\ 0 \\ 0 \\ 0 \\ \vdots \\ 0 \\ 0 \end{pmatrix}$$

which will be abbreviated with

$$G^T \circ D + g^0 \geq 0 \ . \tag{4.16}$$

For the objective function of (4.15),

$$\begin{aligned} F(D) :&= 2D^T \circ E \circ D + D^T \circ D - 2D^T \circ Y + Y^T \circ Y \\ &= D^T(2E + I) \circ D - 2D^T \circ Y + Y^T \circ Y \end{aligned}$$

is true where I denotes the $(m \times m)$-unit matrix and $Y := (y_1, y_2, \ldots, y_m)^T$. Here, $E^* := 2E + I$ is a positive definite matrix. $L \circ L^T = E^*$ represents the CHOLESKY decomposition of E^*. Since the term $Y^T \circ Y$ has no meaning for the determination of the minimum of $F(D)$, $F^*(D) := D^T \circ E^* \circ D - 2D^T \circ Y$ is reformulated. So one has

$$\begin{aligned} F^*(D) :&= D^T \circ (L \circ L^T) \circ D - 2D^T \circ (L \circ L^{-1}) \circ Y \\ &= \{D^T \circ (L \circ L^T) \circ D - 2D^T \circ (L \circ L^{-1}) \circ Y \\ &\quad + Y^T \circ (L^{-1})^T \circ L^{-1} \circ Y\} - Y^T \circ (L^{-1})^T \circ L^{-1} \circ Y \\ &= \{\langle L^T \circ D, L^T \circ D\rangle - 2\langle L^T \circ D, L^{-1} \circ Y\rangle + \langle L^{-1} \circ Y, L^{-1} \circ Y\rangle\} \\ &\quad - Y^T \circ (L^{-1})^T \circ L^{-1} \circ Y \\ &= ||L^T \circ D - L^{-1} \circ Y||^2 - Y^T \circ (L^{-1})^T \circ L^{-1} \circ Y \ . \end{aligned}$$

With the substitutions

$$\begin{aligned} &b := L^T \circ D \\ &\bar{y} := L^{-1} \circ Y \\ &\bar{G} := L^{-1} \circ G, \end{aligned}$$

$$F^*(D) = F^{**}(b) = ||b - \bar{y}||^2 - \bar{y}^T \circ \bar{y}$$

results.

When one observes that $\bar{y}^T \circ \bar{y}$ only determines the size of the value of the objective function F^{**}, but not the state of the optimal vector b^*, then the previous quadratic optimization problem (4.15) can be transformed into a simple norm minimization task with affine linear constraints:

$$\left.\begin{array}{l}\text{Search for the uniquely determined } b \in \mathbb{R}^m\,, \quad \text{for which}\\ \|b - \bar{y}\|^2 \text{ becomes minimal for all vectors that fulfill the}\\ \text{conditions } \bar{G}^T \circ b + g^0 \geq 0\ .\end{array}\right\} \tag{4.17}$$

Before the solution to (4.17) can be discussed any further, several outcomes of the theory of convex optimization problems need of be mentioned. In order to derive useful optimization procedures, the so called SLATER condition is assumed (see JARRE and STOER (2003)). For (4.17) this means the existence of a b^0 for which $\bar{G}^T \circ b^0 + g^0 > 0$ is true. This is equivalent for the demand that $k_i > 0$ for all i from 1 to m.
With this assumption the validity of a LAGRANGE multiplier rule for the optimal b^* is not only a necessary but also a sufficient condition on the optimum. When applied to (4.17) this results in:
b^* is the optimal solution to (4.17) exactly then when a vector $\alpha^* \geq 0$ exists in $\mathbb{R}^{2m}$ for which

$$\left.\begin{array}{l}1.\ b^* - \bar{y} - \bar{G} \circ \alpha^* = 0\\ 2.\ \bar{G}^T \circ b^* + g^0 \geq 0 \text{ and}\\ 3.\ \alpha_i^*(e^{iT} \circ (\bar{G}^T \circ b^* + g^0)) = 0,\ i = 1, 2, \ldots, 2m,\end{array}\right\} \tag{4.18}$$

is true.
Here, e^{iT} is the notation for the unit vector of $\mathbb{R}^{2m}$ that possesses a one at the ith coordinate and otherwise consists of zeros. Since Y fulfills the constraints of (MKO), the vector $b = L^T \circ Y$ from $\mathbb{R}^m$ can serve as a starting point for the algorithm for the solution to (4.17) (cf. SPELUCCI (1993), Page 304 et seqq.). Because of 3. in (4.18) for b in the case $\alpha_i^* = 0$, it is true that $e^{iT} \circ (\bar{G}^T \circ b + g^0) > 0$. That is why in the first equation in (4.18) only the columns g^i of $\bar{G}$ are taken into account for which $\bar{G}^T \circ b^* + g^0 = 0$ is true. $q := q(b)$ is the number of columns, $N := N(b)$ is the $(m \times q)$ matrix that contains g^i as its columns and $A := A(b)$ is the corresponding index set of the original column numbers in $\bar{G}$. To fulfill the SLATER condition it was assumed that all $k_i > 0$, $i = 1, 2, \ldots, m$. These assumptions also secure that N possesses full column rank and that the linear system of equations

$$\begin{pmatrix} I_m & N \\ N^T & 0 \end{pmatrix} \circ \begin{pmatrix} s \\ w \end{pmatrix} = \begin{pmatrix} -(b - \bar{y}) \\ 0 \end{pmatrix} \tag{4.19}$$

has a unique solution. This is why the following Theorem (4.4), which describes a possible solution to (4.17), is true. It should be pointed out again that the SLATER condition or the unique solvability of (4.19) and therefore the validity of Theorem 4.4 are only fulfilled when all k_i, $i = 1, 2, \ldots, m$, are greater than zero!

Theorem 4.4. *Let b be a permissible solution to (4.17) and (s, w) the solution to (4.19). Then:*

(1) *If $s = 0$ and $w_i \leq 0$, $i = 1, 2, \ldots, q$, then $b^* = b$ is the solution to (4.17).*

(2) *When $s \neq 0$, then $b + s$ solves the minimum task $||b' - \bar{y}||^2 = \min!$ for all b', that fulfill the constraints $(g^i)^T \circ b' + (g^0)_i = 0$, $i \in A(b)$. Furthermore, $b + \sigma s$ is a permissible solution to (4.17) and $||(b + \sigma s) - \bar{y}||^2 < ||b - \bar{y}||^2$,*

$$\begin{aligned}\sigma :&= \max\{1 \geq \tau > 0 : b + \tau s \text{ is a permissible solution to } (4.17)\} \\ &= \min\{1, ((g^i)^T \circ b + (g^0)_i)/(-(g^i)^T \circ s) : i \in \{1, 2, \ldots, 2m\} \backslash A(b), \\ &\quad (g^i)^T s < 0\}.\end{aligned}$$

(3) *$s = 0$ and $w_{i_o} > 0$ for an $i_o \in \{1, 2, \ldots, q\}$. $\tilde{N}$ denotes the matrix that is produced from N by the elimination of column i_0. Then the system*

$$\begin{pmatrix} I_m & \tilde{N} \\ \tilde{N}^T & 0 \end{pmatrix} \circ \begin{pmatrix} \hat{s} \\ \hat{w} \end{pmatrix} = \begin{pmatrix} -(b - \bar{y}) \\ 0 \end{pmatrix}$$

defines a direction $\hat{s}$ for which point 2 of this theorem holds true. $A(b) \backslash \{j_{i_o}\}$ is to be used instead of $A(b)$.

(4) *The uniquely determined optimum b^* from (4.17) is obtained by a finite iterative application of Steps (1) to (3).*

Proof. The technical description of the proof can be found in, for example, SPELUCCI (1993), pages 307-309. □

The solution to the previous optimization problem (MKO) is obtained by first transforming b^* back. $D = (L^T)^{-1} \circ b^*$ contains $D_i = s(x_i)$, $i = 1, 2, \ldots, m$, of the natural cubic spline function that solves (MKO). This is constructed by solving (4.2) and determining the remaining spline coefficients from the B_i.

Example 4.3. Given are the 20 points $(0, -0.45)$, $(1, 1.14)$, $(2, 1.02)$, $(3, 39)$, $(4, -0.55)$, $(5, -0.99)$, $(6, 0.42)$, $(7, 1.72)$, $(8, 2.47)$, $(9, 0.58)$,

(10, 0.78), (11, 0.98), (12, 1.08), (13, 3.28), (14, 2.18), (15, 1.12), (16, 0.58), (17, 1.03), (18, 2.22), (20, −0.45).
The same constraints $k_i = 0.3$ were set for all knots x_i. The vector D^* was determined with the algorithm from Theorem 4.4:
(−0.15, 0.84, 0.917741345, 0.24521133, −0.570955043, −0.69, 0.431271341, 1.844218357, 2.17, 0.88, 0.48, 0.68, 1.38, 2.98, 2.48, 1.362822182, 0.88, 1.33, 1.92, −0.15). The uniquely determined natural interpolating spline function $s(x)$ that solves (MKO) is illustrated in Figure 4.6. With regard to the given y-values, the spline function possesses the sum of the squared errors 1.3664. As a comparison, problem (R) for $k_i = 1$, $i = 1, 2, \ldots, 20$, and $S = 1.3664$ was solved for the same data with the REINSCH algorithm. The result is demonstrated in Figure 4.7. It can be clearly seen at several knots that the value of $|s(x_i) - y_i|$ exceeds 0.3.

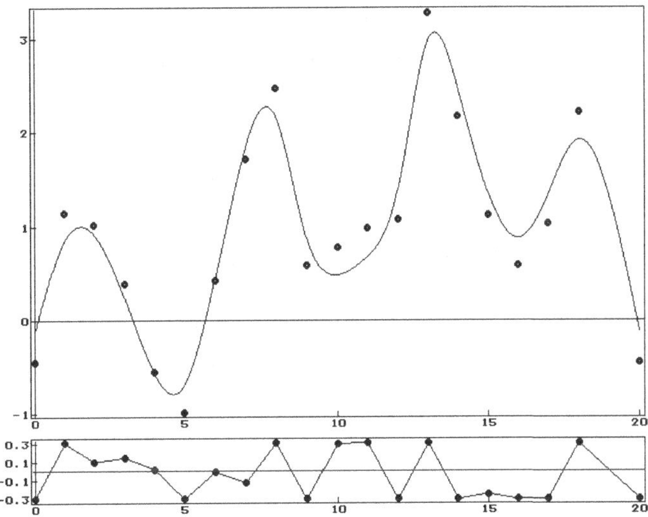

Fig. 4.6: The unique spline solution $s(x)$ to problem (MKO) with the constraints $|s(x_i) - y_i| \leq 0.3$, $i = 1, \ldots, 20$, see Example 4.3), below: the residuals.

Another problem will be studied:
A function from $C^2([x_1, x_m])$ is searched for which runs through the given points (x_i, y_i), $i = 1, 2, \ldots, m$, with minimal weighted sum of the squared errors. Its total curvature should not exceed a given constant T.

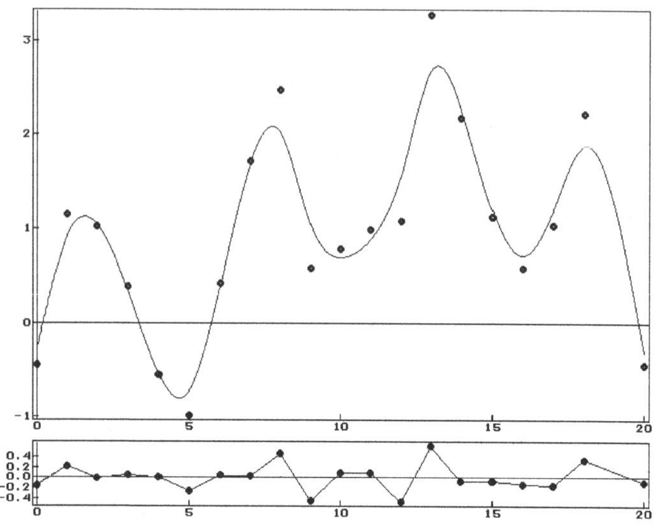

Fig. 4.7: The uniquely determined solution $s(x)$ to the problem (R), see Example 4.3 for the data. Below: the residuals.

Given are the points (x_i, y_i), constants $k_i > 0$, i from 1 to m, and $T \geq 0$. A twice continuously differentiable function $f(x)$ with $\int_{x_1}^{x_m} [f''(x)]^2 \mathrm{d}x \leq T$ is searched for so that $\sum_{i=1}^{m} \left(\frac{f(x_i) - y_i}{k_i} \right)^2$ approaches a minimum.	(GK3)

If (GK3) is now observed in the set of all natural cubic spline functions with regard to the knots x_i then it is possible to transform the constraint into the form

$$\frac{2}{3} \sum_{i=1}^{m-1} \Delta x_i \left[B_i^2 + (B_i + B_{i+1})^2 + B_{i+1}^2 \right] + h^2 - T = 0$$

with the help of the slack variable h (see (4.3)).
If this equation with the LAGRANGE multiplier p and the spline function defining equations (4.2) is built into the objective function with the multi-

pliers V_i, $i = 2, 3, \ldots, m-1$, then the following is obtained:

$$L^{**}(B_2, \ldots, B_{m-1}, D_1, \ldots, D_m, V_2, \ldots, V_{m-1}, p, h) := \sum_{i=1}^{m} \left(\frac{D_i - y_i}{k_i} \right)^2$$

$$+ p \left[\frac{2}{3} \sum_{i=1}^{m-1} \Delta x_i \left[B_i^2 + (B_i + B_{i+1})^2 + B_{i+1}^2 \right] + h^2 - T \right]$$

$$+ \sum_{i=2}^{m-1} V_i \left[\Delta x_{i-1} B_{i-1} + 2(\Delta x_{i-1} + \Delta x_i) B_i + \Delta x_i B_{i+1} \right.$$

$$\left. - 3 \left(\frac{(D_{i+1} - D_i)}{\Delta x_i} - \frac{(D_i - D_{i-1})}{\Delta x_{i-1}} \right) \right] = \min!$$

The necessary conditions $\partial L^{**}/\partial B_i = 0$, $\partial L^{**}/\partial V_i = 0$, $i = 2, 3, \ldots, m-1$, and $\partial L^{**}/\partial D_i = 0$, $i = 1, 2, \ldots, m$, for a local minimum lead to the relationships

$$V = -\frac{4}{3} p B \ , \tag{4.20}$$

$$Z \circ B = Q \circ D \tag{4.21}$$

and

$$2K^{-2} \circ [D - Y] = 3Q^T \circ V \ . \tag{4.22}$$

With (4.20), $K^{-2} \circ [D - Y] = -2pQ^T \circ B$ results from (4.22) and so

$$D = Y - 2pK^2 \circ Q^T \circ B \ . \tag{4.23}$$

$Z \circ B = Q \circ [Y - 2pK^2 \circ Q^T \circ B]$ results due to (4.21), which is

$$[Z + 2pQ \circ K^2 \circ Q^T] \circ B = Q \circ Y \ . \tag{4.24}$$

The matrix of coefficients from (4.24) is positive definite for all $p \geq 0$ and therefore invertible. From this it follows that

$$B = [Z + 2pQ \circ K^2 \circ Q^T]^{-1} \circ Q \circ Y \ . \tag{4.25}$$

If (4.25) is substituted into (4.23) then

$$D = Y - 2pK^2 \circ Q^T \circ [Z + 2pQ \circ K^2 \circ Q^T]^{-1} \circ Q \circ Y \ . \tag{4.26}$$

The following is defined in order to calculate p,

$$\begin{aligned} GK(p) : &= \int_{x_1}^{x_m} [s''(x)]^2 \mathrm{d}x = 2B^T \circ Z \circ B \qquad \text{(Lemma 4.2)} \\ &= 2\{[Z + 2pQ \circ K^2 \circ Q^T]^{-1} \circ Q \circ Y\}^T \circ Z \circ \\ &\quad [Z + 2pQ \circ K^2 \circ Q^T]^{-1} \circ Q \circ Y \qquad \text{(see (4.25))} \\ &= 2Y^T \circ Q^T \circ [Z + 2pQ \circ K^2 \circ Q^T]^{-1} \circ Z \circ \\ &\quad [Z + 2pQ \circ K^2 \circ Q^T]^{-1} \circ Q \circ Y \\ &= T - h^2 \ . \end{aligned}$$

$ph = 0$ is obtained from the necessary condition $\partial L^{**}/\partial h = 0$.

Case 1 $p = 0$. This results that $GK(0) = 2Y^T \circ Q^T \circ Z^{-1} \circ Q \circ Y$ is the total curvature of the interpolating natural cubic spline function through the given (x_i, y_i), $i = 1, 2, \ldots, m$.

Case 2 $p \neq 0$. In this case $h = 0$ and the p that is searched for is determined by the equation $GK(p) = T$.

An initial possibility for the construction of the unique solution $s(x)$ in the set of the natural cubic splines is as follows:

1. If $T \geq 2Y^T \circ Q^T \circ Z^{-1} \circ Q \circ Y$, then the interpolating natural cubic spline function with $s(x_i) = y_i$, $i = 1, 2, \ldots, m$, is the solution.

2.1. Otherwise determine p from the equation $GK(p) = 2Y^T \circ Q^T \circ [Z + 2pQ \circ K^2 \circ Q^T]^{-1} \circ Z \circ [Z + 2pQ \circ K^2 \circ Q^T]^{-1} \circ Q \circ Y = T$,

2.2. Calculate $D_i = s(x_i)$, $i = 1, 2, \ldots, m$, from the relationship $D = Y - 2pK^2 \circ Q^T \circ [Z + 2pQ \circ K^2 \circ Q^T]^{-1} \circ Q \circ Y$ and $B_i = s''(x_i)/2$, $i = 2, 3, \ldots, m-1$, from $B = [Z + 2pQ \circ K^2 \circ Q^T]^{-1} \circ Q \circ Y$.

2.3. Determine the spline coefficients A_i and C_i with (3.3) and (3.4).

Equation (4.22) can be used as a starting point for a second possibility to calculate $s(x)$. If (4.20) is inserted and B is eliminated with the help of (4.21) then

$$K^{-2} \circ [D - Y] = -2pQ^T \circ Z^{-1} \circ Q \circ D \ , \text{ so}$$
$$Y = [I + 2pK^2 \circ Q^T \circ Z^{-1} \circ Q] \circ D \ . \tag{4.27}$$

The matrix of coefficients of (4.27) is positive definite for all $p \geq 0$ (Lemma 4.2) such that this linear system of equations for the determination of the D_i has a unique solution.

From this it follows that

$$\begin{aligned} GK(p) &= 2B^T \circ Z \circ B = 2(Z^{-1} \circ Q \circ D)^T \circ Z \circ Z^{-1} \circ Q \circ D \\ &= 2D^T \circ Q^T \circ Z^{-1} \circ Q \circ D \\ &= 2Y^T \circ [I + 2pK^2 \circ Q^T \circ Z^{-1} \circ Q]^{-1} \circ Q^T \circ Z^{-1} \circ Q \\ &\quad \circ [I + 2pK^2 \circ Q^T \circ Z^{-1} \circ Q]^{-1} \circ Y \ . \end{aligned}$$

The construction algorithm is as follows:

1*. If $T \geq 2Y^T \circ Q^T \circ Z^{-1} \circ Q \circ Y$, then the interpolating natural cubic spline function $s(x_i) = y_i$, $i = 1, 2, \ldots, m$, is the solution.

2*.1. Otherwise determine p by means of the equation

$$\begin{aligned} GK(p) = 2Y^T &\circ [I + 2pK^2 \circ Q^T \circ Z^{-1} \circ Q]^{-1} \circ Q^T \circ Z^{-1} \circ Q \\ &\circ [I + 2pK^2 \circ Q^T \circ Z^{-1} \circ Q]^{-1} \circ Y = T \ , \end{aligned}$$

2*.2. calculate $D = [I + 2pK^2 \circ Q^T \circ Z^{-1} \circ Q]^{-1} \circ Y$, $B = Z^{-1} \circ Q \circ Y$ and
2*.3. the remaining spline coefficients with (3.3) and (3.4).

The solutions to 2.1 and 2*.1 are identical.
Unique solvability is observed for 2.1. This allows proceeding analogously to the problem (RS). Set $G(p) := GK(p)$, $N := Q \circ K^2 \circ Q^T$ and $M := Z$. Since Z and $Q \circ K^2 \circ Q^T$ are positive definite, the total curvature as a function of p over $[0, \infty)$ is differentiable, monotone decreasing and strongly convex with $\lim_{p \to \infty} GK(p) = 0$. That is why a unique positive solution to the equation $GK(p) = T$ for all T between zero and the total curvature of the interpolating natural spline exists that also, due to Kuhn and Tucker (1951), determines the solution $s_o(x)$ to (GK3) in the set of cubic splines. It remains to be examined if other continuously differentiable functions $g(x)$ exist that also fulfill $\int_{x_1}^{x_m} [g''(x)]^2 \mathrm{d}x \leq T$ and whose weighted sum of the squared errors with regard to y_i is less than or equal to that of $s_o(x)$.

Theorem 4.5. *Let* $T_0 := 2Y^T \circ Q^T \circ Z^{-1} \circ Q \circ Y$ *be and* $T \leq T_0$. *The problem* (GK3) *then has a unique solution and the solution is a natural cubic spline function* $s_o(x)$ *for which* $\int_{x_1}^{x_m} [s_o''(x)]^2 \mathrm{d}x = T$.

Proof. If $T > T_0$, then T is greater than the total curvature of the interpolating natural cubic spline function $s(x)$ with $s(x_i) = y_i$, $i = 1, 2, \ldots, m$. The uniqueness of the solution cannot be secured because it is possible that other interpolating functions, such as the interpolating polynomial, fulfill the constraint.
Let $T \leq T_0$ and $s_o(x)$ be the uniquely determined solution to (GK3) in the subset of natural cubic splines for which $\int_{x_1}^{x_m} [s_o''(x)]^2 \mathrm{d}x = T$.
Assume there exists a $g(x) \in C^2([x_1, x_m])$ that isn't a natural cubic spline and that $g(x)$ solves (GK3). The uniquely determined interpolating natural cubic spline function $s_D(x)$ is then constructed for $D_i := g(x_i)$, $i = 1, 2, \ldots, m$. Two cases are observed.

Case 1 $\int_{x_1}^{x_m} [s_D''(x)]^2 \mathrm{d}x = T$.

With Theorem 3.1, $T = \int_{x_1}^{x_m} s_D''(x)^2 \mathrm{d}x < \int_{x_1}^{x_m} [g''(x)]^2 \, \mathrm{d}x$ results. With that $g(x)$ is not a permissible solution to (GK3) because it does not fulfill the constraint.

Case 2 $\int_{x_1}^{x_m} [s_D''(x)]^2 \mathrm{d}x < T$.

In this case $s_D(x) \neq s_o(x)$ would be a natural cubic spline function which minimizes the objective function of (GK3). However, this contradicts the already known fact that $s_0(x)$ is the uniquely determined solution to (GK3) in the set of natural cubic splines with regard to the given system of knots.
Consequently, the unique solvability in the set of natural cubic spline functions is extended to the unique solvability of those in $C^2([x_1, x_m])$.
□

Example 4.4. Given are the eight points $(1, 0)$, $(2, 1)$, $(3, 4)$, $(4, 3)$, $(5, 1)$, $(6, 2)$, $(8, 2)$ and $(11, 3.5)$. With the help of MATHEMATICA®, the values

$$
\begin{aligned}
p_1 &= -25.54274 \ , \\
p_2 &= -22.31966 \ , \\
p_3 &= -6.34666 \ , \\
p_4 &= -1.04606 - 0.68801\mathrm{i} \ , \\
p_5 &= -1.04606 + 0.68801\mathrm{i} \ , \\
p_6 &= -0.18318 - 0.01017\mathrm{i} \ , \\
p_7 &= -0.18318 + 0.01017\mathrm{i} \ , \\
p_8 &= -0.11102 - 0.17359\mathrm{i} \ , \\
p_9 &= -0.11102 + 0.17359\mathrm{i} \ , \\
p_{10} &= -0.02833 - 0.00249\mathrm{i} \ , \\
p_{11} &= -0.02833 + 0.00249\mathrm{i} \text{ and} \\
p_{12} &= 1.69623
\end{aligned}
$$

were determined as the solution to the equation $GK(p) = 1$ (see also Part III). There are four real and eight complex results. The respective spline function was constructed with the four real numbers. The total curvature of each of these splines is one (see Figure 4.9).

A numeric solution for the direct determination of the LAGRANGE multiplier p could be approached similarly to the problem (RS). However, the REINSCH (1967) procedure can be used in the following form:
The iterative change of p is completed first. For every $p \geq 0$, the unique solution to

$$
\frac{p}{2} \int_{x_1}^{x_m} [s''(x)]^2 \mathrm{d}x + \sum_{i=1}^{m} \left(\frac{y_i - s(x_i)}{k_i} \right)^2 = \min!
$$

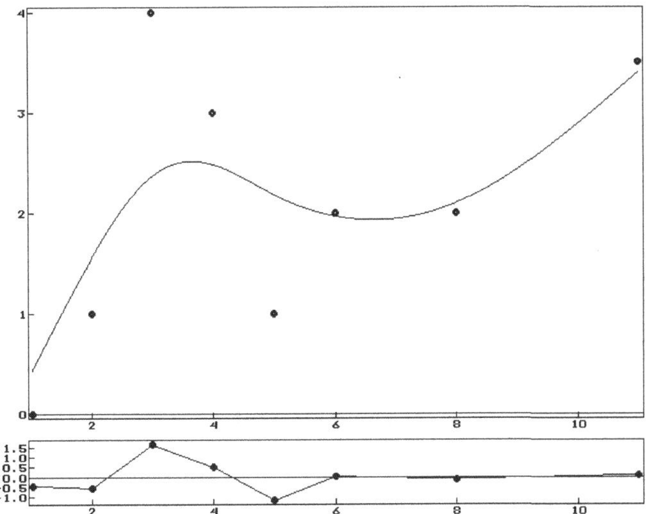

Fig. 4.8: The uniquely determined natural cubic spline with minimal sum of the squared errors and total curvature $T = 1, (p = 1.6962341)$, see Example 4.4 for the data. Below: the residuals.

is calculated. The total curvature $GK(p)$ of $s_p(x)$ results from

$$GK(p) = \frac{4}{3}\sum_{i=1}^{m-1} \Delta x_i(B_i^2 + B_iB_{i+1} + B_{i+1}^2) \ .$$

The interpolating natural cubic spline is obtained for $p = 0$ and the regression line is obtained for p approaching infinity. The LAGRANGE multiplier p must be determined from the equation $GK(p) - T = 0$ for choosen T in $[0, GK(0)]$. For example, the starting value for the applicable NEWTON process is $p_0 = 0$. The iteration sequence then is $p_{j+1} = p_j - (GK(p_j) - T)/GK'(p_j)$.

The required derivative $GK'(p_j)$ can be easily approximated by a respective difference quotient and the monotone characteristic of $GK(p)$ guarantees convergence.

Many variations of the discussed tasks of approximation via natural cubic splines are possible. The clue is the formulation of the constraints. So, for example, for the modeling of growth curves it is a meaningful demand that $s(x_i) \leq s(x_{i+1})$, $i = 1, 2, \ldots, m-1$, should be true. It can also be made a condition that $s'(x_i) \geq 0$ for all i from 1 to m should be true.

Following, the influence on the shape of the resulting quadratic optimization problem is examined when constraining conditions are formulated

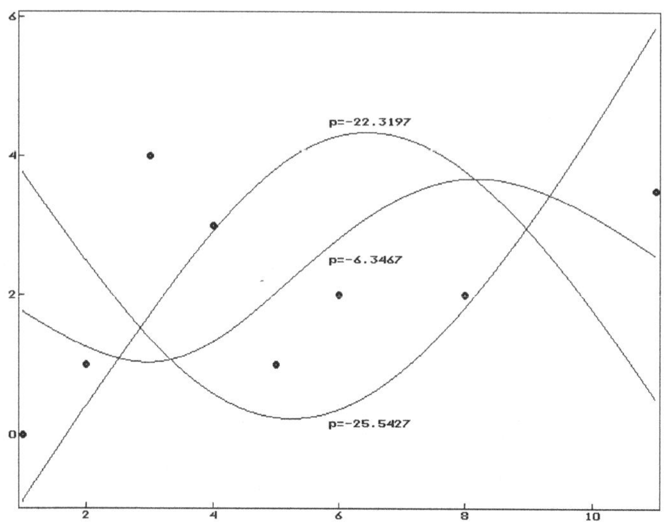

Fig. 4.9: Splines with total curvature $T = 1$ for the real valued negative solutions to $GK(p) = 1$, see Example 4.4.

for the spline coefficients and therefore also for the derivatives $s^{(k)}(x_i)$, $k = 0, 1, 2, 3$, $i = 1, 2, \ldots, m$.

The starting point is the representation $s(x) = A_i(x - x_i)^3 + B_i(x - x_i)^2 + C_i(x - x_i) + D_i$ for all x over the interval $[x_i, x_{i+1}]$. Then the relationships

$$s(x_i) = D_i \ ,$$

$$s'(x_i) = C_i = (D_{i+1} - D_i)/\Delta x_i - 1/3\Delta x_i(2B_i + B_{i+1}) \ , \tag{4.28}$$

$$s''(x_i)/2 = B_i \text{ and}$$

$$s'''(x_i)/6 = A_i = 1/3(B_{i+1} - B_i)/\Delta x_i \tag{4.29}$$

are true.

With this, the vector $B := (B_2, B_3, \ldots, B_{m-1})^T$ is the uniquely determined solution of the linear system of equations $Z \circ B = Q \circ D$ and $B_1 = B_m = 0$. Furthermore, the $(m-1) \times (m-1)$ diagonal matrix ΔX with Δx_i on the main diagonal and the $(m-1) \times (m-2)$ matrix

$$H := \begin{pmatrix} 1 & 0 & 0 \ldots & 0 & 0 \\ -1 & 1 & 0 \ldots & 0 & 0 \\ 0 & -1 & 1 \ldots & 0 & 0 \\ \vdots & \vdots & \vdots \ddots & \vdots & \vdots \\ 0 & 0 & 0 \ldots & -1 & 1 \\ 0 & 0 & 0 \ldots & 0 & -1 \end{pmatrix}$$

are defined.
Then due to (4.29) the following representation is true for the vector $A := (A_1, A_2, \ldots, A_{m-1})^T$,

$$A = 1/3\Delta X^{-1} \circ H \circ B = 1/3\Delta X^{-1} \circ H \circ Z^{-1} \circ Q \circ D \ .$$

The following results for $C := (C_1, C_2, \ldots, C_{m-1})^T$ from (4.28),

$$\begin{aligned} C &= \Delta X^{-1} \circ H' \circ D - 1/3\Delta X \circ G \circ B \\ &= \Delta X^{-1} \circ H' \circ D - 1/3\Delta X \circ G \circ Z^{-1} \circ Q \circ D \\ &= [\Delta X^{-1} \circ H' - 1/3\Delta X \circ G \circ Z^{-1} \circ Q] \circ D =: P \circ D \ . \end{aligned}$$

Here, H' is a $(m-1) \times m$ matrix that results from H by appending the column $(-1, 0, 0, \ldots, 0)^T$ at the front and the column $(0, 0, \ldots, 0, 1)^T$ at the end of this matrix.

$$G := \begin{pmatrix} 1\,0\,0 \ldots 0\,0 \\ 2\,1\,0 \ldots 0\,0 \\ 0\,2\,1 \ldots 0\,0 \\ \vdots\ \vdots\ \vdots\ \ddots\ \vdots\ \vdots \\ 0\,0\,0 \ldots 2\,1 \\ 0\,0\,0 \ldots 0\,2 \end{pmatrix}$$

is of type $(m-1) \times (m-2)$. Therefore, for a given system of knots $x_1, x_2, \ldots, x_m$, it is possible to linearly determine the spline coefficients from the vector D. Every linear constraint of the spline coefficients, and therefore the derivatives $s^{(k)}(x_i)$, $k = 0, 1, 2, 3$, $i = 1, 2, \ldots, m$, is with that a linear constraint for the vector D. Here there is always a need to solve a quadratic optimization problem in $\mathbb{R}^m$ with linear constraints.
Thus, for example, the optimization problem in $C^2([x_1, x_m])$,

$$\left. \begin{array}{l} \int\limits_{x_1}^{x_m} [f''(x)]^2 \mathrm{d}x + \sum\limits_{i=1}^{m} (y_i - f(x_i))^2 = \min! \\ \text{with } f'(x_i) \geq 0 \text{ for } i \text{ from 1 to } m \end{array} \right\}$$

parallels the problem:
Minimize $F(D) := 2D^T \circ Q^T \circ Z^{-1} \circ Q \circ D + D^T \circ D - 2Y^T \circ D$ in $\mathbb{R}^m$ under all vectors with the constraint $[\Delta X^{-1} \circ H' - 1/3\Delta X \circ G \circ Z^{-1} \circ Q] \circ D =: P \circ D \geq 0$.
The end of this section should highlight the most important aspects again. With Theorem 2.1, there exists a unique representation for every natural spline functions $s(x)$ in the following form:

$$s(x) = \sum_{j=1}^{k} c_j x^{j-1} + \sum_{i=1}^{m} c_{k+i} (x - x_i)_+^{2k-1} \ .$$

Thus, for fixed knots x_i and fixed k, there exists a function $H(*)$ that maps every interpolating natural spline function on the vector $C := (c_1, c_2, \ldots, c_{k+m}) \in \mathbb{R}^{k+m}$.
On the other hand, from Theorem 2.3 it follows that there exists a unique function $G(Y)$ that maps every $Y := (y_1, y_2, \ldots, y_m) \in \mathbb{R}^m$ on the interpolating natural spline function $s_Y(x)$. The apparent discrepancy between the two maps $G(*)$ and $H(*)$ is eliminated via Theorem 2.3 and Lemma 2.1. It is shown there that $H(*)$ is not a surjection on $\mathbb{R}^{k+m}$. More so, $H(*)$ is a map on the m-dimensional subspace $\mathbb{R}^{k+m}$, which is determined by the relationships $\sum_{i=1}^{m} c_i x_i^j = 0$, $j = 0, 1, \ldots, k-1$. As a result, $H(*)$ is a one-to-one map of this subspace.
These statements are also particularly true for natural cubic spline functions. If the system of knots x_i, $i = 1, 2, \ldots, m$, is fixed, then the spline is uniquely determined by the vector Y. Similarly to the minimization of the sum of the squared errors for a linear equation, only knowledge about the unknown function is built into the given knots to aid in the construction of the spline function. The course of the function that is to be approximated between the x_i is therefore determined.
In Chapter 4, smoothing natural cubic splines were constructed based on the knowledge about interpolating natural cubic splines. For each of the four problems that were dealt with, the starting point is an optimization problem in the set $C^2([a, b])$ of the twice continuously differentiable functions over the interval $[a, b]$. Since the total curvature effects the respective objective function or the constraint, it results that every problem can be solved by a natural cubic spline function in a unique fashion. Their construction can take place in two steps. In the first step the $D_i := s(x_i)$, $i = 1, 2, \ldots, m$, of the optimum are calculated. Following this, the interpolating natural cubic spline function is defined by the points (x_i, D_i) that are obtained. This is possible because each of the observed optimization problems can be carried over into an equivalent quadratic optimization problem in $\mathbb{R}^m$ in an elementary way. Since all spline coefficients are linearly obtained via the D_i it is possible, with linear constraints for the derivatives $s^{(j)}(x_i)$, $j = 1, 2, 3$, $i = 1, 2, \ldots, m$, to always formulate a quadratic optimization problem with linear constraints. In this case, the respective numerical procedures can be applied.
In the described problems regarding the smoothing splines, the situation where the spline function is determined not only by the data is made more clear. With one exception (in problem (MKO)), the construction of the unique solution of the observed optimization task is dependent on the spec-

ification of one real number. For the optimization task (OP3) it is μ, for (R) it is the constant S and for (GK3) it is the limit T.
Once these numbers are chosen, the respective problem solving spline function is uniquely and completely determined. All knowledge about the unknown functional dependency or about the accuracy of measurement must flow into these constants.
The measuring precision can be taken into account particularly well at the problem (R). If it is assumed that estimations S_{y_i}, $i = 1, 2, \ldots, m$, are known for the standard deviation of y_i then the random variables $(y_i - f(x_i))/S_{y_i}$ have a standard normal distribution when the measurement error of y_i has a normal distribution and an expected value of zero. The unknown function value $f(x_i)$ is then equal to the expected value of the measurement at x_i.
With these requirements the random variable $X := \sum_{i=1}^{m} \left(\frac{y_i - f(x_i)}{S_{y_i}}\right)^2$ is χ^2-distributed with $(m-1)$ degrees of freedom. The constant S in (R) should be chosen near the expectation $E(X)$.
That the solution of each of the discussed smoothing problems is determined by the specification of a number is also mirrored in the defining equations of D_i. They read:

$$D = [I + 2\mu K^2 \circ Q^T \circ Z^{-1} \circ Q]^{-1} \circ Y \text{ for } (OP3) \text{ (Equation (4.10))} ,$$
$$D = [I + 2/p_R K^2 \circ Q^T \circ Z^{-1} \circ Q]^{-1} \circ Y \text{ for } (R) \text{ (Equation (4.12)) and}$$
$$D = [I + 2pK^2 \circ Q^T \circ Z^{-1} \circ Q]^{-1} \circ Y \text{ for } (GK3) \text{ (Equation (4.27))}$$

or

$$D = Y - K^2 \circ Q^T \circ [1/(2\mu)Z + Q \circ K^2 \circ Q^T]^{-1} \circ Q \circ Y \text{ for } (OP3) ,$$
$$D = Y - K^2 \circ Q^T \circ [p_R/2Z + Q \circ K^2 \circ Q^T]^{-1} \circ Q \circ Y \text{ for } (R),$$
$$D = Y - K^2 \circ Q^T \circ [1/(2p)Z + Q \circ K^2 \circ Q^T]^{-1} \circ Q \circ Y \text{ for } (GK3) .$$

In the required matrices only expressions which are defined by the knots and the weights are used. The three problems are in close connection. Suppose $s^*(x)$ is the unique solution to (OP3) for given μ. Then a sum of squared errors S is associated with this solution. $s^*(x)$ again results as the solution of (R) if this S serves as the constraint in (R). An analogous statement is true concerning the total curvature T of $s^*(x)$.

4.1 Estimating the smoothing parameters

As already shown, the parameter μ in (OP3) is a smoothing parameter that steers the equalization of the given points. The interpolating natural cubic spline results for $\mu = 0$ and for $\mu \to \infty$ the regression line with regard to (x_i, y_i) is obtained in the asymptotic process. For problem (R) smoothing is controlled by the parameter S. For $S = 0$, the interpolating spline is also obtained as a solution. When S is chosen between zero and the sum of squared errors of the regression line then (for $k_i = 1,\ i = 1, 2, \ldots, m$) the solution to (R) is a natural cubic spline. Its sum of the squared errors with respect to the y_i, $i = 1, 2, \ldots, m$, is exactly S. In (GK3) smoothing is steered by the bound T of the total curvature. If T lies between zero and the total curvature of the interpolating spline then the solution is a natural cubic spline function that also takes on the bound T.
A great problem is the right choice of the respective smoothing parameter. Depending on the relationship between the sum of the squared errors and total curvature for a given spline, visible smoothing effects appear for very different parameters. In each case the specification is somewhat arbitrary which is not satisfactory.
The cross-validation method for the estimation of smoothing parameters will be illustrated followingly. The problem (OP3) and the choice of μ are observed.
Up to now the smoothing splines were seen as solutions to optimization tasks. The probabilistic model

$$Y = g(X) + e \tag{4.30}$$

is taken into account in order to make statements about the choice of μ. The error e should fulfill $E(e) = 0$ and $V(e) = \sigma^2$. Equation (4.30) states that the relationship $E(Y|X = x_0) = g(x_0)$ is true for arbitrary x_0. Smoothing of the given data $y_i = g(x_i) + e_i$, $i = 1, 2, \ldots, m$, therefore corresponds with an estimation $\hat{g}(x)$ of $g(x)$. It remains to be observed that the errors e_i are presupposed as stochastically independent.
The following statements are true for general linear estimators. However, to follow the topic of this chapter, they are especially formulated for smoothing natural splines (see Hastie and Tibshirani (1990)).
Given is a fixed $\mu \geq 0$. This parameter is used to construct the uniquely determined smoothing natural cubic spline $s_\mu(x) = \hat{g}_\mu(x)$ for the points

(x_i, y_i) which minimizes

$$\mu \int_{x_1}^{x_m} [g''(x)]^2 \mathrm{d}x + \sum_{i=1}^{m} \left(\frac{g(x_i) - y_i}{k_i} \right)^2 . \tag{OP3}$$

This $s_\mu(x)$ is therefore seen as an estimator for the unknown function g. We define the expected mean-square error

$$MSE(\mu) = \frac{1}{m} \sum_{i=1}^{m} E[(s_\mu(x_i) - g(x_i))^2] \tag{4.31}$$

and the expected value of the average predictive squared error

$$PSE(\mu) = \frac{1}{m} \sum_{i=1}^{m} E[(Y_i^* - s_\mu(x_i))^2]. \tag{4.32}$$

Here, Y_i^* are random variables with new realizations $y_i^* = g(x_i) + e_i^*$. The y_i^* did not effect the construction of $s_\mu(x)$.
A simple relationship between $MSE(\mu)$ and $PSE(\mu)$ will be demonstrated next. For every summand of $PSE(\mu)$,

$$\begin{aligned} E[(Y_i^* - s_\mu(x_i))^2] &= E[(g(x_i) + e_i^* - s_\mu(x_i))^2] \\ &= E[(g(x_i) - s_\mu x_i))^2] + 2(g(x_i) - s_\mu(x_i))E[e_i^*] + E[e_i^{*2}] \\ &= E[(g(x_i) - s_\mu(x_i))^2] + \sigma^2 \end{aligned}$$

is true since $E[e_i^{*2}] = E[(e_i^* - 0)^2] = V[e_i^*]$. This results in

$$PSE(\mu) = MSE(\mu) + \sigma^2 \ . \tag{4.33}$$

A successful method for automatically determining a smoothing parameter is the cross-validation method. In this process exactly one point (x_i, y_i) is removed from the data set and the smoothing spline $s_\mu^{-i}(x)$ is calculated for the remaining $(m-1)$ points. The distance between the removed point (x_i, y_i) and the calculated smoothing spline $s_\mu^{-i}(x)$ is expressed by $(y_i - s_\mu^{-i}(x_i))^2$. When every point is removed one after the other one obtains

$$CV(\mu) = CV_{(OP3)}(\mu) = \frac{1}{m} \sum_{i=1}^{m} \left(y_i - s_\mu^{-i}(x_i)\right)^2 \ . \tag{4.34}$$

What can be said about the expected value $E[CV(\mu)]$?
The following is calculated for the random variable $Y_i = g(x_i) + e_i$,

$$\begin{aligned} E[\left(Y_i - s_\mu^{-i}(x_i)\right)^2] &= E[\left(Y_i - g(x_i) + g(x_i) - s_\mu^{-i}(x_i)\right)^2] \\ &= E[(Y_i - g(x_i))^2] + 2E[(Y_i - g(x_i)) \cdot \left(g(x_i) - s_\mu^{-i}(x_i)\right)] \\ &\quad + E[\left(g(x_i) - s_\mu^{-i}(x_i)\right)^2] \\ &= E[\left(g(x_i) - s_\mu^{-i}(x_i)\right)^2] + \sigma^2 \ , \end{aligned}$$

because $E[(Y_i - g(x_i))^2] = E[(g(x_i) - e_i - g(x_i))^2] = \sigma^2$.
$E[Z_1 \cdot Z_2] = E[Z_1] \cdot E[Z_2]$ is true for two independent random variables Z_1 and Z_2. Since $s_\mu^{-i}(x_i)$ is independent of Y_i, $E[(Y_i - g(x_i)) \cdot (g(x_i) - s_\mu^{-i}(x_i))] = E[e_i] \cdot E[(g(x_i) - s_\mu^{-i}(x_i))] = 0$ follows.
With the assumption $s_\mu^{-i}(x_i) \approx s_\mu(x_i)$ we get

$$E[CV(\mu)] \approx PSE(\mu) \ . \tag{4.35}$$

This motivates the following statement:

Use the minimum of $CV(\mu)$ with regard to μ for the smoothing parameter in (OP3).

At first glance we could also use $\frac{1}{m}\sum_{i=1}^{m}(y_i - s_\mu(x_i))^2$ as an estimator for $PSE(\mu)$ and minimize it with regard to μ. This does not make any sense though because the minimum always results for $\mu = 0$. But this corresponds with the interpolating spline.
At first the calculation of $CV(\mu)$ is arduous. m smoothing splines $s_\mu^{-i}(x)$ must be constructed to determine one function value of $CV(\mu)$. A large amount of function values are necessary to determine the minimum. However, it turns out that $s_\mu^{-i}(x_i)$ result from $s_\mu(x_i)$, $i = 1, 2, \ldots, m$. This will be formulated in the following theorem.
$H_\mu = H = (h_{ij})_{i,j=1,\ldots,m}$ denotes the matrix that uniquely determines the smoothing spline over $s_\mu = (s_\mu(x_1), \ldots, s_\mu(x_m))^T = H \circ Y$ (see (4.10)).

Theorem 4.6. (HASTIE and TIBSHIRANI (1990)) *For fixed knots x_i, $i = 1, 2, \ldots, m$, and $\mu > 0$*

$$CV(\mu) = \frac{1}{m}\sum_{i=1}^{m}\left(y_i - s_\mu^{-i}(x_i)\right)^2 = \frac{1}{m}\sum_{i=1}^{m}\left(\frac{y_i - s_\mu(x_i)}{1 - h_{ii}}\right)^2$$

is true.

Proof. When $s_\mu^{-i}(x)$ solves the problem (OP3) for the dataset without the point (x_i, y_i) then $s_\mu^{-i}(x)$ is also the solution for the dataset in which (x_i, y_i) is replaced by the point $(x_i, s_\mu^{-i}(x_i))$. A point that lies directly on the smoothing function does not change the minimum of (OP3) and therefore also does not change the solution. On the other hand, $H = H_\mu$ is the matrix that generates the solution to (OP3) for all m-dimensional vectors Y. Hence $s_\mu^{-i}(x_i) = \sum_{j \neq i} h_{ij} y_j + h_{ii} s_\mu^{-i}(x_i)$ or $s_\mu^{-i}(x_i) = \sum_{j \neq i}\left(\frac{h_{ij}}{1 - h_{ii}}\right) y_j$.

Hence $y_i - s_\mu^{-i}(x_i) = y_i - \sum\limits_{j \neq i} \left(\frac{h_{ij}}{1 - h_{ii}} \right) y_j$ is true. If $\left(\frac{h_{ii}}{1 - h_{ii}} \right) y_i$ is added and at the same time subtracted

$$y_i - s_\mu^{-i}(x_i) = \left(1 + \frac{h_{ii}}{1 - h_{ii}}\right) y_i - \left[\sum_{j \neq i} \left(\frac{h_{ij}}{1 - h_{ii}} \right) y_j + \left(\frac{h_{ii}}{1 - h_{ii}} \right) y_i \right]$$

$$= \frac{y_i - s_\mu(x_i)}{1 - h_{ii}}$$

results and the theorem has been proven. □

Note 4.3. The matrix H^{-i} with elements h_{jk}^{-i} results from H via $h_{jk}^{-i} = \frac{h_{jk}}{1 - h_{ii}}$, $j, k \neq i$. Further, $\sum\limits_{k=1, k \neq i}^{m-1} \frac{h_{jk}}{1 - h_{ii}} = 1$ for all $j = 1, 2, \ldots, m-1$, $j \neq i$, because $\sum\limits_{k=1}^{m} h_{jk} = 1$ for all j.

Similarly, the error function $CV_{(R)}(S) = \sum_{i=1}^{m} \left(y_i - s_S^{-i}(x_i)\right)^2$ can be defined for the optimization problems (R) and the error function $CV_{(GK3)}(T) = \sum_{i=1}^{m} (y_i - s_T^{-i}(x_i))^2$ can be defined for the optimization problem (GK3).
Note that all error functions may possess multiple local minima.
Using the data from Example 4.4, $CV_{(OP3)}(\mu)$ possesses a unique local minimum at $\mu_{\min} = 0.318931$. Figure 4.10 shows a part of this error function and Figure 4.11 shows the corresponding solution to (OP3) for $\mu_{\min}$.
A different situation is represented in Figure 4.12. The data were chosen such that the minimum of $CV_{(OP3)}(\mu)$ is a very large number. The smoothing spline is near the regression line. When corresponding iterative algorithms for the determination of the minimum of $CV_{(OP3)}(\mu)$ do not converge then (OP3) can also be reformulated:
Minimize the convex linear combination

$$\mu \int_{x_1}^{x_m} [f''(x)]^2 \mathrm{d}x + (1 - \mu) \sum_{i=1}^{m} \left(\frac{f(x_i) - y_i}{k_i} \right)^2 \quad \text{for } 0 \leq \mu < 1 \ .$$

Figure 4.13 shows the error function $CV_{(R)}(S)$ for the data used in Example 4.4 as a function of the smoothing parameter $S = SSE$ ($k_i = 1, i = 1, 2, \ldots, m$). The solutions to (R) related to the three local minima $SSE = 1.00787$, $SSE = 5.1193$ and $SSE = 8.96825$ of the function $CV_{(R)}(S)$ are illustrated in Figure 4.14.

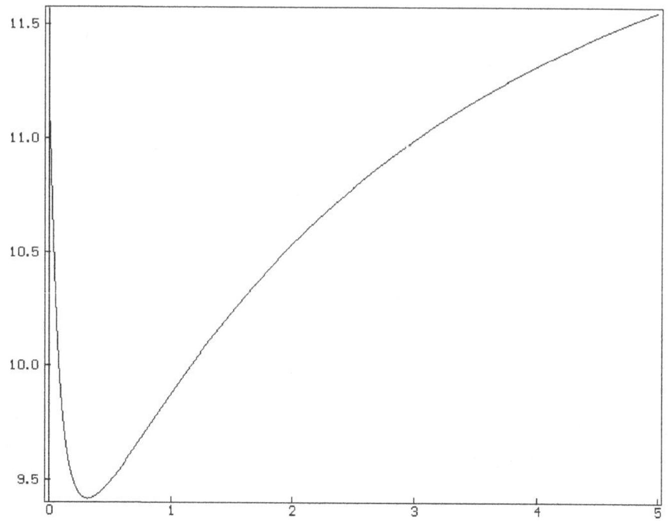

Fig. 4.10: Part of $CV_{(OP3)}(\mu)$ for the data from Example 4.4.

Figure 4.15 shows the error function $CV_{(GK3)}(T) = CV_{(GK3)}(GK)$. The minimum lies at $T = GK_{\min} = 3.124514$. The respective solution to (GK3) for $T = GK_{\min}$ and $k_i = 1$, $i = 1, 2, \ldots, m$, is illustrated in Figure 4.16.

Example 4.5. A fictitious data set is used to document the efficiency of the estimation procedure. An unknown functional process is described by the points in Figure 4.17. No knowledge exists with regard to the type of a related function and the accuracy of the measurement also cannot be verified. Due to this the approach used in Example 4.2 cannot be taken. Nevertheless, a structure within the data should be sought. The natural cubic spline function constructed for the local minimum of $CV_{(OP3)}(\mu)$ is illustrated in Figure 4.18 (see Part III). The data stem from the function $\sin(x)/x$. From these function, 500 points between 0.1 and 10 were taken and distorted by a $N(0, (0.7)^2)$-distributed error. Not only the respective oscillation of the model function, but also the decrease in amplitude is clarified through the spline function.

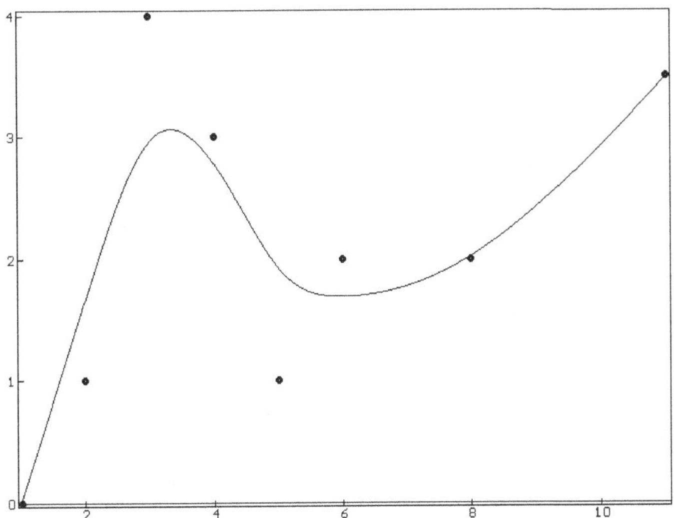

Fig. 4.11: Solution to (OP3) where $\mu{=}\mu_{\min} = 0.318931$, data from Example 4.4.

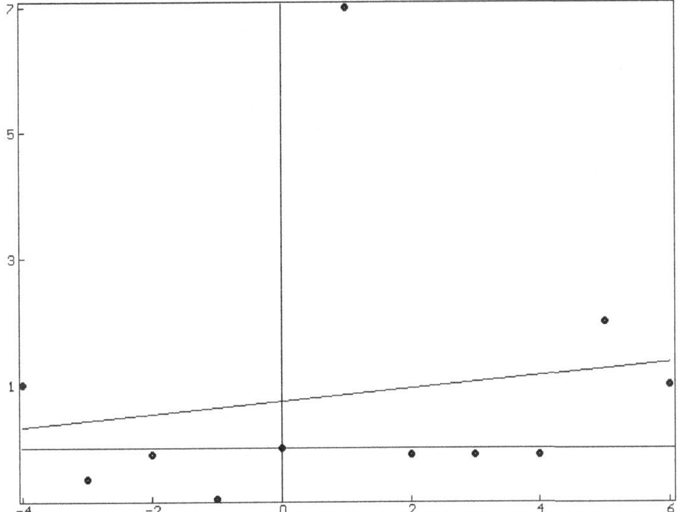

Fig. 4.12: The regression line as a solution to (OP3), data such that $CV_{(OP3)}(\mu)$ is very large.

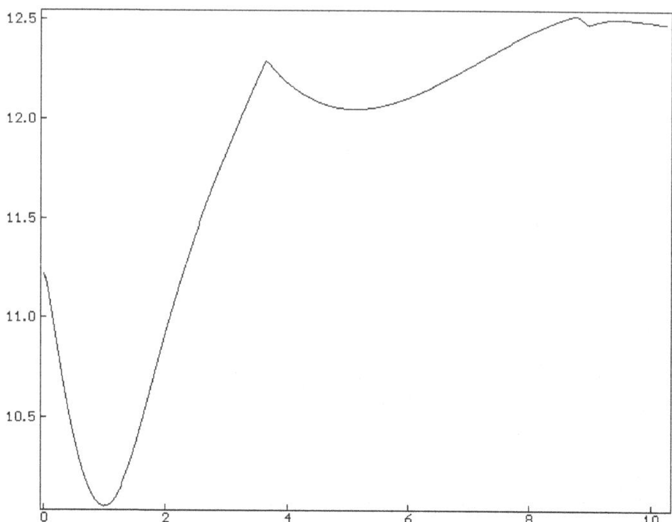

Fig. 4.13: Error function $CV_{(R)}(S)$, data from Example 4.4.

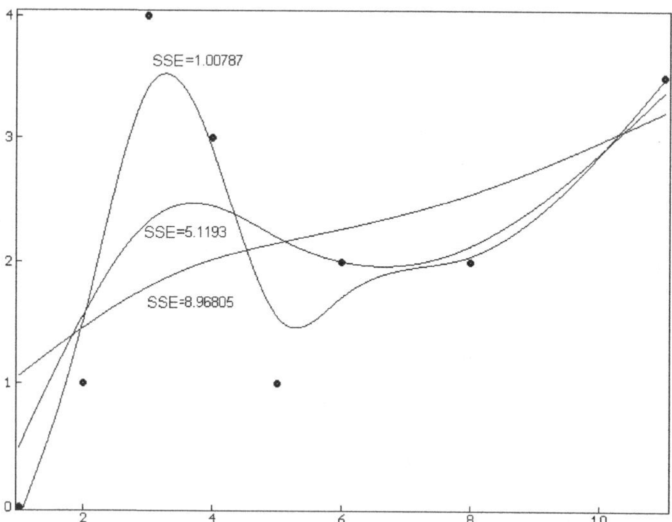

Fig. 4.14: The solutions to (R) related the local minima of $CV_{(R)}(S)$, data from Example 4.4.

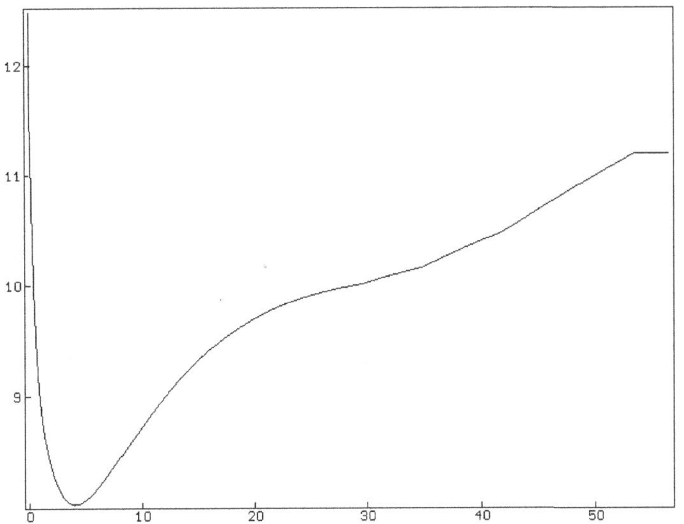

Fig. 4.15: The error function $CV_{(GK3)}(T)$, data from Example 4.4.

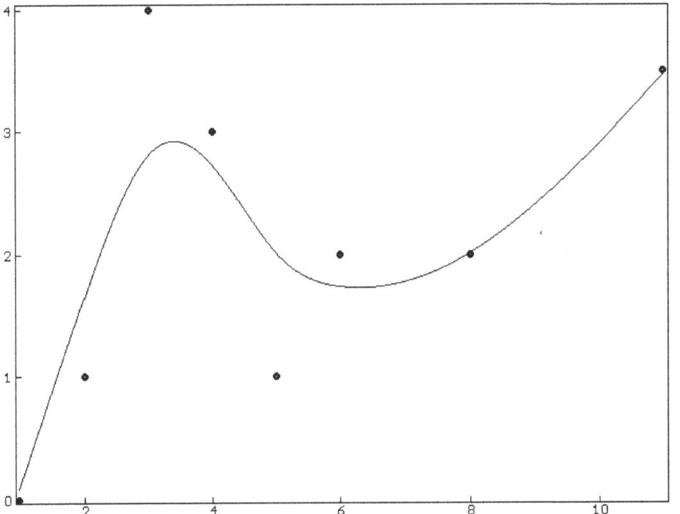

Fig. 4.16: The solution to (OP3) with $T = GK_{\min} = 3.124514$, data from Example 4.4.

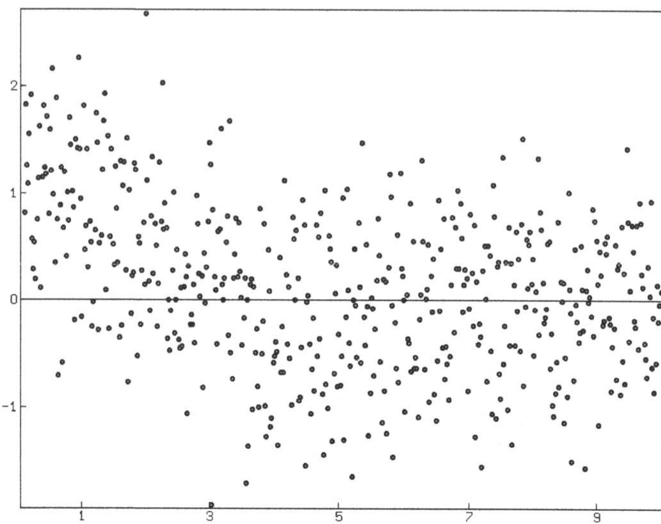

Fig. 4.17: Raw data of a process, see data from Example 4.5.

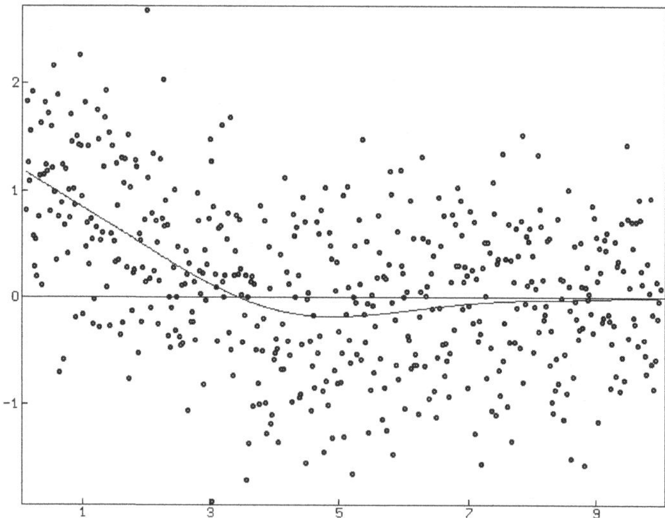

Fig. 4.18: Regressing natural cubic spline with estimated smoothing parameter μ, see data from Example 4.5.

Chapter 5

Interpolating quadratic splines

The class of quadratic splines is interesting because it does not contain natural splines and because the numeric calculation of the interpolating problem requires little effort. Interpolating quadratic splines with minimal total curvature are developed. Further optimization criteria for the construction of interpolating quadratic splines are observed. In addition, the following question is posed: when is a polynomial of the degree 2 exactly reconstructed by a quadratic spline? More interesting results can be seen here than for cubic splines.
Similarly to cubic splines, this chapter will discuss quadratic spline functions. The discussion in this Chapter is based on the general definition of a spline function of degree n, the spline functions that are to be observed are introduced.

Definition 5.1 (quadratic spline). *A function $s_2 : \mathbb{R} \to \mathbb{R}$ is called a spline function of degree 2 or a quadratic spline function with the knots $x_1 < x_2 < \ldots < x_m$ if and only if*

a) *In each interval $[x_i, x_{i+1}]$, $i = 1, 2, \ldots, m-1$, $(-\infty, x_1]$ and $[x_m, +\infty)$, $s_2(x)$ is given by some polynomial of degree 2 or less.*
b) *$s_2(x)$ is continuously differentiable everywhere.*

For the task of interpolation a quadratic spline function is searched for that fulfills the conditions $s_2(x_i) = y_i$ for the given points (x_i, y_i), $i = 1, 2, \ldots, m$, where $x_1 < x_2 < \ldots < x_m$. According to Definition 5.1 we proceed with $s_2(x) = A_i(x-x_i)^2 + B_i(x-x_i) + C_i$ over the interval $[x_i, x_{i+1}]$.

$$C_i = y_i, \qquad i = 1, 2, \ldots, m-1\,, \tag{5.1}$$

result due to the interpolation conditions.

In each case, $s_2'(x)$ is a straight line on the intervals between the knots. That is why the two point equation is

$$s_2'(x) = 2A_i(x - x_i) + B_i = \frac{s_2'(x_{i+1}) - s_2'(x_i)}{x_{i+1} - x_i}(x - x_i) + s_2'(x_i)\,.$$

From that $s_2''(x) = s_2''(x_i) = 2A_i = (s_2'(x_{i+1}) - s_2'(x_i))/(x_{i+1} - x_i)$ results for all $x \in [x_i, x_{i+1})$. One obtains

$$A_i = \frac{1}{2}(B_{i+1} - B_i)/\Delta x_i, \qquad i = 1, 2, \ldots, m-1 \tag{5.2}$$

with the short form $\Delta x_i := x_{i+1} - x_i$.
An interpolating spline function of degree 2 is therefore uniquely determined by the first derivatives in the knots $s_2'(x_i) = B_i$, $i = 1, 2, \ldots, m$.
To calculate these derivatives it is used that $s_2(x)$ is continuous. In particular, for all knots it is true that

$$A_i(x_{i+1} - x_i)^2 + B_i(x_{i+1} - x_i) + C_i = y_{i+1}, \qquad i = 1, 2, \ldots, m-1\,.$$

If the obtained expressions for A_i and C_i are inserted in these relationships and similarly $\Delta y_i := y_{i+1} - y_i$ is introduced then

$$B_i + B_{i+1} = 2\left(\frac{\Delta y_i}{\Delta x_i}\right), \qquad i = 1, 2, \ldots, m-1\,, \tag{5.3}$$

is obtained.

Theorem 5.1. *Given are the m points (x_i, y_i) where $x_1 < x_2 < \ldots < x_m$. The interpolation task can then uniquely be solved by a quadratic spline in each of the following described situations:*

a) *The first derivative $s_2'(x_k) = f_k$ is given for one arbitrary k in $\{1, 2, \ldots, m\}$.*
b) *The second derivative $s_2''(x_k) = g_k$ is given for one arbitrary k in $\{1, 2, \ldots, m-1\}$.*
c) *The relationship $u \cdot s_2'(x_k) = s_2'(x_{k+1})$ is true for one certain k in $\{1, 2, \ldots, m-1\}$ and $u \neq -1$.*
d) *When m is an even number, $s_2(x_1) = s_2(x_m)$ and $s_2'(x_1) = s_2'(x_m)$ are true. The spline is constructed as a periodic function with the period $x_m - x_1$.*
   When m is an odd number, the antiperiodicity condition $s_2'(x_1) = -s_2'(x_m)$ is true.

Proof. a) The quadratic spline is determined by the $B_i = s_2'(x_i)$, $i = 1, 2, \ldots, m$. These values can successively be calculated from the given f_k due to (5.3).
If with b) $s_2''(x_k) = g_k = 2A_k = (B_{k+1} - B_k)/\Delta x_k$ is true, then $B_{k+1} = g_k \Delta x_k + B_k$ is obtained. With this one has $B_k = \frac{\Delta y_k}{\Delta x_k} - \frac{g_k \Delta x_k}{2}$. This situation leads back to a).
The same is true for c), $B_k + B_{k+1} = B_k + uB_k = 2\Delta y_k/\Delta x_k$ leads to $B_k = 2/(1+u)\Delta y_k/\Delta x_k$.
Concerning d) it is first noted that the $y_1 = y_m$ is not directly needed for the proof. However, if it is not true, we cannot speak of a periodic spline anymore. The condition $B_1 = B_m$ or $B_1 - B_m = 0$ is appended to the linear system of equations (5.3) as the mth equation. With this the following matrix of coefficients results

$$A := \begin{pmatrix} 1\,1\,0\,0 \ldots 0 & 0 \\ 0\,1\,1\,0 \ldots 0 & 0 \\ 0\,0\,1\,1 \ldots 0 & 0 \\ \vdots\;\vdots\;\vdots\;\vdots \;\ddots\; \vdots & \vdots \\ 0\,0\,0\,0 \ldots 1 & 1 \\ 1\,0\,0\,0 \ldots 0 & -1 \end{pmatrix} .$$

To calculate the determinant of A, this is developed with respect to the last row. Both of the remaining subdeterminants are of simple structure. This means that the subdeterminants each can be calculated from the product of the elements of the main diagonal and are therefore equal to one.
For odd m, the signs that are to be observed in the development are both positive such that $\det(A) = 1 \cdot 1 \cdot 1 + 1 \cdot (-1) \cdot 1 = 0$. For even m, different signs appear such that $\det(A) = -1 \cdot 1 \cdot 1 + 1 \cdot (-1) \cdot 1 = -2$ is true. With this, the periodic quadratic spline is only determined for even values of m. The course of the proof is analogous for uneven values of m and $B_1 + B_m = 0$.
□

Example 5.1. Given are the 10 points $(1, 1)$, $(2, 2)$, $(2.5, 2)$, $(3, 2.3)$, $(4, 1.8)$, $(4.2, 1.5)$, $(4.5, 1.1)$, $(5, 1.3)$, $(6, 1.5)$ and $(7, 1)$.
Constructed are the uniquely determined interpolating quadratic splines related to the conditions $s_2'(x_2) = s_2'(2) = 0$, $s_2''(x_2) = s_2''(2) = 10$ and $s_2'(x_2) = s_2'(x_1) = s_2'(2) = s_2'(1)$, respectively (see Figure 5.1).

Example 5.2. For the given condition $y_1 = y_m$, the uniquely determined interpolating periodic quadratic spline, where $s_2'(x_1) = s_2'(x_m)$, was constructed for the points of Example 5.1 (see Figure 5.2).

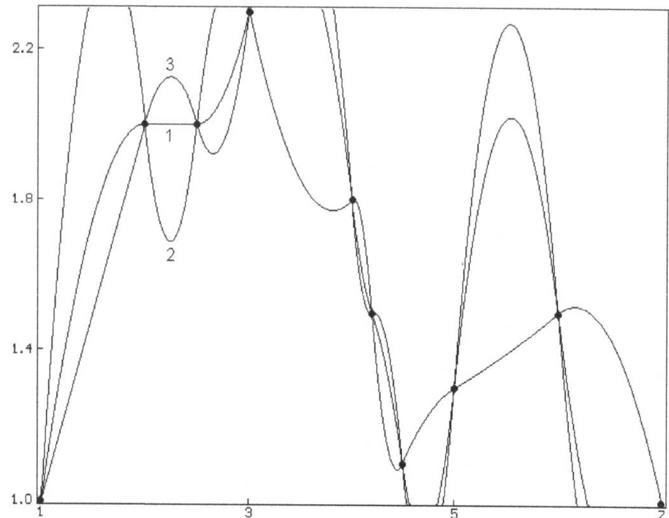

Fig. 5.1: Interpolating quadratic splines, data from Example 5.1:
1: $s_2'(x_2) = s_2'(2) = 0$, 2: $s_2''(x_2) = s_2''(2) = 10$, 3: $s_2'(x_2) = s_2'(x_1)$; $u = 1$.

From the theory regarding cubic splines it is known that the interpolating natural cubic spline possesses minimal total curvature in the set of all twice continuously differentiable functions that also solve the task of interpolation.

In Figure 5.1 it can be seen that the quadratic spline can possess large oscillations when the constraints that need to be specified are chosen inadequately. That is why an attempt should be made to construct an interpolating spline function of degree two that fulfills a peculiar condition of optimization. The following three criteria (O1), (O2) and (O3) are observed:

$$GK := \int_{x_1}^{x_m} [s_2''(x)]^2 \mathrm{d}x = \min! \tag{O1}$$

$$AQ := \int_{x_1}^{x_m} [s_2'(x)]^2 \mathrm{d}x = \min! \tag{O2}$$

$$LK := \int_{x_1}^{x_m} \frac{[s_2''(x)]^2}{(1 + s_2'(x)^2)^3} \mathrm{d}x = \min! \tag{O3}$$

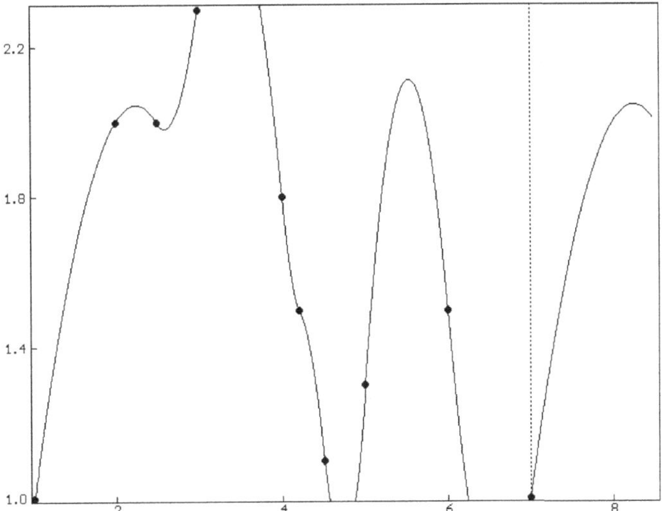

Fig. 5.2: Interpolating periodic quadratic spline, data from Example 5.2.

Since $s_2(x)$ is comprehensively described by the points (x_i, y_i) and the B_i, the objective functions of (O1), (O2) and (O3) can also be described by these values.

For GK,

$$\begin{aligned}\int_{x_1}^{x_m} [s_2''(x)]^2 \mathrm{d}x &= \sum_{i=1}^{m-1} \int_{x_i}^{x_{i+1}} (2A_i)^2 \mathrm{d}x = \sum_{i=1}^{m-1} 4A_i^2 \Delta x_i = \sum_{i=1}^{m-1} \frac{1}{\Delta x_i} [B_{i+1} - B_i]^2 \\ &= \sum_{i=1}^{m-1} \frac{1}{\Delta x_i} [B_{i+1} + B_i - 2B_i]^2 = 4 \sum_{i=1}^{m-1} \frac{1}{\Delta x_i} \left[\frac{\Delta y_i}{\Delta x_i} - B_i\right]^2 \\ &= 4 \sum_{i=1}^{m-1} \frac{1}{\Delta x_i} [D_i - B_i]^2 ,\end{aligned}$$

where $D_i := \Delta y_i / \Delta x_i$, is true.

For AQ,

$$\begin{aligned}\int_{x_1}^{x_m} [s_2'(x)]^2 \mathrm{d}x &= \sum_{i=1}^{m-1} \int_{x_i}^{x_{i+1}} [2A_i(x - x_i) + B_i]^2 \mathrm{d}x \\ &= \sum_{i=1}^{m-1} \left[\frac{4}{3} A_i^2 \Delta x_i^3 + 2A_i B_i \Delta x_i^2 + B_i^2 \Delta x_i\right]\end{aligned}$$

is true. Due to the fact that

$$A_i = \frac{B_{i+1} - B_i}{2\Delta x_i} = \frac{B_{i+1} + B_i - 2B_i}{2\Delta x_i} = \frac{\frac{\Delta y_i}{\Delta x_i} - B_i}{\Delta x_i} = \frac{D_i - B_i}{\Delta x_i}$$

we obtain

$$\begin{aligned}\int_{x_1}^{x_m} [s_2'(x)]^2 \,\mathrm{d}x &= \sum_{i=1}^{m-1} \Delta x_i \left[\frac{4}{3}(D_i - B_i)^2 + 2(D_i - B_i)B_i + B_i^2\right] \\ &= \sum_{i=1}^{m-1} \Delta x_i [\frac{1}{3}B_i^2 - \frac{2}{3}B_i D_i + \frac{4}{3}D_i^2] \\ &= \frac{1}{3}\sum_{i=1}^{m-1} \Delta x_i (D_i - B_i)^2 + \sum_{i=1}^{m-1} \Delta x_i D_i^2 \,.\end{aligned}$$

For LK, the minimization of the squared total local curvature

$$\int_{x_1}^{x_m} \frac{[s_2''(x)]^2}{(1 + s_2'(x)^2)^3} \mathrm{d}x$$

falls back on MAESZ (1984). To simplify the problem, in each case, the first derivative of $s_2(x)$ over the interval $[x_i, x_{i+1}]$ is approximated by D_i. With this

$$\int_{x_1}^{x_m} \frac{[s_2''(x)]^2}{(1 + s_2'(x)^2)^3} \mathrm{d}x \approx 4\sum_{i=1}^{m-1} \frac{\Delta x_i}{(1 + D_i^2)^3} A_i^2 = 4\sum_{i=1}^{m-1} \frac{1}{\Delta x_i \, (1 + D_i^2)^3} (D_i - B_i)^2 \,.$$

Based on Theorem 5.1 it is known that B_1 needs to be given for the unique construction of $s_2(x)$. With this, the above optimization problems (O1), (O2) and (O3) can be formulated as follows:

$$\left.\begin{array}{l} F = \sum\limits_{i=1}^{m-1} w_i (D_i - B_i)^2 = \min\limits_{B_1}! \text{ with the constraints} \\ w_i > 0 \text{ and } B_i + B_{i+1} = 2D_i, \; i = 1, 2, \ldots, m-1 \,. \end{array}\right\}$$

Thereby for (O1) $w_i = 1/\Delta x_i$, for (O2) $w_i = \Delta x_i$, since $\sum_{i=1}^{m-1} \Delta x_i D_i^2$ is no longer dependent on B_i, and in the case of (O3) $w_i = 1/\Delta x_i \left(1 + D_i^2\right)^3$. First the objective function F is represented as a function of $t := B_1$. With that one receives the representations from the constraints for the B_i,
$i = 1$, $B_1 = t$,
$i = 2$, $B_2 = 2D_1 - t$,

$i = 3$, $B_1 = 2D_2 - B_2 = 2D_2 - 2D_1 + t$.
In general,

$$B_i = 2\sum_{k=1}^{i-1}(-1)^{i+k+1}D_k + (-1)^{i+1}t\,. \tag{5.4}$$

If these expressions are inserted into the objective function $F(t)$, the following is obtained:

$$F(t) = \sum_{i=1}^{m-1} w_i[D_i - 2\sum_{k=1}^{i-1}(-1)^{i+k+1}D_k - (-1)^{i+1}t]^2\,.$$

Necessary for a local minimum is

$$\begin{aligned} F'(t) &= 2\sum_{i=1}^{m-1} w_i \left[D_i - 2\sum_{k=1}^{i-1}(-1)^{i+k+1}D_k - (-1)^{i+1}t\right](-1)^i \\ &= 2\sum_{i=1}^{m-1} w_i \left[(-1)^i D_i - 2\sum_{k=1}^{i-1}(-1)^{k+1}D_k + t\right] = 0\,. \end{aligned}$$

From this it follows that

$$B_1 = t = \frac{\sum\limits_{i=1}^{m-1} w_i\left((-1)^{i+1}D_i + 2\sum\limits_{k=1}^{i-1}(-1)^{k+1}D_k\right)}{\sum\limits_{i=1}^{m-1} w_i} \tag{5.5}$$

or, after rearranging,

$$B_1 = \frac{\sum\limits_{i=1}^{m-1}(-1)^{i+1}\left(w_i + 2\sum\limits_{k=i+1}^{m-1} w_k\right)D_i}{\sum\limits_{i=1}^{m-1} w_i}\,. \tag{5.6}$$

From $F''(t) = 2\sum_{i=1}^{m-1} w_i > 0$ follows that (5.6) is a local minimum. According to the goal of construction, only the w_i need to be made particular.

Example 5.3. Figure 5.3 shows the interpolating quadratic splines for the five points (1.0, 10.0), (2.0, 5.0), (3.0, 4.0), (6.0, 2.0) and (8.0, 1.0), that solve (O1), (O2) and (O3), respectively.

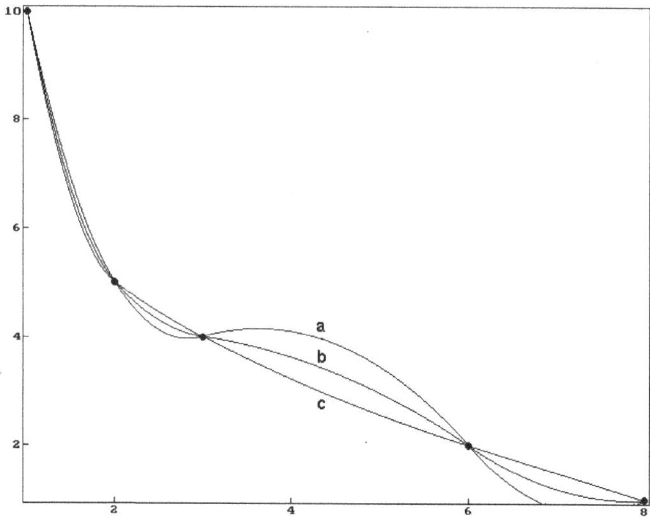

Fig. 5.3: Interpolating quadratic splines (a) as a solution to (O1), (b) as a solution to (O2) and (c) as a solution to (O3), data from Example 5.3.

Remark 5.1. In the case of equidistant knots, (O1) and (O2) are equivalent. $\Delta x_i = h$ is then true for all i from 1 to $(m-1)$. In the calculation of (O1), the constraints $w_i = 1/\Delta x_i$ are set. The relationship (5.6) leads to

$$B_1 = \frac{\frac{1}{h}\sum_{i=1}^{m-1}(-1)^{i+1}\left(1+2\sum_{k=i+1}^{m-1}1\right)\frac{\Delta y_i}{h}}{\sum_{i=1}^{m-1}\frac{1}{h}},$$

$$B_1 = \frac{\sum_{i=1}^{m-1}(-1)^{i+1}\left(1+2\sum_{k=i+1}^{m-1}1\right)\Delta y_i}{(m-1)h},$$

$$B_1 = \frac{\sum_{i=1}^{m-1}(-1)^{i+1}(2(m-i)-1)\Delta y_i}{(m-1)h}.$$

For the problem (O2), the $w_i = \Delta x_i = h$, $i = 1, 2, \ldots, m-1$, need to be set so that the same initial condition B_1 results.

The uniquely determined interpolating quadratic spline function $s_2(x)$ with minimal total curvature over $[x_1, x_m]$, constructed for the data in Example

5.1, is represented together with the interpolating natural cubic spline $s_3(x)$ in Figure 5.4.

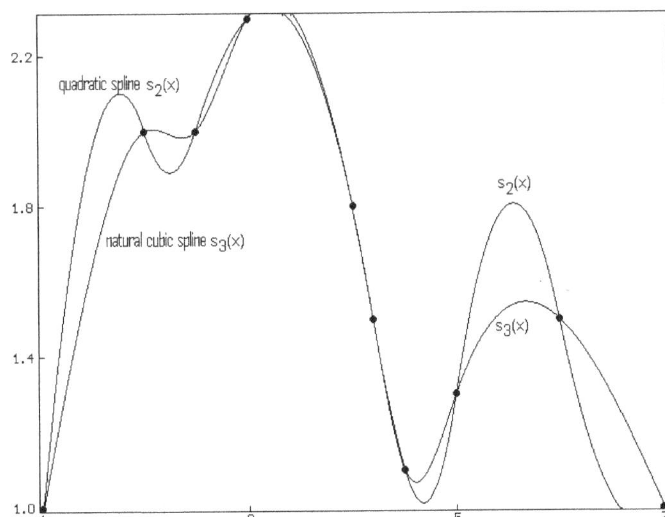

Fig. 5.4: Interpolating quadratic and cubic spline with minimal total curvature, data from Example 5.1.

It can be seen clearly that the interpolating natural cubic spline function $s_3(x)$ has an optically smoother course than the quadratic spline with minimal total curvature. In fact, the total curvature for $s_3(x)$ is 21.09571 units and for $s_2(x)$ it is 54.489697. This brings up the question if this result is coincidental. With regard to this, are the interpolating natural cubic splines predominant over the interpolating quadratic splines with minimal total curvature? HOLLADAY's theorem cannot be used directly to answer this question because the quadratic spline functions in general are only once continuously differentiable. The main idea of the proof however will be applied in the following lemma.

Lemma 5.1. *Given are m points (x_i, y_i), $i = 1, 2, \ldots, m$, where $x_1 < x_2 < \ldots < x_m$, the respective interpolating natural cubic spline function $s_3(x)$ and an arbitrary interpolating spline $s_2(x)$ of degree 2.*
Then $\int_a^b [s_3''(x)]^2 \, \mathrm{d}x \leq \int_a^b [s_2''(x)]^2 \, \mathrm{d}x$ for all $a \leq x_1$ and $b \geq x_m$. The equality is only true when $s_2(x) \equiv s_3(x) = \alpha x + \beta$.

Proof.

$$\int_a^b [s_2''(x)]^2 \, dx = \int_a^b [s_3''(x) + s_2'' - s_3''(x)]^2 \, dx$$

$$= \int_a^b [s_3''(x)]^2 \, dx + \int_a^b [s_2''(x) - s_3''(x)]^2 \, dx + 2\int_a^b s_3''(x)[s_2''(x) - s_3''(x)]dx.$$

After partial integration, the evaluation of the third integral results in

$$\int_a^b s_3''(x)[s_2''(x) - s_3''(x)]dx = \sum_{i=1}^{m-1} \int_{x_i}^{x_{i+1}} s_3''(x)[s_2''(x) - s_3''(x)]dx$$

$$= \sum_{i=1}^{m-1} s_3''(x)[s_2'(x) - s_3'(x)]|_{x_i}^{x_{i+1}} - \sum_{i=1}^{m-1} \int_{x_i}^{x_{i+1}} s_3'''(x)[s_2'(x) - s_3'(x)]dx$$

$$= -s_3''(x_1)[s_2''(x_1) - s_3''(x_1)] + s_3''(x_m)[s_2''(x_m) - s_3''(x_m)]$$

$$- \sum_{i=1}^{m-1} s_3'''(x_i)[s_2(x_{i+1}) - s_3(x_{i+1}) - s_2(x_i) + s_3(x_i)] = 0 \, ,$$

because the second derivative $s_3''(x) \equiv 0$ over $\mathbb{R}\backslash(x_1, x_m)$, $s_3'''(x) = s_3'''(x_i)$ is constant over $[x_i, x_{i+1})$ and $s_3(x_i) = s_2(x_i)$ is true for all i from 1 to m. With this the inequality of the lemma is proven.
To prove the second statement of the lemma, $\int_a^b [s_2''(x) - s_3''(x)]^2 dx = 0$ is evaluated.
If x is an arbitrary value in $[a, b]$ then there exists a $j \in \{0, 1, 2, \ldots, m\}$ so that $x \in [x_j, x_{j+1})$. With this, $x_0 = a$ and $x_{m+1} = b$. $s_2''(x) = \alpha_j$ is true over $[x_j, x_{j+1})$ so that $s_3''(x) = \alpha_j$ must also be true for all $x \in [x_j, x_{j+1})$. From the continuity of $s_3''(x)$ over $\mathbb{R}$, $s_2''(x) = s_3''(x) = \alpha$ is obtained for all $x \in [a, b]$.
However, since $s_3(x)$ is a natural cubic spline function, $s_3''(x) = 0$ must be true for all $x \in [a, x_1]$ and $x \in [x_m, b]$. In total, $s_2''(x) \equiv s_3''(x) \equiv 0$ results over $[a, b]$. Then $s_2(x)$ and $s_3(x)$ are each a straight line. $s_2(x) \equiv s_3(x) = \alpha x + \beta$ follows from the shared conditions of interpolation. □

Remark 5.2. Lemma 5.1 can be generalized insofar as $s_3(x)$ does not need to be a natural cubic spline. Only $s_3'(x_1) = s_2'(x_1)$ and $s_3'(x_m) = s_2'(x_m)$ need to be true for the cases $a = x_1$ and $b = x_m$.

Since the assertion from Lemma 5.1 is true for arbitrary interpolating splines of degree 2, it is naturally also true for the quadratic splines with

minimal total curvature. With this it is clear that, with regard to total curvature, the interpolating quadratic splines $s_2(x)$ can never produce better results than the interpolating natural cubic splines. When the interpolation curve is not a straight line, then the total curvature of the natural cubic $s_3(x)$ is always less than the total curvature of $s_2(x)$.
Another interesting characteristic of interpolating quadratic splines with equidistant knots is described in the following theorem.

Theorem 5.2. *Given are m points (x_i, y_i), $i = 1, 2, \ldots, m$, with an equidistant system of knots $h := x_{i+1} - x_i$ for all i from one to $(m-1)$ and $x_1 < x_2 < \ldots < x_m$.*
Furthermore, $\alpha = s_2'(x_1)$ denotes the first derivative of an arbitrary interpolating quadratic spline $s_2(x)$ at the knot x_1.
Then the following is true
for **even** $m = 2j + 2$:

$$\int_{x_1}^{x_m} s_2(x)\mathrm{d}x = \frac{h}{3}\left[2y_1 + 2y_2 + 4y_3 + 2y_4 + 4y_5 + \ldots + 2y_{m-2} + 4y_{m-1} \right.$$
$$\left. + y_m + \frac{h}{2}\alpha\right]$$

and for **odd** $m = 2j + 1$:

$$\int_{x_1}^{x_m} s_2(x)\mathrm{d}x = \frac{h}{3}[y_1 + 4y_2 + 2y_3 + 4y_4 + 2y_5 + \ldots + 4y_{m-3} + 2y_{m-2}$$
$$+ 4y_{m-1} + y_m] \, .$$

Proof. With the known attempt $s_2(x) = A_i(x-x_i)^2 + B_i(x-x_i) + y_i$ over $[x_i, x_{i+1}]$, $B_2 = \frac{2}{h}[\,-y_1 + y_2 - \frac{1}{2}h\alpha\,]$, $B_3 = \frac{2}{h}[y_1 - 2y_2 + y_3 + \frac{1}{2}h\alpha]$ is true due to $B_i + B_{i+1} = 2\Delta y_i/\Delta x_i = 2(y_{i+1} - y_i)/h$ and $B_1 = \alpha$. Generally, for all $i = 2, 3, \ldots, m$, $B_i = \frac{2}{h}[\,(-1)^{i+1}y_1 + 2\sum_{k=2}^{i-1}(-1)^{i+k}y_k + y_i + \frac{1}{2}(-1)^{i+1}h\alpha\,]$ is true.
$A_1 = \frac{1}{h^2}[\,-y_1 + y_2 - h\alpha\,]$ results due to $A_i = (B_{i+1} - B_i)/(2h)$ and the following is true for i from 2 to $(m-1)$:

$$A_i = \frac{1}{h^2}[(-1)^i y_1 + 2\sum_{k=2}^{i}(-1)^{i+k+1}y_k + y_{i+1} + \frac{1}{2}(-1)^i h\alpha$$
$$+ (-1)^i y_1 + 2\sum_{k=2}^{i-1}(-1)^{i+k+1}y_k - y_i + \frac{1}{2}(-1)^i h\alpha]$$
$$A_i = \frac{1}{h^2}[\,2(-1)^i y_1 + 4\sum_{k=2}^{i-1}(-1)^{i+k+1}y_k - 3y_i + y_{i+1} + (-1)^i h\alpha]\,,$$

$$\int_{x_1}^{x_m} s_2(x)\mathrm{d}x = \int_{x_1}^{x_2} s_2(x)\mathrm{d}x + \sum_{i=2}^{m-1} \int_{x_i}^{x_{i+1}} s_2(x)\mathrm{d}x$$

$$\int_{x_1}^{x_m} s_2(x)\mathrm{d}x = h\left[\frac{1}{3}A_1h^2 + \frac{1}{2}B_1h + y_1\right] + h\sum_{i=2}^{m-1}\left[\frac{1}{3}A_ih^2 + \frac{1}{2}B_ih + y_i\right]$$

$$\int_{x_1}^{x_m} s_2(x)\mathrm{d}x = h\left[-\frac{1}{3}y_1 + \frac{1}{3}y_2 - \frac{1}{3}h\alpha + \frac{1}{2}h\alpha + y_1\right] + h\sum_{i=2}^{m-1}\left[\frac{2}{3}(-1)^i y_1\right.$$

$$+\frac{4}{3}\sum_{k=2}^{i-1}(-1)^{i+k+1}y_k - y_i + \frac{1}{3}y_{i+1} + \frac{1}{3}(-1)^i h\alpha - (-1)^i y_1$$

$$\left. -2\sum_{k=2}^{i-1}(-1)^{i+k+1}y_k + y_i - (-1)^i h\alpha + y_i\right]$$

$$\int_{x_1}^{x_m} s_2(x)\mathrm{d}x = h\left[\frac{2}{3}y_1 + \frac{1}{3}y_2 + \frac{1}{6}h\alpha\right] + h\sum_{i=2}^{m-1}\left[\frac{1}{3}(-1)^{i+1}y_1\right.$$

$$\left. +\frac{2}{3}\sum_{k=2}^{i-1}(-1)^{i+k}y_k + y_i + \frac{1}{3}y_{i+1} + \frac{1}{6}(-1)^{i+1}h\alpha\right]$$

$$\int_{x_1}^{x_m} s_2(x)\mathrm{d}x = \frac{h}{3}\left[2y_1 + y_2 + \frac{1}{2}h\alpha\right] + \frac{h}{3}\sum_{i=2}^{m-1}\left[(-1)^{i+1}y_1 + 2\sum_{k=2}^{i-1}(-1)^{i+k}y_k\right.$$

$$\left. + 3y_i + y_{i+1} + \frac{1}{2}(-1)^{i+1}h\alpha\right].$$

Case 1 m is even and $m-1 = 2j+1$. Then

$$\int_{x_1}^{x_m} s_2(x)\mathrm{d}x = \frac{h}{3}\left[2y_1 + y_2 + \frac{1}{2}h\alpha + \sum_{k=1}^{j}(y_{2k} + 4y_{2k+1} + y_{2k+2})\right]$$

$$= \frac{h}{3}\left[2y_1 + 2y_2 + 4y_3 + \ldots + 2y_{m-2} + 4y_{m-1} + y_m + \frac{1}{2}h\alpha\right].$$

Case 2 m is odd and $m-1 = 2j$. Then with case 1:

$$\int_{x_1}^{x_m} s_2(x)\mathrm{d}x = \int_{x_1}^{x_{m-1}} s_2(x)\mathrm{d}x + \int_{x_{m-1}}^{x_m} s_2(x)\mathrm{d}x$$

$$\int_{x_1}^{x_m} s_2(x)\mathrm{d}x = \frac{h}{3}\left[2y_1 + 2y_2 + 4y_3 + \ldots + 2y_{m-3} + 4y_{m-2} + y_{m-1}\right.$$

$$\left.+\frac{1}{2}h\alpha\right] + \frac{h}{3}\left[(-1)^m y_1 + 2\sum_{k=2}^{m-2}(-1)^{m-1+k}y_k + 3y_{m-1}\right.$$

$$\left.+ y_m + \frac{1}{2}(-1)^m h\alpha\right]$$

$$\int_{x_1}^{x_m} s_2(x)\mathrm{d}x = \frac{h}{3}\left[2y_1 + 2y_2 + 4y_3 + \ldots + 2y_{m-3} + 4y_{m-2} + y_{m-1}\right.$$

$$\left.+ \frac{1}{2}h\alpha\right] + \frac{h}{3}\left[-y_1 + 2\sum_{k=2}^{m-2}(-1)^k y_k + 3y_{m-1} + y_m - \frac{1}{2}h\alpha\right]$$

$$\int_{x_1}^{x_m} s_2(x)\mathrm{d}x = \frac{h}{3}\left[y_1 + 4y_2 + 2y_3 + 4y_4 + 2y_5 + \ldots + 4y_{m-3} + 2y_{m-2}\right.$$

$$\left.+ 4y_{m-1} + y_m\right] .$$

□

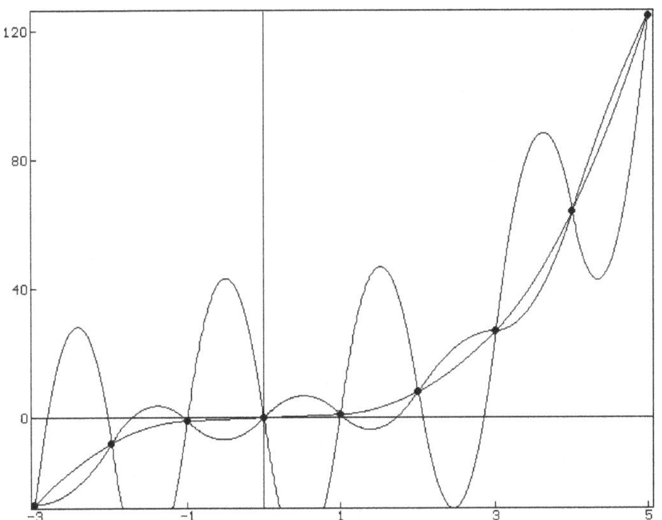

Fig. 5.5: Different interpolating quadratic splines with the same integral.

With this theorem, with odd m, fixed data points and equidistant knots,

all interpolating quadratic spline possess the same integral over $[x_1, x_m]$. This is calculated via the SIMPSON'S rule.
In Figure 5.5, three interpolating quadratic splines with identical integrals over $[-3, 5]$ are illustrated for $m = 9$.

Chapter 6

Interpolating quadratic splines and parabolas

Observed was an interpolating task using quadratic splines. This has a unique solution when an additional condition is fulfilled (Theorem 5.1). Parabolas are special quadratic spline functions. In this chapter the question of how a given parabola can be reconstructed by interpolating quadratic splines is examined. Initial possibilities become apparent when the necessary additional condition, due to Theorem 5.1, is suitable formulated. Another approach is offered by the different attempts at optimization. All of these problems of optimization can be solved by the realization of an initial condition. The examination of this condition under the assumption that the interpolation points stem from a parabola produces additional possibilities for reconstruction. These can be optimization conditions in connection with the number of points or conditions for the knots of the points of interpolation. If the formulation of a special optimization problem is foregone, it is still possible to find initial conditions that can, using the given points, reproduce every quadratic polynomial.
We begin with the conditions formulated in Theorem 5.1.

Theorem 6.1. (x_i, y_i), $i = 1, 2, \ldots, m > 1$, *are points of a quadratic polynomial* $P_2(x) = \alpha x^2 + \beta x + \gamma$. *Then in each of the following situations, the interpolating quadratic spline* $s_2(x)$ *constructed for these points is identical to* $P_2(x)$.

a) $s_2'(x_k) = P_2'(x_k)$ *is true for at least one* $k \in \{1, 2, \ldots, m\}$.
b) $B_i = P_2'(x_i)$ *for all* $i = 1, 2, \ldots, m$, *and* $y_k = P_2(x_k)$ *for at least one* $k \in \{1, 2, \ldots, m\}$.
c) $s_2''(x_k) = P_2''(x_k)$ *for at least one* $k \in \{1, 2, \ldots, m\}$.
d) The relationship $u \cdot s_2'(x_k) = s_2'(x_{k+1})$ *is true for at least one* $k \in \{1, 2, \ldots, m-1\}$, *where* $u = P_2'(x_{k+1})/P_2'(x_k)$ *and* $P_2'(x_k) \neq 0$.

Proof. Situation a) is a consequence of Theorem 5.1. In this case, the interpolating quadratic spline is uniquely determined. Since $P_2(x)$ is also an interpolating quadratic spline, $P_2(x) \equiv s_2(x)$ follows. The argument for b) is similar. By the specification of all B_i and a special y_k, the missing y_i can also be determined from (5.3).
The cases c) and d) lead back to a). By the specification of a second derivative, the relationship $s_2'(x_{k+1}) = s_2'(x_k) + P_2''(x_k)\Delta x_k$ results from $P_2''(x_k) = s_2''(x_k) = (s_2'(x_{k+1}) - s_2'(xk))/\Delta x_k$. If this is inserted into (5.3), $s_2'(x_k) = \Delta y_k/\Delta x_k - \frac{1}{2}P_2''(x_k)\Delta x_k$ results. Because of the conditions of interpolation, $\Delta y_k = P_2(x_{k+1}) - P_2(x_k) = \alpha(x_{k+1}^2 - x_k^2) + \beta(x_{k+1} - x_k)$. Equation $s_2'(x_k) = \alpha(x_{k+1} + x_k) + \beta - \alpha(x_{k+1} - x_k) = 2\alpha x_k + \beta = P_2'(x_k)$ results.
If a condition of the form $u \cdot s_2'(x_k) = s_2'(x_{k+1})$ is given for a certain $k \in \{1, 2, \ldots, m-1\}$, the following results from (5.3) with $u \neq -1$:

$$s_2'(x_k) = \frac{2}{(1+u)}\frac{\Delta y_k}{\Delta x_k} = \frac{2\alpha(x_{k+1}^2 - x_k^2) + 2\beta\Delta x_k}{(1+u)\Delta x_k} = \frac{2\alpha x_{k+1} + 2\alpha x_k + 2\beta}{\dfrac{P_2'(x_k) + P_2'(x_{k+1})}{P_2'(x_k)}},$$

$s_2'(x_k) = P_2'(x_k)$.
The condition $s_2'(x_k) = -s_2'(x_{k+1})$ is only to be fulfilled when $y_k = y_{k+1}$, and therefore $\Delta y_i = 0$, is true. □

Equation (5.5) (and (5.6)) allow access to additional conditions under which parabolas can be created by means of quadratic splines. The following is true:

Theorem 6.2. *Given are points (x_i, y_i), $i = 1, 2, \ldots, m > 1$, of a parabola $P_2(x) = \alpha x^2 + \beta x + \gamma$ and w_i, $i = 1, 2, \ldots, m-1$, where $\sum_{i=1}^{m-1} w_i \neq 0$. The interpolating quadratic spline $s_2(x)$ constructed with the initial condition*

$$B_1 = \frac{\sum\limits_{i=1}^{m-1} w_i \left((-1)^{i+1}D_i + 2\sum\limits_{k=1}^{i-1}(-1)^{k+1}D_k\right)}{\sum\limits_{i=1}^{m-1} w_i}$$

possesses the first derivative

$$s_2'(x_1) = B_1 = P_2'(x_1) + \frac{\alpha\left(\sum\limits_{i=1}^{m-1}(-1)^{i+1}w_i\Delta x_i\right)}{\sum\limits_{i=1}^{m-1} w_i}.$$

Proof. By the assumption, $D_i := \Delta y_i/\Delta x_i = \alpha(x_i + x_{i+1}) + \beta$ for all i from 1 to $(m-1)$. Then (5.5) turns into

$$B_1 = \frac{\alpha \sum_{i=1}^{m-1} w_i \left((-1)^{i+1}(x_i + x_{i+1}) + 2\sum_{k=1}^{i-1} (-1)^{k+1}(x_k + x_{k+1}) \right)}{\sum_{i=1}^{m-1} w_i} + \frac{\beta \sum_{i=1}^{m-1} w_i \left((-1)^{i+1} + 2\sum_{k=1}^{i-1} (-1)^{k+1} \right)}{\sum_{i=1}^{m-1} w_i} .$$

When x_1 is added and subtracted at the same time in the brackets of the first summand in the case $i = 1$, the following results:

$$B_1 = \frac{\alpha \sum_{i=1}^{m-1} w_i \left(2x_1 + (-1)^{i+1}(x_{i+1} - x_i)\right)}{\sum_{i=1}^{m-1} w_i} + \frac{\beta \sum_{i=1}^{m-1} w_i}{\sum_{i=1}^{m-1} w_i}$$

$$B_1 = 2\alpha x_1 + \frac{\alpha \left(\sum_{i=1}^{m-1} (-1)^{i+1} w_i \Delta x_i \right)}{\sum_{i=1}^{m-1} w_i} + \beta$$

$$B_1 = P_2'(x_1) + \frac{\alpha \left(\sum_{i=1}^{m-1} (-1)^{i+1} w_i \Delta x_i \right)}{\sum_{i=1}^{m-1} w_i} .$$

□

Theorem 6.2 states that the interpolating spline produces the quadratic polynomial in the case

$$\delta := \frac{\alpha \left(\sum_{i=1}^{m-1} (-1)^{i+1} w_i \Delta x_i \right)}{\sum_{i=1}^{m-1} w_i} = 0 . \tag{6.1}$$

This δ can be regarded as an error term.

Theorem 6.3. *The* (x_i, y_i), $i = 1, 2, \ldots, m > 1$, *are points of a parabola* $P_2(x) = \alpha x^2 + \beta x + \gamma$ *and* $w_i \in \mathbb{R}$ *where* $\sum_{i=1}^{m-1} w_i \neq 0$. *The interpolating* *quadratic spline* $s_2(x)$ *constructed with*

$$B_1 = \frac{\sum\limits_{i=1}^{m-1} w_i \left((-1)^{i+1} D_i + 2 \sum\limits_{k=1}^{i-1} (-1)^{k+1} D_k \right)}{\sum\limits_{i=1}^{m-1} w_i}$$

is identical to $P_2(x)$ *for both of the following situations:*

a) $\alpha = 0$ *is true.*

b) $\alpha \neq 0$ *and* $\sum\limits_{i=1}^{m-1} (-1)^{i+1} w_i \Delta x_i = 0$ *are true. In the case* (O1) *it means the same as* m *odd.*

Proof. The error

$$\frac{\alpha \left(\sum\limits_{i=1}^{m-1} (-1)^{i+1} w_i \Delta x_i \right)}{\sum\limits_{i=1}^{m-1} w_i}$$

needs to be observed due to Theorem 6.2. This expression is zero in the case where $\alpha = 0$ or when $\sum_{i=1}^{m-1} (-1)^{i+1} w_i \Delta x_i = 0$. For the weights, $w_i = \frac{1}{\Delta x_i}$ are true for optimization problem (O1). With this, $\sum_{i=1}^{m-1} (-1)^{i+1} w_i \Delta x_i = \sum_{i=1}^{m-1} (-1)^{i+1}$. This sum is zero when m is odd, otherwise it is one. □

If the problem (O2) is solved, $\sum_{i=1}^{m-1} (-1)^{i+1} \Delta x_i^2 = 0$ has to be true so that $s_2(x)$ and $P_2(x)$ are identical. This can be reached for different constellations of Δx_i. For example, let $\Delta x_{2k-1} = \Delta x_{2k}$, $k = 1, 2, \dots, j$, $(m = 2j+1)$ be. Another example is given with an even number of knots: $x_1 = -12$, $x_2 = -9$, $x_3 = -4$, $x_4 = 0$, $x_5 = 3$, $x_6 = 8$, $x_7 = 12$, $x_8 = 15$, $x_9 = 20$ and $x_{10} = 24$.

When the requirements $w_i > 0$, $i = 1, 2, \dots, m-1$, are avoided, initial conditions for B_1 can be found that reproduce all quadratic polynomials with regard to a system of knots. Only $\sum_{i=1}^{m-1} w_i \neq 0$ is required in Theorem 6.2. If for example $w_i = (-1)^{i+1}$ is set for all i from 1 to $(m-2)$, then $\sum_{i=1}^{m-1} (-1)^{i+1} w_i \Delta x_i = \sum_{i=1}^{m-2} \Delta x_i + (-1)^m w_{m-1} \Delta x_{m-1} = 0$ results exactly when

$$w_{m-1} = \frac{(-1)^{m+1} \sum\limits_{i=1}^{m-2} \Delta x_i}{\Delta x_{m-1}} = \frac{(-1)^{m+1} (x_{m-1} - x_1)}{\Delta x_{m-1}}$$

is true (see (6.1)).

The following is true for this approach:

$$\sum_{i=1}^{m-1} w_i = \sum_{i=1}^{m-2}(-1)^{i+1} + \frac{(-1)^{m+1}(x_{m-1}-x_1)}{\Delta x_{m-1}}$$

$$= \begin{cases} \dfrac{-(x_{m-1}-x_1)}{\Delta x_{m-1}} < 0 & \text{for even } m \\ 1 + \dfrac{x_{m-1}-x_1}{\Delta x_{m-1}} > 0 & \text{for odd } m \ . \end{cases}$$

By (6.1) $\delta = 0$ is true and that the polynomial of degree 2 is represented exactly. However, a connection to an optimization task does not exist anymore.

Even the specification $w'_i = (-1)^i$ for i from 1 to $(m-2)$ and respectively

$$w'_{m-1} = \frac{(-1)^m \sum\limits_{i=1}^{m-2} \Delta x_i}{\Delta x_{m-1}} = \frac{(-1)^m(x_{m-1}-x_1)}{\Delta x_{m-1}}$$

is possible. The following is then true:

$$\sum_{i=1}^{m-1} w'_i = \sum_{i=1}^{m-2}(-1)^{i} + \frac{(-1)^{m}(x_{m-1}-x_1)}{\Delta x_{m-1}}$$

$$= \begin{cases} \dfrac{(x_{m-1}-x_1)}{\Delta x_{m-1}} > 0 & \text{for even } m \\ -1 - \dfrac{x_{m-1}-x_1}{\Delta x_{m-1}} < 0 & \text{for odd } m. \end{cases}$$

Therefore, for every constellation of Δx_i, it is possible to find a set of weights with $\sum_{i=1}^{m-1} w_i > 0$ such that "some sort" of minimal spline can be spoken of. These splines that are each constructed for a fixed set of data with these two sets of weights w_i or w'_i are always identical, independent of m. Equation $w_i = -w'_i$ is true for all i. The same B_1 therefore results.

Finally, the values of the objective function from (O1) and (O2) are considered. Given are points $(x_i, y_i = P_2(x_i))$, $i = 1, 2, \ldots, m$, from a parabola $P_2(x)$. Let $s_2(x)$ be an arbitrary interpolating quadratic spline with respect to these points. The values of the objective functions for $P_2(x)$ and $s_2(x)$ will be compared. Theorem 6.3 already produces $\int_{x_1}^{x_m}[s''_2(x)]^2\mathrm{d}x = \int_{x_1}^{x_m} P''_2(x)]^2\mathrm{d}x$ for (O1) and an odd value of m.

Lemma 6.1. *(x_i, y_i), $i = 1, \ldots, m$, are points of the parabola $P_2(x) = \alpha x^2 + \beta x + \gamma$ and $s_2(x)$ is the uniquely determined interpolating spline function of degree 2 with regard to (x_i, y_i), $i = 1, 2, \ldots, m$, where*

$$s'_2(x_1) = P'_2(x_1) + \delta = (2\alpha x_1 + \beta) + \delta \ .$$

Then $(s''_2(x_i) - P''_2(x_i))\Delta x_i = 2\delta(-1)^i$ is true for all i from 1 to $(m-1)$.

Proof. Since $s_2'(x)$ is a straight line over the interval $[x_i, x_{i+1}]$, then

$$G := [s_2''(x_i) - P_2''(x_i)]\Delta x_i = \left[\frac{s_2'(x_{i+1}) - s_2'(x_i)}{\Delta x_i} - P_2''(x_i)\right]\Delta x_i$$
$$= s_2'(x_{i+1}) - s_2'(x_i) - 2\alpha\Delta x_i.$$

The following is obtained using (5.4):

$$G = \sum_{k=1}^{i} 2(-1)^{i+k+2}D_k + (-1)^i s_2'(x_1)$$
$$-\sum_{k=1}^{i-1} 2(-1)^{i+k+1}D_k - (-1)^{i+1}s_2'(x_1) - 2\alpha\Delta x_i$$
$$G = 4\sum_{k=1}^{i-1}(-1)^{i+k}D_k + 2D_i + 2(-1)^i s_2'(x_1) - 2\alpha\Delta x_i.$$

Since (x_i, y_i) are the points of the parabola, $D_k = \alpha x_{k+1} + \alpha x_k + \beta$ is true and due to the assumption, $s_2'(x_1) = 2\alpha x_1 + \beta + \delta$. From this it follows that

$$G = 4\alpha\sum_{k=1}^{i-1}(-1)^{i+k}(x_{k+1} + x_k) + 4\beta\sum_{k=1}^{i-1}(-1)^{i+k} + 2\alpha x_{i+1} + 2\alpha x_i + 2\beta$$
$$+ 4\alpha(-1)^i x_1 + 2(-1)^i\beta + 2(-1)^i\delta - 2\alpha x_{i+1} + 2\alpha x_i,$$
$$G = 4\alpha(-1)^{i+1}x_1 + 4\alpha(-1)^{2i-1}x_i + 4\beta\sum_{k=1}^{i-1}(-1)^{i+k} + 2\beta + 2\beta(-1)^i$$
$$+ 4\alpha(-1)^i x_1 + 4\alpha x_i + (-1)^i 2\delta = (-1)^i 2\delta.$$

When i is an odd number, $4\beta\sum_{k=1}^{i-1}(-1)^{i+k} = 0$ and $2\beta(-1)^i = -2\beta$ follow. When i is an even number, $4\beta\sum_{k=1}^{i-1}(-1)^{i+k} = -4\beta$ and $2\beta(-1)^i = 2\beta$ results. □

This lemma geometrically proves that the areas defined by the constant functions $P_2''(x) = 2\alpha$ and $s_2''(x_i)$, respectively, over the intervals $[x_i, x_{i+1}]$ are equal for each of these intervals. They lie alternatingly over and under 2α. Figure 6.1 illustrates the situation for 20 points of the parabola $P_2(x) = -1.234x^2 + 5.678x - 0.9$ and a respective interpolating quadratic spline function.

The lemma can be used to verify the difference in the total curvatures of $P_2(x)$ and $s_2(x)$.

Theorem 6.4. *(x_i, y_i), $i = 1, \ldots, m$, are points of the parabola $P_2(x) = \alpha x^2 + \beta x + \gamma$ and $s_2(x)$ is the uniquely determined interpolating spline* *function of degree 2 where $s_2'(x_1) = P_2'(x_1) + \delta$ with regard to these points. Then*

$$\int_{x_1}^{x_m} [s_2''(x)]^2 \mathrm{d}x = \int_{x_1}^{x_m} [P_2''(x)]^2 \mathrm{d}x + \begin{cases} 4\delta^2 \sum\limits_{i=1}^{m-1} \dfrac{1}{\Delta x_i} - 8\delta\alpha & \textit{when } m \textit{ is even} \\ 4\delta^2 \sum\limits_{i=1}^{m-1} \dfrac{1}{\Delta x_i} & \textit{when } m \textit{ is odd.} \end{cases}$$

Proof. $\int_{x_1}^{x_m} [s_2''(x)]^2 \mathrm{d}x = \sum_{i=1}^{m-1} \int_{x_i}^{x_{i+1}} [s_2''(x)]^2 \mathrm{d}x = \sum_{i=1}^{m-1} [s_2''(x_i)]^2 \Delta x_i$ and with the result of the preceding lemma, $s_2''(x_i) = (-1)^i 2\delta/\Delta x_i + 2\alpha$. From this it follows that

$$\begin{aligned} \int_{x_1}^{x_m} [s_2''(x)]^2 \mathrm{d}x &= \left[\sum_{i=1}^{m-1} \left(\frac{(-1)^i 2\delta}{\Delta x_i} + 2\alpha \right)^2 \Delta x_i \right] \\ &= 4\delta^2 \sum_{i=1}^{m-1} \frac{1}{\Delta x_i} + 8\delta\alpha \sum_{i=1}^{m-1} (-1)^i + 4\alpha^2 \sum_{i=1}^{m-1} \Delta x_i \\ &= 4\alpha^2 (x_m - x_1) + 4\delta^2 \sum_{i=1}^{m-1} \frac{1}{\Delta x_i} + 8\delta\alpha \sum_{i=1}^{m-1} (-1)^i \\ &= \int_{x_1}^{x_m} [P_2''(x)]^2 \mathrm{d}x + \begin{cases} 4\delta^2 \sum\limits_{i=1}^{m-1} \dfrac{1}{\Delta x_i} - 8\delta\alpha & \text{when } m \text{ is even} \\ 4\delta^2 \sum\limits_{i=1}^{m-1} \dfrac{1}{\Delta x_i} & \text{when } m \text{ is odd.} \end{cases} \end{aligned}$$

□

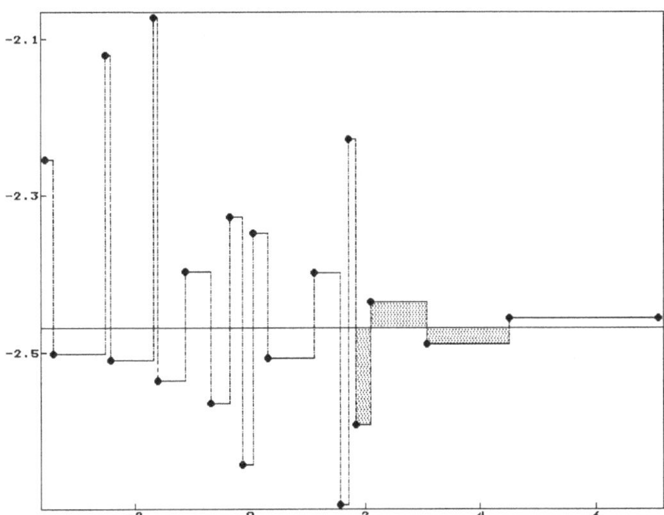

Fig. 6.1: $s_2''(x)$ and $P_2''(x) \equiv -2.468$, explanation see text.

Given is an odd number of points of a parabola $P_2(x) = \alpha x^2 + \beta x + \gamma$. This parabola can then be reconstructed by the construction of an interpolating quadratic spline function which solves (O1). The error δ (see (6.1)) is zero. For an even number of points m, $s_2(x)$ is not identical to $P_2(x)$. For the error of the first derivative $s_2'(x_1)$ at point x_1, the relationship $\delta = \alpha / \sum_{i=1}^{m-1} \frac{1}{\Delta x_i}$ is true due to $w_i = 1/\Delta x_i$. From Theorem 6.4 it follows that

$$\int_{x_1}^{x_m} [s_2''(x)]^2 \mathrm{d}x = \int_{x_1}^{x_m} [P_2''(x)]^2 \mathrm{d}x + \frac{4\alpha^2}{\sum\limits_{i=1}^{m-1} \frac{1}{\Delta x_i}} - \frac{8\alpha^2}{\sum\limits_{i=1}^{m-1} \frac{1}{\Delta x_i}}$$

$$= \int_{x_1}^{x_m} [P_2''(x)]^2 \mathrm{d}x - \frac{4\alpha^2}{\sum\limits_{i=1}^{m-1} \frac{1}{\Delta x_i}} .$$

In the special case $\Delta x_i = h$ for all i,

$$GK := \int_{x_1}^{x_m} [s_2''(x)]^2 \mathrm{d}x = \int_{x_1}^{x_m} [P_2''(x)]^2 \mathrm{d}x - \frac{4\alpha^2 h}{m-1}$$

is true. One recognize that the differences between $s_2(x)$ and $P_2(x)$ decrease with growing m.

Theorem 6.5. *(x_i, y_i), $i = 1, 2, \ldots, m$, are points of the parabola $P_2(x) = \alpha x^2 + \beta x + \gamma$ and $s_2(x)$ is the uniquely determined interpolating spline function of degree 2 with regard to (x_i, y_i), $i = 1, 2, \ldots, m$, where $s_2'(x_1) = P_2'(x_1) + \delta$. Then*

$$\int_{x_1}^{x_m} [s_2'(x)]^2 \mathrm{d}x = \int_{x_1}^{x_m} [P_2'(x)]^2 \mathrm{d}x - \frac{2}{3}\alpha\delta \sum_{i=1}^{m-1} (-1)^{i+1} \Delta x_i^2 + \frac{1}{3}\delta^2 (x_m - x_1)$$

is true.

Proof. The following is true for the polynomial $P_2(x) = \alpha x^2 + \beta x + \gamma$:

$$\int_{x_1}^{x_m} [P_2'(x)]^2 \mathrm{d}x = \sum_{i=1}^{m-1} \int_{x_i}^{x_{i+1}} (2\alpha x + \beta)^2 \mathrm{d}x = \sum_{i=1}^{m-1} \int_{x_i}^{x_{i+1}} (4\alpha^2 x^2 + 4\alpha\beta x + \beta^2) \mathrm{d}x,$$

$$\int_{x_1}^{x_m} [P_2'(x)]^2 \mathrm{d}x = \frac{4}{3}\alpha^2 \sum_{i=1}^{m-1} (x_{i+1}^3 - x_i^3) + 2\alpha\beta \sum_{i=1}^{m-1} (x_{i+1}^2 - x_i^2) + \beta^2 \sum_{i=1}^{m-1} \Delta x_i .$$

The starting point of the proof is the relationship

$$\int_{x_1}^{x_m} [s'_2(x)]^2 \mathrm{d}x = \frac{1}{3}\sum_{i=1}^{m-1} \Delta x_i (D_i - B_i)^2 + \sum_{i=1}^{m-1} \Delta x_i D_i^2 .$$

Remember the abbreviation $D_i := \frac{\Delta y_i}{\Delta x_i}$. For a parabola, $D_i = \alpha(x_{i+1} + x_i) + \beta$ is true. If B_i is replaced by (5.4), the following is obtained for the first sum

$$S_1 := \frac{1}{3}\sum_{i=1}^{m-1} \Delta x_i (D_i - B_i)^2 \ ,$$

$$S_1 = \frac{1}{3}\sum_{i=1}^{m-1} \Delta x_i \left(\alpha(x_{i+1} + x_i) + \beta + 2\sum_{k=1}^{i-1} (-1)^{i+k} \left[\alpha(x_{k+1} + x_k) + \beta\right] \right.$$
$$\left. + (-1)^i B_1 \right)^2 ,$$

$$S_1 = \frac{1}{3}\sum_{i=1}^{m-1} \Delta x_i \left((-1)^{i+1}(2\alpha x_1 + \beta) + \alpha \Delta x_i + (-1)^i P_2'(x_1) + (-1)^i \delta\right)^2 ,$$

$$S_1 = \frac{1}{3}\sum_{i=1}^{m-1} \Delta x_i \left(\alpha \Delta x_i + (-1)^i \delta\right)^2 ,$$

$$S_1 = \frac{1}{3}\alpha^2 \sum_{i=1}^{m-1} \Delta x_i^3 + \frac{2}{3}\alpha\delta \sum_{i=1}^{m-1} (-1)^i \Delta x_i^2 + \frac{1}{3}\delta^2 \sum_{i=1}^{m-1} \Delta x_i \ .$$

In a small additional calculation, the equality

$$\Delta x_i^3 = (x_{i+1} - x_i)^3 = (x_{i+1}^3 - x_i^3) - 3(x_{i+1}^2 x_i - x_{i+1} x_i^2)$$

can be seen. Therefore,

$$S_1 = \frac{1}{3}\alpha^2 \sum_{i=1}^{m-1} (x_{i+1}^3 - x_i^3) + \frac{2}{3}\alpha\delta \sum_{i=1}^{m-1} (-1)^i \Delta x_i^2 + \frac{1}{3}\delta^2 \sum_{i=1}^{m-1} \Delta x_i$$
$$- \alpha^2 \sum_{i=1}^{m-1} (x_{i+1}^2 x_i - x_{i+1} x_i^2)$$

is true. The following is obtained for the second sum:

$$S_2 := \sum_{i=1}^{m-1} \Delta x_i D_i^2 = \sum_{i=1}^{m-1} \Delta x_i \left[\alpha(x_{i+1} + x_i) + \beta\right]^2 ,$$

$$S_2 = \alpha^2 \sum_{i=1}^{m-1} \Delta x_i (x_{i+1} + x_i)^2 + 2\alpha\beta \sum_{i=1}^{m-1} \Delta x_i (x_{i+1} + x_i) + \beta^2 \sum_{i=1}^{m-1} \Delta x_i \ .$$

$$\Delta x_i(x_{i+1}+x_i)^2 = (x_{i+1}-x_i)(x_{i+1}+x_i)^2 = (x_{i+1}^3-x_i^3)+(x_{i+1}^2x_i-x_{i+1}x_i^2)$$

and

$$\Delta x_i(x_{i+1}+x_i) = (x_{i+1}^2 - x_i^2)$$

leads to

$$S_2 = \alpha^2\sum_{i=1}^{m-1}(x_{i+1}^3-x_i^3) + 2\alpha\beta\sum_{i=1}^{m-1}(x_{i+1}^2-x_i^2) + \beta^2\sum_{i=1}^{m-1}\Delta x_i$$
$$+\alpha^2\sum_{i=1}^{m-1}(x_{i+1}^2x_i - x_{i+1}x_i^2)\ .$$

In total,

$$\int_{x_1}^{x_m}[s_2'(x)]^2\mathrm{d}x = S_1+S_2 = \frac{4}{3}\alpha^2\sum_{i=1}^{m-1}(x_{i+1}^3-x_i^3) + 2\alpha\beta\sum_{i=1}^{m-1}(x_{i+1}^2-x_i^2)$$
$$+\beta^2\sum_{i=1}^{m-1}\Delta x_i + \frac{2}{3}\alpha\delta\sum_{i=1}^{m-1}(-1)^i\Delta x_i^2 + \frac{1}{3}\delta^2\sum_{i=1}^{m-1}\Delta x_i$$

and so

$$\int_{x_1}^{x_m}[s_2'(x)]^2\mathrm{d}x = \int_{x_1}^{x_m}[P_2'(x)]^2\mathrm{d}x - \frac{2}{3}\alpha\delta\sum_{i=1}^{m-1}(-1)^{i+1}\Delta x_i^2 + \frac{1}{3}\delta^2(x_m-x_1)\ .$$

□

If $s_2(x)$ is constructed with minimal $\int_{x_1}^{x_m}[s_2'(x)]^2\mathrm{d}x$, the following results from Theorem 6.2 and with $w_i = \Delta x_i$ for the error:

$$\delta = \frac{\alpha\left(\sum\limits_{i=1}^{m-1}(-1)^{i+1}\Delta x_i^2\right)}{\sum\limits_{i=1}^{m-1}\Delta x_i} = \frac{\alpha\left(\sum\limits_{i=1}^{m-1}(-1)^{i+1}\Delta x_i^2\right)}{x_m - x_1}\ .$$

The following is obtained if this errorterm is inserted into Theorem 6.5:

$$\int_{x_1}^{x_m}[s_2'(x)]^2\mathrm{d}x = \int_{x_1}^{x_m}[P_2'(x)]^2\mathrm{d}x - \frac{1}{3}\alpha^2\frac{\left(\sum\limits_{i=1}^{m-1}(-1)^{i+1}\Delta x_i^2\right)^2}{x_m-x_1}\ .$$

If equidistant knots exist, then $\Delta x_i = h$ for all i from 1 to $(m-1)$ and

$$\int_{x_1}^{x_m}[s_2'(x)]^2\mathrm{d}x = \int_{x_1}^{x_m}[P_2'(x)]^2\mathrm{d}x - \frac{1}{3}\alpha^2h^4\frac{\left(\sum\limits_{i=1}^{m-1}(-1)^{i+1}\right)^2}{(m-1)h}$$
$$= \int_{x_1}^{x_m}[P_2'(x)]^2\mathrm{d}x - \begin{cases} 0 & \text{for odd } m \\ \dfrac{\alpha^2h^3}{3(m-1)} & \text{for even } m. \end{cases}$$

Remark 6.1. Lemma 6.1 is not required to determine the difference between $P_2(x)$ and $s_2(x)$ with regard to GK. The total curvature of $s_2(x)$ can be calculated analogously to the remodeling of S_1 in Theorem 6.5. The following is true with (5.4):

$$\begin{aligned}\int_{x_1}^{x_m} [s_2''(x)]^2 \mathrm{d}x &= 4\sum_{i=1}^{m-1} \frac{1}{\Delta x_i}(D_i - B_i)^2 \\ &= 4\sum_{i=1}^{m-1} \frac{1}{\Delta x_i} \Bigg(\alpha(x_{i+1} + x_i) + \beta \\ &\quad + 2\sum_{k=1}^{i-1}(-1)^{i+k}[\alpha(x_{k+1} + x_k) + \beta] + (-1)^i B_1\Bigg)^2 \\ &= 4\sum_{i=1}^{m-1} \frac{1}{\Delta x_i} \left(\alpha\Delta x_i + (-1)^i\delta\right)^2 \\ &= 4\alpha^2(x_m - x_1) + 4\delta^2 \sum_{i=1}^{m-1} \frac{1}{\Delta x_i} + 8\delta\alpha \sum_{i=1}^{m-1}(-1)^i \ .\end{aligned}$$

Chapter 7

Smoothing quadratic splines

The function space $\sum^2(\Delta X)$ of all spline functions of degree 2 is looked at with regard to a fixed system of knots $\Delta X := \{x_1 < x_2 < \ldots < x_m\}$. Optimization problems are formulated for these spline functions that are analogous to (OP3), (R) and (GK3), see Chapter 4. A unique solution can be proven for each of these problems. The relationships between the solutions are observed. For each of the examples, it is proven that the value of the objective functions for the quadratic splines can never be better than those of the respective solutions to (OP3), (R) and (GK3). Making use of the optimization task in the set of all once continuously differentiable functions over $[x_1, x_m]$ therefore does not make sense because the solution is definitely not a quadratic spline. The optimization problem (OP2) is now formulated similarly to problem (OP3) in $C^2([x_1, x_m])$.

Given are the points (x_i, y_i), $x_1 < x_2 < \ldots < x_m$, the weights $k_i > 0$, $i = 1, 2, \ldots, m$, and a real number $\mu \geq 0$. A quadratic spline function $s_2(x)$ in $\sum^2(\Delta X)$ is sought after so that (OP2)

$$F(s_2) := \mu \int_{x_1}^{x_m} [s_2''(x)]^2 \mathrm{d}x + \sum_{i=1}^{m} \left(\frac{s_2(x_i) - y_i}{k_i} \right)^2 \text{ becomes minimal.}$$

The initial considerations should apply to the existence and uniqueness of a solution to (OP2) in $\sum^2(\Delta X)$.
For each of the intervals $[x_i, x_{i+1}]$, $s_2(x) = A_i(x - x_i)^2 + B_i(x - x_i) + C_i$, $i = 1, 2, \ldots, m-1$, is defined. If a B_m and a C_m are also introduced, then $s_2(x_i) = C_i$ and $s_2'(x_i) = B_i$, $i = 1, 2, \ldots, m$, are true. With knowledge of the vector $C := (C_1, C_2, \ldots, C_m)^T$, the spline $s_2(x)$ is completely determined because of (5.1) to (5.3). Since (OP2) can only be solved by a

spline function from $\sum^2(\Delta X)$ with minimum total curvature, the following is true due to (5.6) and $w_i = 1/\Delta x_i$:

$$B_1 = B_1(C_1, \ldots, C_m) = \frac{\sum\limits_{i=1}^{m-1} \frac{(-1)^{i+1}}{\Delta x_i} \left(\frac{1}{\Delta x_i} + 2 \sum\limits_{k=i+1}^{m-1} \frac{1}{\Delta x_k} \right) (C_{i+1} - C_i)}{\sum\limits_{i=1}^{m-1} \frac{1}{\Delta x_i}}.$$

If this value is inserted into (5.4) the total curvature GK results:

$$\begin{aligned} GK &= GK(C_1, \ldots, C_m) \\ &= 4 \sum_{i=1}^{m-1} \frac{1}{\Delta x_i} \left[\frac{(C_{i+1} - C_i)}{\Delta x_i} + \sum_{k=1}^{i-1} (-1)^{i+k} 2 \frac{(C_{k+1} - C_k)}{\Delta x_k} + (-1)^i B_1 \right]^2 . \end{aligned}$$

The total curvature GK is therefore dependent on the vector C.
(OP2) can be replaced in $\mathbb{R}^m$ by the following equivalent quadratic optimization problem (OPQ) since the equations (5.3) that produce $s_2(x)$ have already been introduced into the calculation of the total curvature.

Determine $C := (C_1, C_2, \ldots, C_m) \in \mathbb{R}^m$ so that the objective function

$$F_{\mathbb{R}}(C) = 4\mu \sum_{i=1}^{m-1} \frac{1}{\Delta x_i} \left[\frac{C_{i+1} - C_i}{\Delta x_i} + \sum_{k=1}^{i-1} 2(-1)^{i+k} \frac{C_{k+1} - C_k}{\Delta x_k} + (-1)^i B_1 \right]^2 + \sum_{i=1}^{m} \left(\frac{C_i - y_i}{k_i} \right)^2 \text{ becomes minimal} \qquad \text{(OPQ)}$$

for given(x_i, y_i), k_i, $i = 1, 2, \ldots, m$, and for a given $\mu \geq 0$.

Theorem 7.1. *For each system of points* (x_i, y_i), $i = 1, 2, \ldots, m$, $x_1 < x_2 < \ldots < x_m$, *and* $0 \leq \mu$, *the optimization problem* (OPQ) *and therefore* (OP2) *has a unique solution in* $\sum^2(\Delta X)$. *This solution is determined by the uniquely solvable linear system of equations*

$$\frac{\partial F_{\mathbb{R}}(C)}{\partial C_n} = 0, \; n = 1, 2, \ldots, m \; .$$

Proof. Observe the following compact and not empty set

$$E := \{(C_1, C_2, \ldots, C_m) \in \mathbb{R}^m : \sum_{i=1}^{m} \left(\frac{s_2(x_i) - y_i}{k_i} \right)^2 = \sum_{i=1}^{m} \left(\frac{C_i - y_i}{k_i} \right)^2 \leq S_g\}$$

where S_g is the sum of the squared errors of the regression line $g(x)$ corresponding with the weights k_i. When $s_2(x)$ solves problem (OP2) then the sum of the squared errors

$$\sum_{i=1}^{m}\left(\frac{s_2(x_i)-y_i}{k_i}\right)^2$$

cannot be larger than S_g. Otherwise $g(x) \in \sum^2(\Delta X)$ would be a spline of degree 2 with a total curvature of zero that possesses a smaller objective function value than $s_2(x)$. Then $C_g := (g(x_1), g(x_2), \ldots, g(x_m)) \in \mathbb{R}^m$ would solve problem (OPQ).
Due to the quadratic terms in which only linear expressions of the C_i are used, $F_{\mathbb{R}}(*)$ is the sum of a continuous convex and a strongly convex function over $\mathbb{R}^m$ and therefore it is itself strongly convex and continuous. This is why $F_{\mathbb{R}}(*)$, as a continuous function over all $\mathbb{R}^m$, approaches a uniquely determined minimum over the compact and convex set E.
A necessary condition for a local minimum is $\partial F_{\mathbb{R}}(C)/\partial C_n = 0$, $n = 1, 2, \ldots, m$. This is a linear system of equations because the unknown values of the C_i, $i = 1, 2, \ldots, m$, only appear linearly in the quadratic summands. The given conditions are also sufficient for strongly convex functions. That is why the linear system of equations $\partial F_{\mathbb{R}}(C)/\partial C_n = 0$, $n = 1, 2, \ldots, m$, possesses exactly one solution. Via (5.3), together with B_1, it determines the complete function $s_2(x)$. □

In matrix notation the objective function of (OPQ) has the form

$$F(C) = \mu C^T \circ G^T \circ \begin{pmatrix} \frac{4}{\Delta x_1} & 0 & \ldots & 0 \\ 0 & \frac{4}{\Delta x_2} & \ldots & 0 \\ \vdots & \vdots & \ddots & \vdots \\ 0 & 0 & \ldots & \frac{4}{\Delta x_{m-1}} \end{pmatrix} \circ G \circ C + ||K^{-1} \circ (C - Y)||^2.$$

K is a diagonal matrix with the diagonal elements k_i. G is an $(m-1) \times m$ matrix that is defined such that

$$C^T \circ G^T \circ \begin{pmatrix} \frac{4}{\Delta x_1} & 0 & \ldots & 0 \\ 0 & \frac{4}{\Delta x_2} & \ldots & 0 \\ \vdots & \vdots & \ddots & \vdots \\ 0 & 0 & \ldots & \frac{4}{\Delta x_{m-1}} \end{pmatrix} \circ G \circ C$$

results in the total curvature. Its rows are defined by the coefficients of

$$\frac{(C_{i+1} - C_i)}{\Delta x_i} + \sum_{k=1}^{i-1} 2\,(-1)^{i+k}\,\frac{(C_{k+1} - C_k)}{\Delta x_k} + (-1)^i\,B_1,$$

$i = 1, 2, \ldots, m-1$, with regard to the C_j. The abbreviation

$$H := G^T \circ \begin{pmatrix} \frac{4}{\Delta x_1} & 0 & \ldots & 0 \\ 0 & \frac{4}{\Delta x_2} & \ldots & 0 \\ \vdots & \vdots & \ddots & \vdots \\ 0 & 0 & \ldots & \frac{4}{\Delta x_{m-1}} \end{pmatrix} \circ G$$

is set. The following notation is then determined by the necessary conditions for a local minimum:

$$\begin{aligned} 2\mu H \circ C + 2K^{-2} \circ [C - Y] = 0 \text{ or} \\ [I + \mu K^2 \circ H] \circ C = Y \ . \end{aligned} \tag{7.1}$$

H is a positive semi-definite and symmetric matrix. It is used to calculate the total curvature of the quadratic spline with minimal total curvature which interpolates the (x_i, C_i), $i = 1, 2, \ldots, m$.
Consequently, $[I + \mu K^2 \circ H]$ is positive definite for all $\mu \geq 0$ and therefore invertible. With this one has a further proof for the unique solvability of (OP2).
To obtain an explicit representation of H, (5.6) is applied to (5.3) as the first equation. $F \circ B = D \circ C$ is obtained with $B := (B_1, B_2, \ldots, B_m)^T$,

$$F := \begin{pmatrix} 1 & 0 & 0 & \ldots & 0 & 0 \\ 1 & 1 & 0 & \ldots & 0 & 0 \\ 0 & 1 & 1 & \ldots & 0 & 0 \\ \vdots & \vdots & \vdots & \ddots & \vdots & \vdots \\ 0 & 0 & 0 & \ldots & 1 & 1 \end{pmatrix} \text{ and } D := \begin{pmatrix} d_{11} & d_{12} & d_{13} & \ldots & d_{1m-1} & d_{1m} \\ -\frac{2}{\Delta x_1} & \frac{2}{\Delta x_1} & 0 & \ldots & 0 & 0 \\ 0 & -\frac{2}{\Delta x_2} & \frac{2}{\Delta x_2} & \ldots & 0 & 0 \\ \vdots & \vdots & \vdots & \ddots & \vdots & \vdots \\ 0 & 0 & 0 & \ldots & -\frac{2}{\Delta x_{m-1}} & \frac{2}{\Delta x_{m-1}} \end{pmatrix} .$$

F is invertible and $B = F^{-1} \circ D \circ C$ is true. The inverse of F can also be given. This is

$$F^{-1} = \begin{pmatrix} 1 & 0 & 0 & \ldots & 0 & 0 \\ -1 & 1 & 0 & \ldots & 0 & 0 \\ 1 & -1 & 1 & \ldots & 0 & 0 \\ \vdots & \vdots & \vdots & \ddots & \vdots & \vdots \\ (-1)^{m-1} & (-1)^{m-2} & (-1)^{m-3} & \ldots & -1 & 1 \end{pmatrix} .$$

To calculate the first row of D, the first step is to break down the numerator of

$$B_1 = \frac{\sum_{i=1}^{m-1} \frac{(-1)^{i+1}}{\Delta x_i} \left(\frac{1}{\Delta x_i} + 2 \sum_{k=i+1}^{m-1} \frac{1}{\Delta x_k} \right) (C_{i+1} - C_i)}{\sum_{i=1}^{m-1} \frac{1}{\Delta x_i}}$$

into the individual summands

$$\begin{array}{lllll}
i=1: & \frac{1}{\Delta x_1}(\frac{1}{\Delta x_1}+2[\frac{1}{\Delta x_2}+ & \frac{1}{\Delta x_3}+ & \frac{1}{\Delta x_4}+\ldots+\frac{1}{\Delta x_{m-1}}])(C_2-C_1) \\
i=2: & -\frac{1}{\Delta x_2}(\frac{1}{\Delta x_2}+2[& \frac{1}{\Delta x_3}+ & \frac{1}{\Delta x_4}+\ldots+\frac{1}{\Delta x_{m-1}}])(C_3-C_2) \\
i=3: & \frac{1}{\Delta x_3}(\frac{1}{\Delta x_3}+2[& & \frac{1}{\Delta x_4}+\ldots+\frac{1}{\Delta x_{m-1}}])(C_4-C_3) \\
\vdots & & & \\
\end{array}$$

$i=m-1$: $\frac{(-1)^m}{\Delta x_{m-1}}(\frac{1}{\Delta x_{m-1}})(C_m-C_{m-1})$.

The respective coefficients of the C_k, $k=1,2,\ldots,m$, are summarized.

$$d_{1k}=(-1)^k\left\{\frac{1}{\Delta x_{k-1}^2}+\frac{1}{\Delta x_k^2}+\frac{1}{\Delta x_{k-1}}\sum_{j=k}^{m-1}\frac{2}{\Delta x_j}+\frac{1}{\Delta x_k}\sum_{j=k+1}^{m-1}\frac{2}{\Delta x_j}\right\}/\alpha$$

result for k from 1 to $(m-1)$ (for $k=1$ set $1/\Delta x_{k-1}=1/\Delta x_0=0$) and

$$d_{1m}=\frac{(-1)^m}{\Delta x_{m-1}^2\alpha},\ \text{where } \alpha=\sum_{i=1}^{m-1}\frac{1}{\Delta x_i}\ .$$

The following is true for the total curvature:

$$\begin{aligned}
GK:&=\int_{x_1}^{x_m}[s_2''(x)]^2\mathrm{d}x=\sum_{i=1}^{m-1}\int_{x_i}^{x_{i+1}}(2A_i)^2\mathrm{d}x=4\sum_{i=1}^{m-1}\left(\frac{B_{i+1}-B_i}{2\Delta x_i}\right)^2\Delta x_i \\
&=\sum_{i=1}^{m-1}\frac{B_{i+1}^2-2B_{i+1}B_i+B_i^2}{\Delta x_i}\ .
\end{aligned}$$

With the $(m\times m)$ matrix

$$Z^*:=\begin{pmatrix}
\frac{1}{\Delta x_1} & -\frac{1}{\Delta x_1} & 0 & \ldots & 0 & 0 & 0 \\
-\frac{1}{\Delta x_1} & \frac{1}{\Delta x_1}+\frac{1}{\Delta x_2} & -\frac{1}{\Delta x_2} & \ldots & 0 & 0 & 0 \\
\vdots & \vdots & \vdots & \ddots & \vdots & \vdots & \vdots \\
0 & 0 & 0 & \ldots & -\frac{1}{\Delta x_{m-2}} & \frac{1}{\Delta x_{m-2}}+\frac{1}{\Delta x_{m-1}} & -\frac{1}{\Delta x_{m-1}} \\
0 & 0 & 0 & \ldots & 0 & -\frac{1}{\Delta x_{m-1}} & \frac{1}{\Delta x_{m-1}}
\end{pmatrix}$$

$GK=B^T\circ Z^*\circ B=(F^{-1}\circ D\circ C)^T\circ Z^*\circ F^{-1}\circ D\circ C=C^T\circ D^T\circ(F^{-1})^T\circ Z^*\circ F^{-1}\circ D\circ C$ results so that $H=D^T\circ(F^{-1})^T\circ Z^*\circ F^{-1}\circ D$ is true.

Observe that the elements of H are exclusively produced from the Δx_i. That is why the same statements are valid here for the "smoothing matrix" $[I+\mu K^2\circ H]^{-1}$ as are true for cubic splines.

The construction algorithm for the solution to (OP2) is similar to the one of the solution to (OP3). The system of knots is fixed. That is why the former optimization task in the function space $\sum^2(\Delta X)$ can be carried over into a quadratic optimization problem in $\mathbb{R}^m$. In a first step its solution produces the unknown $C_i = s_2(x_i)$, $i = 1, 2, \ldots, m$. In a following step, the uniquely determined interpolating quadratic spline function with minimal total curvature is calculated. With Theorem 5.1 and (5.6) this is the function $s_2^o(x)$ that is constructed with the condition

$$B_1 = \frac{\sum_{i=1}^{m-1} \frac{(-1)^{i+1}}{\Delta x_i} \left(\frac{1}{\Delta x_i} + 2 \sum_{k=i+1}^{m-1} \frac{1}{\Delta x_k} \right) (C_{i+1} - C_i)}{\sum_{i=1}^{m-1} \frac{1}{\Delta x_i}}.$$

In this sense the objective function $F_{\mathbb{R}}(C) = F(s_2)$ is not only a function of $C \in \mathbb{R}^m$, but also a function of μ over the interval $[0, \infty)$. Each μ uniquely defines a vector C. Exactly one interpolating quadratic spline with minimal total curvature is determined in this way. It possesses total curvature $GK(\mu) = GK(C) = GK(s_2)$ and sum of the squared errors $SSE(\mu) = SSE(C) = SSE(s_2)$.
The following statement is proven similarly to Lemma 4.3.

Lemma 7.1. *The functional $F(\mu) = \mu GK(\mu) + SSE(\mu)$ increases strongly monotone for $\mu \geq 0$.*

Example 7.1. The uniquely determined quadratic spline $s_2^o(x)$ is constructed for the points from Example 5.1 (see Figure 7.1). This solves problem (OP2) for $\mu = 1$ and $k_i = 1$, $i = 1, 2, \ldots, m$ (see Part III). Characterizing parameters of the solution are

$$\begin{aligned} \text{squared sum of errors SSE} &= 0.66860 \\ \text{total curvature GK} &= 0.27123 \text{ and} \\ F(s_2^o(x)) = \mu \cdot GK + SSE &= 0.93983. \end{aligned}$$

If problem (OP3) is calculated with $\mu = 1$ and the data from Example 5.1 then the uniquely determined solution is a natural cubic spline $s_3^o(x)$.
Both solutions $s_2^o(x)$ and $s_3^o(x)$ are now compared with respect to the associated objective function. Lemma 5.1 shows that for identical interpolation points, the total curvature of the interpolating quadratic splines can never be larger than those of the natural cubic splines. $s_3^o(x_i)$ and $s_2^o(x_i)$ are generally different. So Lemma 5.1 cannot be used. Nevertheless, the objective

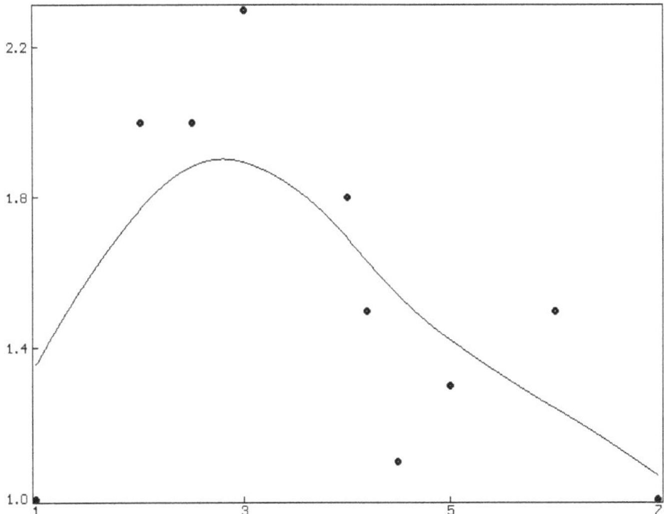

Fig. 7.1: The unique solution $s_2^o(x)$ to (OP2) where $\mu = 1$, see Example 5.1 for the data.

function value of $s_2^o(x)$ can never be better than that of $s_3^o(x)$.

Theorem 7.2. *Given is a system of points (x_i, y_i), $i = 1, 2, \ldots, m$, where $x_1 < x_2 < \ldots < x_m$, and $0 \leq \mu$. Let $s_3^o(x)$ and $s_2^o(x)$ be the unique solutions to* (OP3) *and* (OP2)*, respectively.*
Then for the corresponding objective functions $Z(s_3^o(x)) \leq F(s_2^o(x))$ holds. The equals sign is only true in the case that $s_2(x) = s_3(x) = \alpha x + \beta$.

Proof. Assume there is a constellation for which $Z(s_3^o(x)) > F(s_2^o(x))$ is true for the respective objective functions. The uniquely determined interpolating natural cubic spline function $s_3^*(x)$ can then be constructed for the vector $C_0 := (s_2^o(x_1), s_2^o(x_2), \ldots, s_2^o(x_m))$. This possesses the same weighted sum of the squared errors as $s_2^o(x)$ with regard to the y_i, $i = 1, 2, \ldots, m$. However, with Lemma 5.1 the total curvature of $s_3^*(x)$ is less than or equal to that of $s_2^o(x)$. Consequently, $s_3^o(x)$ does not solve (OP3) and $Z(s_3^o(x)) > F(s_2^o(x))$ cannot be true. The same argument also states that equality only in the case of a straight line applies. $\square$

The solutions to (OP3) and (OP2) for $\mu = 1$ with regard to the data from Example 7.1 are shown in Figure 7.2. The following exist for the solution $s_3^o(x)$ to (OP3):

squared sum of errors SSE	=	0.63210
total curvature GK	=	0.28112 and
$Z(s_3^o(x)) = \mu GK + SSE$	=	0.91318.

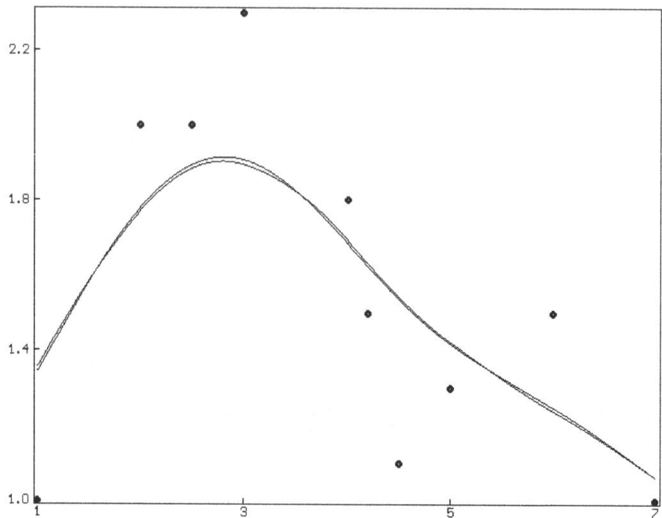

Fig. 7.2: The solution to (OP2) and (OP3) for $\mu = 1$, see Example 7.1 for the data.

When compared with the results of Example 7.1, it can be seen that the total curvature of $s_2^o(x)$ is less than the total curvature of $s_3^o(x)$. The relation is the opposite for the sum of the squared errors. According to Theorem 7.2 it can be seen that $Z(s_3^o(x)) < F(s_2^o(x))$.

Example 7.2. The 10 points $(9.3, 15.9)$, $(10.2, 20.7)$, $(11.4, 18.6)$, $(11.8, 20.4)$, $(13.1, 19.0)$, $(13.4, 18.0)$, $(14.0, 22.5)$, $(14.2, 21.5)$, $(15.3, 23.9)$, $(17.2, 26.2)$ were used and (OP2) and (OP3) were each solved with $\mu = 0.001$. The following is true for $s_2^o(x)$:

squared sum of errors SSE	=	1.36863
total curvature GK	=	1434.95610 and
objective function $\mu GK + SSE$	=	2.80359.

For $s_3^o(x)$ one has:

squared sum of errors SSE	=	0.40773
total curvature GK	=	997.16844 and
value of $\mu GK + SSE$	=	1.40490.

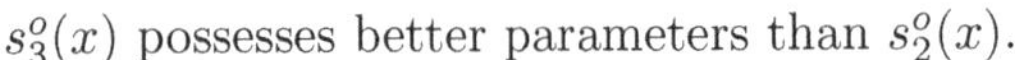

$s_3^o(x)$ possesses better parameters than $s_2^o(x)$.

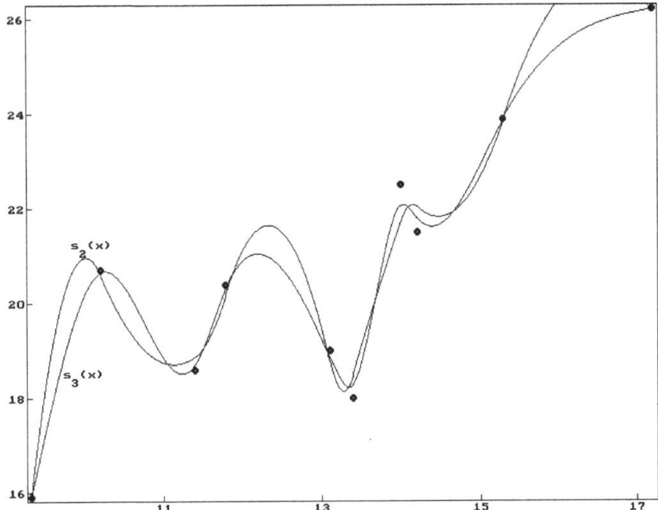

Fig. 7.3: The solution to (OP2) and (OP3), see Example 7.2 for the data.

The next example dealing with $\sum^2(\Delta X)$ is handled with respect to problem (R) by REINSCH, which falls within the set of the twice continuously differentiable functions.

> Given are the points (x_i, y_i), the weights k_i, $i = 1, 2, \ldots, m$, and a $S \geq 0$. Suppose $x_1 < x_2 < \ldots < x_m$. A spline function $s_2(x)$ in $\sum^2(\Delta X)$ is searched for so that $\displaystyle\int_{x_1}^{x_m} [s_2''(x)]^2 \, dx$ becomes minimal. Accepted are all quadratic splines $s_2(x)$ for which $\displaystyle\sum_{i=1}^{m} \left(\frac{s_2(x_i) - y_i}{k_i}\right)^2 \leq S$ is true. (R2)

To begin it is determined that the solution to (R2), in the case that it exists, is realized by an $s_2(x)$ in $\sum^2(\Delta X)$ with minimal total curvature. The first derivative $s'_2(x_1) = B_1$ possesses the representation given by (5.6), where $w_i = 1/\Delta x_i$. Similarly to (OP2), (R2) can be transformed into an equivalent quadratic optimization problem in $\mathbb{R}^m$:

$$4\sum_{i=1}^{m-1}\frac{1}{\Delta x_i}\left[\frac{C_{i+1}-C_i}{\Delta x_i}+\sum_{k=1}^{i-1}2(-1)^{i+k}\frac{C_{k+1}-C_k}{\Delta x_k}+(-1)^i B_1\right]^2 = \min! \text{ where } \sum_{i=1}^{m}\left(\frac{C_i-y_i}{k_i}\right)^2 \leq S\,. \tag{R2Q}$$

A necessary condition for a local extremum is that all partial derivatives of the LAGRANGE-function

$$G(C,p,h)=4\sum_{i=1}^{m-1}\frac{1}{\Delta x_i}\left[\frac{C_{i+1}-C_i}{\Delta x_i}+\sum_{k=1}^{i-1}2(-1)^{i+k}\frac{C_{k+1}-C_k}{\Delta x_k}+(-1)^i B_1\right]^2 +p\left[\sum_{i=1}^{m}\left(\frac{C_i-y_i}{k_i}\right)^2+h^2-S\right]$$

disappear. Here the auxiliary condition is built in with the LAGRANGE multiplier p and the slack variable h. According to (5.6),

$$B_1=B_1(C_1,\ldots,C_m)=\frac{\sum\limits_{i=1}^{m-1}\frac{(-1)^{i+1}}{\Delta x_i}\left(\frac{1}{\Delta x_i}+2\sum\limits_{k=i+1}^{m-1}\frac{1}{\Delta x_k}\right)(C_{i+1}-C_i)}{\sum\limits_{i=1}^{m-1}\frac{1}{\Delta x_i}}\,.$$

Two cases result from $\partial G(C,p,h)/\partial h = 2ph = 0$.

a) $p = 0$.
Here the total curvature is to be minimized. All straight lines that fulfill the auxiliary condition are considered in the solution, in particular, the regression line with minimal weighted sum of the squared errors with regard to the points (x_i, y_i), $i = 1, 2, \ldots, m$.

b) $h = 0$ and $p \neq 0$.
Then the necessary minimum condition reads as

$$\frac{\partial G(C,p,0)}{\partial p}=\sum_{i=1}^{m}\left(\frac{C_i-y_i}{k_i}\right)^2 - S = 0\,.$$

The weighted sum of the squared errors is equal to the predefined S for a solution to (R2Q).

The case b) shall be looked at in greater detail. If the LAGRANGE multiplier p would be known and greater than zero, then the vector C and therefore the entire quadratic spline from the linear system of equations could be calculated

$$\frac{\partial G(C,p,0)}{\partial C_j} = 0, \qquad j = 1,2,\ldots,m\,.$$

It can easily be seen that this system is similar to that of type (OP2) and therefore can uniquely be solved for all $p > 0$. Set $\mu := 1/p$ (Theorem 7.1 or Formula (7.1)).
Then it is true that $[I + \frac{1}{p}K^2 \circ H] \circ C = Y$ or

$$C = [I + \frac{1}{p}K^2 \circ H]^{-1} \circ Y = p[pI + K^2 \circ H]^{-1} \circ Y. \tag{7.2}$$

Due to (7.2) the unknown p could be derived from the equation

$$\begin{aligned} SSE(p) :&= \sum_{i=1}^{m} \left(\frac{C_i - y_i}{k_i}\right)^2 = \|K^{-1} \circ [C - Y]\|^2 \\ &= \| - \frac{1}{p}K \circ H \circ C\|^2 = \frac{1}{p^2}C^T \circ H \circ K^2 \circ H \circ C \\ &= Y^T \circ [pI + K^2 \circ H]^{-1} \circ H \circ K^2 \circ H \circ [pI + K^2 \circ H]^{-1} \circ Y \\ &= S\,. \end{aligned} \tag{7.3}$$

The previous results are summarized now.

Theorem 7.3. *The problem* (R2) *can uniquely be solved for all points* (x_i, y_i), $i = 1,2,\ldots,m$, *and for all* $0 \le S \le S_g$, *where* S_g *is the weighted sum of the squared errors of the regression lines with regard to the* y_i.

Proof. Analogously to the statement made in Chapter 4 regarding the problem (R), the function $SSE(p)$ decreases strong monotone, is convex and is differentiable for all $p > 0$. With this, for every bound $0 < S < S_g$ there exists a unique solution to the equality (7.3) where $p > 0$. This is used for the determination of C in (7.2) because the existence of a positive p under these conditions is also sufficient to solve the optimization problem. For the case $S = S_g$ it remains to be noted that the regression line is defined as the solution. For p approaching infinity, the interpolating quadratic spline with minimal total curvature is the solution. With this, the solution for $S = 0$ can also be uniquely defined. □

It should be examined whether an advantage exists in the size of the objective function $\int_{x_1}^{x_m} [s_2''(x)]^2 \mathrm{d}x$ with quadratic splines in comparison to the natural cubic spline functions for the problem (R). For the problem (OP2) it is not the case. This is to be formulated in greater detail.

Theorem 7.4. *Let $s_3^o(x)$ be the uniquely determined solution to* (R)*, $s_2^o(x)$ the unique solution to* (R2) *for the same data (x_i, y_i), $i = 1, 2, \ldots, m$, and S where $0 \leq S \leq S_g$.*
Then $\int_a^b [s_3^{o\prime\prime}(x)]^2 \mathrm{d}x \leq \int_a^b [s_2^{o\prime\prime}(x)]^2 \mathrm{d}x$ for all $a \leq x_1$ and $b \geq x_m$. The equality is only true when $s_2^o(x) \equiv s_3^o(x) = ax + b$.

Proof. The proof for this assertion is a direct consequence of Lemma 5.1. If it is assumed that there is a situation where the objective function of (R) is greater than that of (R2), then construct the interpolating natural cubic spline for the vector $(s_2^o(x_1), \ldots, s_2^o(x_m))$. In the case that $s_2^o(x) \not\equiv s_3^o(x) = ax + b$, there is a contradiction to the assumption that $s_3^o(x)$ solves problem (R). □

Example 7.3. The uniquely determined quadratic spline $s_2^o(x)$ with minimal total curvature and the sum of the squared errors $SSE = S = 10$ was constructed for the 6 points $(1, 5)$, $(5, 1)$, $(8, 4)$, $(10, 8)$, $(15, 10)$, $(18, 9)$ and $k_i = 1$ for $i = 1, \ldots, 6$ (see Part III).
This function is defined by the LAGRANGE multiplier $p = 1/\mu = 0.070880768$, which is the only positive solution to (7.3). The spline possesses the total curvature $GK = 0.365204$.
Figure 7.4 also shows the splines that result from the real valued negative solutions to (7.3). The respective total curvatures of the functions are:
for $p = 1/\mu = -6.06178$, $GK = 73.06976$,
for $p = 1/\mu = -4.53841$, $GK = 42.72649$ and
for $p = 1/\mu = -0.83735$, $GK = 14.55585$ units.
For the given knots, Figure 7.5 illustrates a part of the sum of the squared errors $SSE(p) = Y^T \circ [pI + K^2 \circ H]^{-1} H \circ K^2 \circ H \circ [pI + K^2 \circ H]^{-1} \circ Y$ as a function of p. The real valued solutions to the equality $SSE(p) = 10$ are marked.

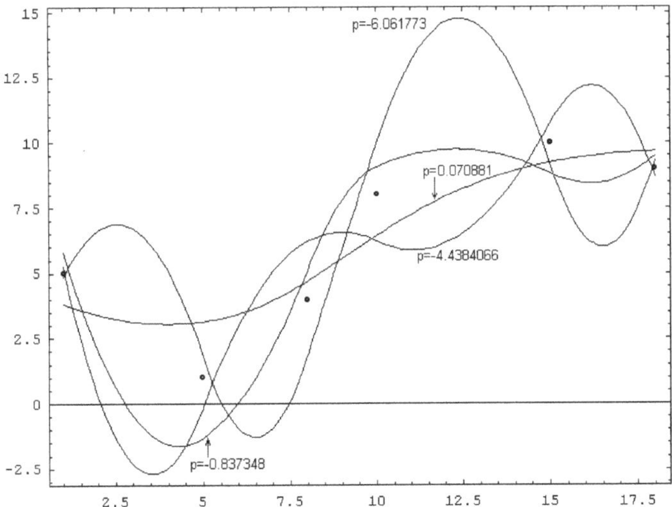

Fig. 7.4: All splines for the solution to $SSE(p) = 10$, see Example 7.3 for the data.

Finally problem (GK3) from Chapter 4 is formulated for quadratic splines.

> Given are the points (x_i, y_i) with $x_1 < x_2 < \ldots < x_m$, the weights k_i, $i = 1, 2, \ldots, m$, and a real number $T \geq 0$. A quadratic spline $s_2(x)$ is searched for so that $\sum_{i=1}^{m} \left(\frac{s_2(x_i) - y_i}{k_i} \right)^2$ becomes minimal. Permissible are all splines of degree 2 where $\int_{x_1}^{x_m} [s_2''(x)]^2 \mathrm{d}x \leq T$. (GK2)

As in problem (R2), the LAGRANGE function

$$H(C, p^*, h^*) = p^* \left(4 \sum_{i=1}^{m-1} \frac{1}{\Delta x_i} \left[\frac{C_{i+1} - C_i}{\Delta x_i} + \sum_{k=1}^{i-1} 2(-1)^{i+k} \frac{C_{k+1} - C_k}{\Delta x_k} \right.\right.$$
$$\left.\left. + \ (-1)^i B_1 \right]^2 + h^{*2} - T \right) + \sum_{i=1}^{m} \left(\frac{C_i - y_i}{k_i} \right)^2$$

is constructed with regard to the integral inequality of the LAGRANGE multiplier p^* and the slack variable h^*. Again, the two cases $h^* = 0$ or $p^* = 0$ are obtained from the necessary condition $\partial H(C, p^*, h^*)/\partial h^* = 2p^* h^* = 0$.

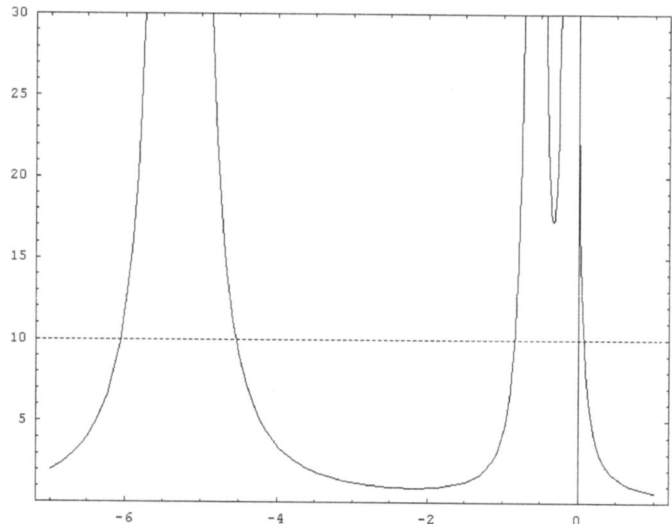

Fig. 7.5: The sum of the squared errors as a function of $p = 1/\mu$, see Example 7.3 for the data.

For $p^* = 0$, only the sum of the squared errors is to be minimized. Each interpolating quadratic spline that fulfills the constraint in (GK2) is then a solution.

When $h^* = 0$ and $p^* \neq 0$, then

$$\frac{\partial H(C, p^*, 0)}{\partial p^*} = 4 \sum_{i=1}^{m-1} \frac{1}{\Delta x_i} \left[\frac{C_{i+1} - C_i}{\Delta x_i} + \sum_{k=1}^{i-1} 2(-1)^{i+k} \frac{C_{k+1} - C_k}{\Delta x_k} \right. \\ \left. + (-1)^i B_1 \right]^2 - T = 0 \,.$$

The solution to (GK2) is then a quadratic spline with minimal total curvature that takes on the given bound T. If p^* were known and greater than zero then the solution to (GK2) could be determined from the linear system of equations $\partial H(C, p^*, 0)/\partial C_j = 0$, $j = 1, 2, \ldots, m$. This has a unique solution for all $p^* \geq 0$ because it corresponds with the equalities to be solved for problem (OP2) when $\mu = p^*$.

According to (7.1), $[I + p^* K^2 \circ H]^{-1} \circ Y = C$ results so that $GK = GK(p^*) = C^T \circ H \circ C = Y^T \circ ([I + p^* K^2 \circ H]^{-1})^T \circ H \circ [I + p^* K^2 \circ H]^{-1} \circ Y$ exists for the total curvature. Determine p^* from the equation

$$GK(p^*) = Y^T \circ ([I + p^* K^2 \circ H]^{-1})^T \circ H \circ [I + p^* K^2 \circ H]^{-1} \circ Y = T \quad (7.4)$$

and use the respective $p^* \geq 0$ in order to obtain the interpolation points of the resulting spline according to (7.1). In a final step, the uniquely determined interpolating quadratic spline with minimal total curvature is constructed.

Theorem 7.5. *Given are (x_i, y_i), $i = 1, 2, \ldots, m$, and a T with $T \leq T_o$. The number T_o denotes the minimal total curvature of an interpolating quadratic spline with respect to the given points. Problem* (GK2) *then has a unique solution. The solution is determined by the only positive p^* which fulfills* (7.4).

Proof. First it is noted that the uniqueness of the solution is not given for $T > T_o$ because the sum of the squared errors is also zero for other interpolating quadratic splines. These splines are also permissible beside the spline with minimal total curvature.
Furthermore it can be shown that the total curvature in dependence of $p^* \geq 0$ is a strong monotone decreasing, convex and differentiable function. With this, (7.4) possesses exactly one positive solution. So $p^* \geq 0$ is also sufficient, the statement of the theorem follows. □

The comparison of the objective functions of (GK3) and (GK2) precipitates again in favor of the cubic splines.

Theorem 7.6. *Let (x_i, y_i), $i = 1, 2, \ldots, m$, be a given data set, k_i, $i = 1, 2, \ldots, m$, given weights and T^* the total curvature of the natural cubic spline interpolating these points. Further, let $T \leq T^*$. $s_3^o(x)$ denotes the unique solution to* (GK3) *and $s_2^o(x)$ denotes the unique solution to* (GK2). *The weighted sum of the squared errors $SSE(s_3^o)$ of $s_3^o(x)$ with regard to the y_i is then always less than or at most equal to the respective sum of the squared errors $SSE(s_2^o)$ of $s_2^o(x)$. The equality is only true when $s_3^o(x) \equiv s_2^o(x) = ax + b$.*

Proof. It is assumed that $SSE(s_2^o) < SSE(s_3^o)$. In this case, the interpolating natural cubic spline $s_{int}(x)$ is constructed for the vector $(s_2^o(x_1), \ldots, s_2^o(x_m))$. According to Lemma 5.1, for the total curvature one has $GK_{int} \leq GK_{s_2} = T$. The equality is only true when both functions are a straight line. With that, $s_{int}(x)$ is a permissible solution to (GK3), whose objective function value is indeed less than that of $s_3^o(x)$. This however conflicts with the assumption that $s_3^o(x)$ is the uniquely determined solution to (GK3). □

Example 7.4. The points from Example 7.3 are also used for an example calculation of problem (GK2). $T = 0.36520$ is used. This is the total curvature of the solution to (R2). In addition, all weights are defined to be one, $k_i = 1$.With these data one obtains

for $p^* = -\,30.84527$, $SSE = 40.50326$,
for $p^* = -\,72.33282$, $SSE = 46.13801$,
for $p^* = -\,116.4269$, $SSE = 76.03800$ and
for $p^* = 14.10820$, $SSE = 10$.

The solutions to (R2) and (GK2) correspond with the respectively chosen bounds and $p = 1/p^*$.
A section of the total curvature as a function

$$GK(p^*) = Y^T \circ \left(\left[I + p^* K^2 \circ H\right]^{-1}\right)^T \circ H \circ \left[I + p^* K^2 \circ H\right]^{-1} \circ Y$$

of the LAGRANGE parameter p^* is illustrated in Figure 7.7.

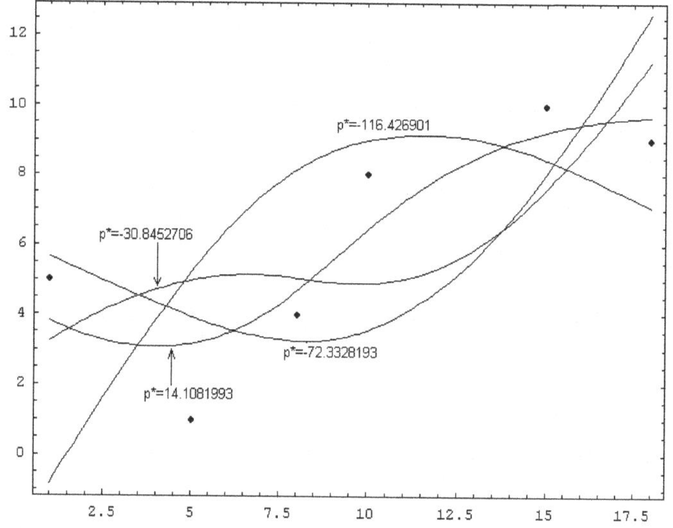

Fig. 7.6: Example 7.4: all quadratic splines for the solution to $GK(p^*) = 0.36520$, see Example 7.3 for the data.

Example 7.5. In order to show that the differences between the solution to (GK3) and (GK2) are often minimal, the data from Example 4.4 from the Chapter 4 were used.
Four real valued solutions p^* resulted from equality $GK(p^*) = 1$. The respective quadratic splines are shown in Figure 7.8. Compare these with the respective natural cubic spline function in Figures 4.8 and 4.9.

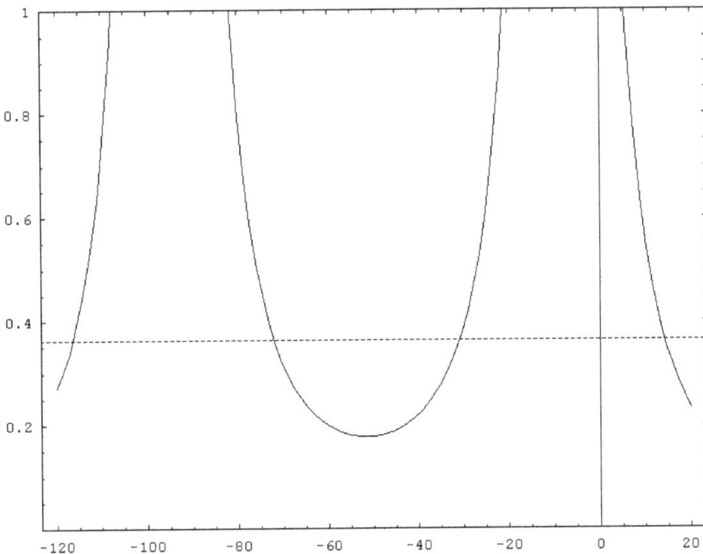

Fig. 7.7: Example 7.4: the total curvature as a function $GK(p^*)$ of the LAGRANGE parameter p^*, see Example 7.3 for the data.

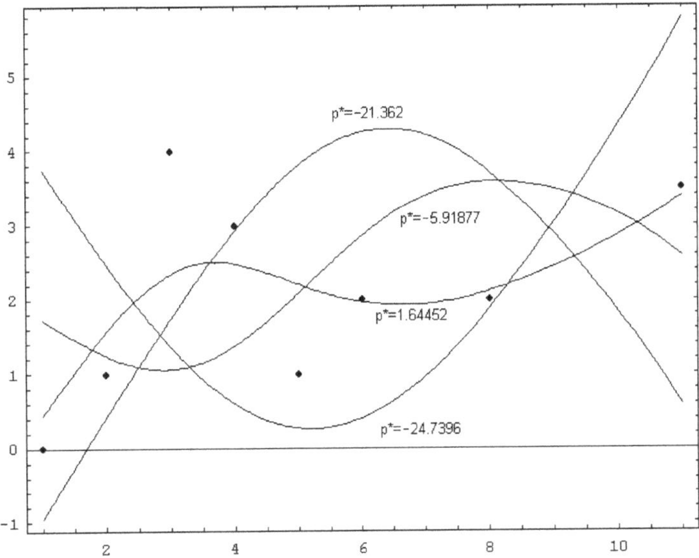

Fig. 7.8: The quadratic splines related to all real valued solutions to (GK2) where $T = 1$, see Example 4.4 for the data.

7.1 Smoothing quadratic splines and the integral of the quadratic first derivative

It is possible to construct interpolating splines of degree 2 with different minimal conditions by the selection of an appropriate initial condition. This makes the calculation of a smoothing quadratic spline with this minimal condition possible.

Given are m points (x_i, y_i) with the condition $x_1 < x_2 < \ldots < x_m$. Then a uniquely determined interpolating quadratic spline $s_2(x)$ exists that minimizes the measure of curvature $\int_{x_1}^{x_m} [s_2'(x)]^2 \mathrm{d}x$. This $s_2(x)$ is defined by the initial condition

$$B_1 = \frac{\sum\limits_{i=1}^{m-1} \frac{(-1)^{i+1}}{\Delta x_i} \left(\Delta x_i + 2 \sum\limits_{k=i+1}^{m-1} \Delta x_k \right) \Delta y_i}{\sum\limits_{i=1}^{m-1} \Delta x_i}. \tag{7.5}$$

The following optimization problem is formulated with respect to a fixed system of knots $x_1 < x_2 < \ldots < x_m$ in the set $\sum^2(\Delta X)$ of all quadratic splines:

Given are the points (x_i, y_i) with $x_1 < x_2 < \ldots < x_m$, the weights k_i, $i = 1, 2, \ldots, m$, and a real number $0 \le \mu$. A spline function $s_2(x)$ is searched for in $\sum^2(\Delta X)$, so that $F(s_2) := \mu \int\limits_{x_1}^{x_m} [s_2'(x)]^2 \mathrm{d}x + \sum\limits_{i=1}^{m} \left(\frac{s_2(x_i) - y_i}{k_i} \right)^2$ becomes a minimum.	(OP2')

The ability to solve problem (OP2') is now examined. It is clear that every solution is a quadratic spline with minimal $\int_{x_1}^{x_m} [s_2'(x)]^2 \mathrm{d}x$.

Due to the definition of a quadratic spline, the known representation $s_2(x) = A_i(x - x_i)^2 + B_i(x - x_i) + C_i$ over $[x_i, x_{i+1}]$ should be used. $C_i = s_2(x_i)$ and $A_i = (B_{i+1} - B_i)/2\Delta x_i$, $i = 1, 2, \ldots, m$, are then true. With this, the $B_i = s_2'(x_i)$, $i = 1, 2, \ldots, m$, are determined by the still unknown C_i from the linear system of equations

$$\left.\begin{aligned} B_1 &= \frac{\sum\limits_{i=1}^{m-1} \frac{(-1)^{i+1}}{\Delta x_i}\left(\Delta x_i + 2\sum\limits_{k=i+1}^{m-1}\Delta x_k\right)(C_{i+1}-C_i)}{\sum\limits_{i=1}^{m-1}\Delta x_i} \\ &\text{and} \\ B_i + B_{i+1} &= 2\frac{C_{i+1}-C_i}{\Delta x_i}, \qquad i = 1, 2, \dots, m-1\,. \end{aligned}\right\} \tag{7.6}$$

Since the spline that is sought after is uniquely defined by the derivative, an expression in terms of the B_i for the new measure of curvature is searched for,

$$\begin{aligned} \int\limits_{x_1}^{x_m} [s_2'(x)]^2 \mathrm{d}x &= \sum_{i=1}^{m-1} \int\limits_{x_i}^{x_{i+1}} [s_2'(x)]^2 \mathrm{d}x = \sum_{i=1}^{m-1} \int\limits_{x_i}^{x_{i+1}} [2A_i(x-x_i)+B_i]^2 \mathrm{d}x \\ &= \sum_{i=1}^{m-1}\left[\frac{4}{3}A_i^2\Delta x_i^3 + 2A_iB_i\Delta x_i^2 + B_i^2\Delta x_i\right] \\ &= \sum_{i=1}^{m-1}\left[\frac{1}{3}(B_{i+1}-B_i)^2 + (B_{i+1}-B_i)B_i + B_i^2\right]\Delta x_i \\ &= \frac{1}{3}\sum_{i=1}^{m-1}\Delta x_i(B_i^2 + B_iB_{i+1} + B_{i+1}^2)\,. \end{aligned} \tag{7.7}$$

With the help of the following matrix, the integral of the quadratic first derivative can be represented in quadratic form $B^T \circ Z' \circ B$, where $B := (B_1, B_2, \dots, B_m)^T$ and

$$Z' := \frac{1}{3}\begin{pmatrix} \Delta x_1 & \frac{\Delta x_1}{2} & 0 & 0 & \dots & 0 & 0 \\ \frac{\Delta x_1}{2} & \Delta x_1 + \Delta x_2 & \frac{\Delta x_2}{2} & 0 & \dots & 0 & 0 \\ 0 & \frac{\Delta x_2}{2} & \Delta x_2 + \Delta x_3 & \frac{\Delta x_3}{2} & \dots & 0 & 0 \\ \vdots & \vdots & \vdots & \vdots & \ddots & \vdots & \vdots \\ 0 & 0 & 0 & 0 & \dots & \Delta x_{m-2} + \Delta x_{m-1} & \frac{\Delta x_{m-1}}{2} \\ 0 & 0 & 0 & 0 & \dots & \frac{\Delta x_{m-1}}{2} & \Delta x_{m-1} \end{pmatrix}.$$

Z' is a symmetrical and invertible matrix because the GAUSSIAN sum of rows criterion is fulfilled. In addition, Z' is positive definite because the matrix determines the measure of curvature. This is only zero when all derivatives $s_2'(x_i)$ disappear.

The system of equations (7.6) is described in the same way as above in this chapter with $F \circ B = D' \circ C$ where

$$F = \begin{pmatrix} 1\,0\,0 \ldots 0\,0 \\ 1\,1\,0 \ldots 0\,0 \\ 0\,1\,1 \ldots 0\,0 \\ \vdots\,\vdots\,\vdots \ddots \vdots\,\vdots \\ 0\,0\,0 \ldots 1\,1 \end{pmatrix}$$

and

$$D' = \begin{pmatrix} d'_{11} & d'_{12} & d'_{13} & \ldots & d'_{1m-1} & d'_{1m} \\ -\frac{2}{\Delta x_1} & \frac{2}{\Delta x_1} & 0 & \ldots & 0 & 0 \\ 0 & -\frac{2}{\Delta x_2} & \frac{2}{\Delta x_2} & \ldots & 0 & 0 \\ \vdots & \vdots & \vdots & \ddots & \vdots & \vdots \\ 0 & 0 & 0 & \ldots & -\frac{2}{\Delta x_{m-1}} & \frac{2}{\Delta x_{m-1}} \end{pmatrix}.$$

Since F is invertible, one obtains $B = F^{-1} \circ D' \circ C$ where

$$F^{-1} = \begin{pmatrix} 1 & 0 & 0 & \ldots & 0\;0 \\ -1 & 1 & 0 & \ldots & 0\;0 \\ 1 & -1 & 1 & \ldots & 0\;0 \\ \vdots & \vdots & \vdots & \ddots & \vdots\;\vdots \\ (-1)^{m-1} & (-1)^{m-2} & (-1)^{m-3} & \ldots & -1\;1 \end{pmatrix}.$$

To calculate the first row of D', the denominator from (7.5) is broken down into the individual summands

$$\begin{aligned} i = 1 :&\quad \tfrac{1}{\Delta x_1}(\Delta x_1 + 2[\Delta x_2 + \Delta x_3 + \Delta x_4 + \cdots + \Delta x_{m-1}])(C_2 - C_1) \\ i = 2 :&\; - \tfrac{1}{\Delta x_2}(\Delta x_2 + 2[\Delta x_3 + \Delta x_4 + \cdots + \Delta x_{m-1}])(C_3 - C_2) \\ i = 3 :&\quad \tfrac{1}{\Delta x_3}(\Delta x_3 + 2[\Delta x_4 + \cdots + \Delta x_{m-1}])(C_4 - C_3) \\ &\;\vdots \\ i = m-1 :&\; (-1)^m (C_m - C_{m-1})\,. \end{aligned}$$

Following the summation of the appropriate coefficients for the C_k,

$$d'_{1k} = (-1)^k \left\{ \frac{\Delta x_{k-1}}{\Delta x_{k-1}} + 1 + \frac{2}{\Delta x_{k-1}} \sum_{j=k}^{m-1} \Delta x_j + \frac{2}{\Delta x_k} \sum_{j=k+1}^{m-1} \Delta x_j \right\} / \alpha'$$

is obtained for k from 1 to $(m-1)$ (here is $\Delta x_{k-1}/\Delta x_{k-1} = 0$ for $k = 1$) and

$$d'_{1m} = \frac{(-1)^m}{\alpha'}, \qquad \text{with } \alpha' = \sum_{i=1}^{m-1} \Delta x_i\,.$$

The following results:

$$\int_{x_1}^{x_m} [s_2'(x)]^2 \mathrm{d}x = (F^{-1} \circ D' \circ C)^T \circ Z' \circ F^{-1} \circ D' \circ C$$

$$= C^T \circ D'^T \circ (F^{-1})^T \circ Z' \circ F^{-1} \circ D' \circ C\,.$$

The matrix $H' := D'^T \circ (F^{-1})^T \circ Z' \circ F^{-1} \circ D'$ is symmetrical and positive semi-definite. It is positive semi-definite because $C^T \circ H' \circ C = 0$ for all vectors C where $C_i = \beta \in \mathbb{R}$, $i = 1, 2, \ldots, m$.
These considerations allow the objective function of (OP2') to be represented in the form $F(C) := \mu C^T \circ H' \circ C + ||K^{-1} \circ (C - Y)||^2$.
$2\mu H' \circ C + 2K^{-2} \circ (C - Y) = 0$ or $[I + \mu K^2 \circ H'] \circ C = Y$ is necessary for a local minimum.
The arguments up to now can be summarized in the following theorem (see also Part III).

Theorem 7.7. *Problem* (OP2') *can uniquely be solved for all* $\mu \geq 0$. *The solution is the quadratic spline for which the* $C_i = s_2(x_i)$, $i = 1, 2, \ldots, m$, *are calculated from the uniquely solvable linear system of equations*

$$[I + \mu K^2 \circ H'] \circ C = Y.$$

The first derivatives $B_i = s_2'(x_i)$ *are obtained from* (7.6).

Proof. The matrix $I + \mu K^2 \circ H'$ is positive definite and therefore invertible for $0 \leq \mu$. The statement of the theorem follows because the interpolating quadratic spline with minimal $\int_{x_1}^{x_m} [s_2'(x)]^2 \mathrm{d}x$ is also uniquely determined. □

Similar problems like (R2) and (GK2) where the total curvature is merely replaced by $AQ := \int_{x_1}^{x_m} [s_2'(x)]^2 \mathrm{d}x$ can also be formulated.
These problems can uniquely be solved. Algorithms result where H is replaced by H' and Z^* is replaced by Z' (compare the construction of the solution to (OPQ)).
The way to construct smoothing quadratic splines suggested at the example of minimal

$$GK = \int_{x_1}^{x_m} [s_2''(x)]^2 \mathrm{d}x \text{ or } AQ = \int_{x_1}^{x_m} [s_2'(x)]^2 \mathrm{d}x$$

can be generalized.

Let $\Psi : \sum^2(\Delta X) \to \mathbb{R}^+$ be a map from the space of all quadratic splines with respect to a fixed system of knots into $\mathbb{R}^+$. The following is assumed:

(1) The interpolation task $s_2^*(x_i) = C_i$, $i = 1, 2, \ldots, m$, with the constraint $\Psi(s_2^*) \le \Psi(s_2)$ for all $s_2 \in \sum^2(\Delta X)$ satisfying $s_2(x_i) = C_i$ has a unique solution.

(2) This solution is determined by B_1 (see Theorem 5.1). Let B_1 have the representation $B_1 = \sum_{i=1}^m \beta_i C_i$ with coefficients independent of the C_i.

(3) A matrix Z_Ψ with $\Psi(s_2) = B^T \circ Z_\Psi \circ B \ge 0$ for all $s_2 \in \sum^2(\Delta X)$ exists independent of the C_i. Here B is defined as $B := (B_1, B_2, \ldots, B_m)^T = (s_2'(x_1), s_2'(x_2), \ldots, s_2'(x_m))^T$.

The integrals GK and AQ can serve as examples of Ψ. It was shown in Chapter 5 that the corresponding interpolating tasks are uniquely solvable. B_1 has the representation

$$B_1 = \frac{\sum\limits_{i=1}^{m-1} (-1)^{i+1} \left(w_i + 2 \sum\limits_{k=i+1}^{m-1} w_k \right) \dfrac{C_{i+1} - C_i}{\Delta x_i}}{\sum\limits_{i=1}^{m-1} w_i}$$

with corresponding w_i. This is a linear combination of the C_i. The coefficients have been determined in the solution to (OP2) and (OP2'), respectively. The matrix demanded in the third assumption is given by Z^* in case GK and by Z' in case AQ. These are also indicated in the solutions to (OP2) and (OP2'), respectively.
The following generalized problem is now examined.

Given are the points $(x_i, y_i), x_1 < x_2 < \ldots < x_m$, the weights $k_i > 0$, $i = 1, 2, \ldots, m$, and a real number $\mu \ge 0$. A quadratic spline function $s_2(x)$ in $\sum^2(\Delta X)$ is sought after so that $F_\Psi(s_2) := \mu\Psi(s_2) + \sum\limits_{i=1}^{m} \left(\dfrac{s_2(x_i) - y_i}{k_i} \right)^2$ becomes minimal.	(OP2_Ψ)

If (OP2_Ψ) has a solution then it must fulfill the equations

$$\left.\begin{aligned} &B_1 = \sum_{i=1}^{m} \beta_i C_i \qquad \text{(assumption (2))} \\ &\text{and} \\ &B_i + B_{i+1} = 2\frac{C_{i+1} - C_i}{\Delta x_i}, \qquad i = 1, 2, \ldots, m-1\,. \end{aligned}\right\} \qquad (*)$$

The $C_i = s_2(x_i)$ are the unknown function values of a solution to OP2_Ψ.

With the matrices

$$F := \begin{pmatrix} 1\,0\,0 \ldots 0\,0 \\ 1\,1\,0 \ldots 0\,0 \\ 0\,1\,1 \ldots 0\,0 \\ \vdots\,\vdots\,\vdots \ddots \vdots\,\vdots \\ 0\,0\,0 \ldots 1\,1 \end{pmatrix}$$

and

$$D_\psi := \begin{pmatrix} \beta_1 & \beta_2 & \beta_3 & \cdots & \beta_{m-1} & \beta_m \\ -\frac{2}{\Delta x_1} & \frac{2}{\Delta x_1} & 0 & \cdots & 0 & 0 \\ 0 & -\frac{2}{\Delta x_2} & \frac{2}{\Delta x_2} & \cdots & 0 & 0 \\ \vdots & \vdots & \vdots & \ddots & \vdots & \vdots \\ 0 & 0 & 0 & \cdots & -\frac{2}{\Delta x_{m-1}} & \frac{2}{\Delta x_{m-1}} \end{pmatrix}$$

the system $(*)$ is seen as $F \circ B = D_\Psi \circ C$ or $B = F^{-1} \circ D_\Psi \circ C$.
The problem ($\mathrm{OP2}_\Psi$) can be reformulated because of the assumption $\Psi(s_2) = B^T \circ Z_\Psi \circ B$ as follows:

$$\left.\begin{array}{l} \text{A vector } C := (C_1, \ldots, C_m)^T \in \mathbb{R}^m \text{ is sought after so that} \\ F_\psi(s_2) = F_\psi(C) = \mu C^T \circ H_\psi \circ C + \|K^{-1} \circ [C - Y]\|^2 \\ \text{becomes minimal.} \end{array}\right\}$$

Here the abbreviation $H_\Psi := (D_\Psi)^T \circ (F^{-1})^T \circ Z_\Psi \circ F^{-1} \circ D_\Psi$ is used. $2\mu H_\Psi \circ C + 2K^{-2} \circ [C - Y]{=}0$ is necessary for a local extreme value. From this follows $[I + \mu K^2 \circ H_\Psi] \circ C = Y$.
Due to the assumptions, the matrix H_Ψ is positive semi-definite and therefore $[I + \mu K^2 \circ H_\Psi]$ is positive definite and invertible. This means ($\mathrm{OP2}_\Psi$) has a unique solution. Use C to construct the uniquely determined interpolating spline $s_2(x)$ that minimizes Ψ.
The map

$$\Psi = \Psi(s_2) := \sum_{i=1}^{m-1} \int_{x_i}^{x_{i+1}} \frac{[s_2''(x)]^2}{(1 + (\Delta y_i/\Delta x_i)^2)^3} \mathrm{d}x$$

will be considered as a new example. One compares this with the total local curvature

$$LK := \int_{x_1}^{x_m} \frac{[s_2''(x)]^2}{(1 + [s_2'(x)]^2)^3} \mathrm{d}x \, .$$

The vector $Y = (y_1, y_2, \ldots, y_m)^T$ here is arbitrary but fixed. It was shown in Chapter 5 that there exists a uniquely determined interpolating quadratic spline $s_2(x)$ with respect to the (x_i, C_i), $i = 1, 2, \ldots, m$, minimizing Ψ. For this function must be true

$$B_1 = \frac{\sum_{i=1}^{m-1} (-1)^{i+1} \left(w_i + 2 \sum_{k=i+1}^{m-1} w_k \right) \frac{C_{i+1} - C_i}{\Delta x_i}}{\sum_{i=1}^{m-1} w_i}$$

with

$$w_i = \frac{1}{\Delta x_i \left(1 + (\Delta y_i / \Delta x_i)^2\right)^3} \,.$$

The denominator of B_1 is broken down into the single summands

$$\begin{array}{ll}
\frac{1}{\Delta x_1}(w_1 + 2[w_2 + w_3 + w_4 + \cdots + w_{m-1}])(C_2 - C_1) & \text{for } i = 1, \\
-\frac{1}{\Delta x_2}(w_2 + 2[w_3 + w_4 + \cdots + w_{m-1}])(C_3 - C_2) & \text{for } i = 2, \\
\frac{1}{\Delta x_3}(w_3 + 2[w_4 + \cdots + w_{m-1}])(C_4 - C_3) & \text{for } i = 3, \\
\vdots & \\
\frac{(-1)^m}{\Delta x_{m-1}}(w_{m-1})(C_m - C_{m-1}) & \text{for } i = m-1 \,.
\end{array}$$

The respective coefficients for C_k, $k = 1, 2, \ldots, m$, are summarized:

$$\beta_1 = \frac{-1}{\Delta x_1} \left(w_1 + 2 \sum_{j=2}^{m-1} w_j \right) / \alpha,$$

$$\beta_k = (-1)^k \left\{ \frac{1}{\Delta x_{k-1}} \left(w_{k-1} + 2 \sum_{j=k}^{m-1} w_j \right) + \frac{1}{\Delta x_k} \left(w_k + 2 \sum_{j=k+1}^{m-1} w_j \right) \right\} / \alpha,$$

$k = 2, 3, \ldots, m-1$, and

$$\beta_m = \frac{(-1)^m}{\Delta x_{m-1}} w_{m-1} / \alpha \text{ with } \alpha := \sum_{j=1}^{m-1} w_j \,.$$

So one has $B_1 = \sum_{i=1}^{m} \beta_i \, C_i$. The representation $\Psi = B^T \circ Z_\Psi \circ B$ with

$$Z_\psi = \begin{pmatrix}
w_1 & -w_1 & 0 & \ldots & 0 & 0 & 0 \\
-w_1 & (w_1 + w_2) & -w_2 & \ldots & 0 & 0 & 0 \\
\vdots & \vdots & \vdots & \ddots & \vdots & \vdots & \vdots \\
0 & 0 & 0 & \ldots & -w_{m-2} & (w_{m-2} + w_{m-1}) & -w_{m-1} \\
0 & 0 & 0 & \ldots & 0 & -w_{m-1} & w_{m-1}
\end{pmatrix}$$

results because of

$$\Psi(s_2) = \sum_{i=1}^{m-1} \frac{4\Delta x_i}{(1+(\Delta y_i/\Delta x_i)^2)^3} A_i^2 = \sum_{i=1}^{m-1} \frac{(B_i^2 - 2B_iB_{i+1} + B_{i+1}^2)}{\Delta x_i\,(1+(\Delta y_i/\Delta x_i)^2)^3}.$$

For given points (x_i, y_i) and $\mu \geq 0$, a function $s_2 \in \sum^2(\Delta X)$ is sought that minimizes

$$F_\Psi(s_2) = \mu \sum_{i=1}^{m-1} \int_{x_i}^{x_{i+1}} \frac{[s_2''(x)]^2}{(1+\Delta y_i^2/\Delta x_i^2)^3}\,\mathrm{d}x + \sum_{i=1}^{m} \left(\frac{s_2(x_i) - y_i}{k_i}\right)^2.$$

This optimization problem has a unique solution, due to the generalization developed above: Solve the linear system of equations

$$[I + \mu K^2 \circ H_\Psi] \circ C = Y$$

with $H_\Psi := (D_\Psi)^T \circ (F^{-1})^T \circ Z_\Psi \circ F^{-1} \circ D_\Psi$. Then construct the uniquely determined interpolating quadratic spline minimizing Ψ with respect to the (x_i, C_i).

Example 7.6. The 11 points $(-4.0, 1.0)$, $(-3.0, -0.5)$, $(-2.0, -0.1)$, $(-1.0, -0.8)$, $(0.0, 0.0)$, $(1.0, 7.0)$, $(2.0, -0.1)$, $(3.0, -0.1)$, $(4.0, -0.1)$, $(5.0, 2.0)$, $(6.0, 1.0)$ were used to construct the uniquely determined quadratic spline $s_2(x)$ that minimizes

$$0.5 \sum_{i=1}^{m-1} \int_{x_i}^{x_{i+1}} \frac{[s_2''(x)]^2}{(1+\Delta y_i^2/\Delta x_i^2)^3}\,\mathrm{d}x + \sum_{i=1}^{m} (s_2(x_i) - y_i)^2.$$

One obtains with the data (see Figure 7.9)

the sum of squared errors	SSE	$= 19.38411$,
the total curvature	GK	$= 97.88917$,
the total local curvature	LK	$= 8.35954$ and
$\int_{x_1}^{x_m} [s_2'(x)]^2 \mathrm{d}x$	AQ	$= 28.47128$.

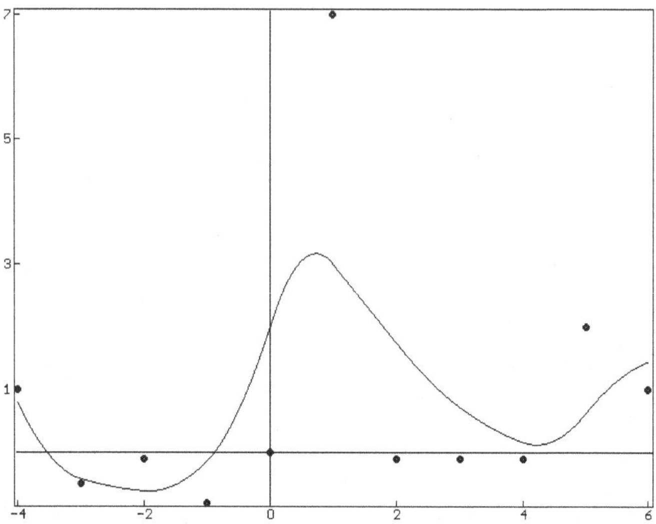

Fig. 7.9: Smoothing quadratic spline $s_2(x)$ with minimal total local curvature, see Example 7.6 for the data and the approximation criterion.

7.2 Quadratic splines smoothing the predefined first derivatives

Another problem concerns the construction of quadratic splines smoothing the predefined first derivatives $b_i = s_2'(x_i)$, $i = 1, 2, \ldots, m$, $x_1 < x_2 < \ldots < x_m$. This problem was examined by KOBZA (1992). According to (5.3) follows $s_2(x)$ is uniquely and completely determined when the b_i, $i = 1, 2, \ldots, m$, are given and for an index $k \in \{1, 2, \ldots, m\}$ is required $s_2(x_k) = y_k$. The respective unknown y_i, $i \neq k$ can then successively be calculated from

$$y_{i+1} - y_i = (b_i + b_{i+1})\frac{\Delta x_i}{2}, \qquad i = 1, 2, \ldots, m-1\,. \tag{7.8}$$

The formulation of a respective smoothing problem is also possible. This is first observed in the set $W_2^2[x_1, x_m]$ of all absolutely continuously differentiable functions $f(x)$ over $[x_1, x_m]$ for which $\int_{x_1}^{x_m} [f''(x)]^2 \mathrm{d}x$ exists and is finite.

A function $f(x)$ in ${W_2}^2[x_1, x_m]$ is searched for so that the objective function

$$F^*(s_2) := \mu \int_{x_1}^{x_m} [f''(x)]^2 \mathrm{d}x + \sum_{i=1}^{m} k_i (b_i - f'(x_i))^2 \tag{S1}$$

becomes a minimum for given $k_i > 0$, b_i, $i = 1, 2, \ldots, m$, and $0 \leq \mu$.

Lemma 7.2. *Let $f(x)$ be from $W_2^2[x_1, x_m]$ where $f'(x_i) = b_i$, $i = 1, 2, \ldots, m$. Then the inequality $\int_{x_1}^{x_m} [s_2''(x)]^2 \mathrm{d}x \leq \int_{x_1}^{x_m} [f''(x)]^2 \mathrm{d}x$ is true for each quadratic spline function $s_2(x)$ that also fulfills the conditions $s_2'(x_i) = b_i$. The equality can only be true when $f(x)$ is a quadratic spline.*

Proof. Again, the starting point is the identity

$$\begin{aligned}
\int_{x_1}^{x_m} [f''(x)]^2 \mathrm{d}x &= \int_{x_1}^{x_m} [s_2''(x) + f''(x) - s_2''(x)]^2 \mathrm{d}x \\
&= \int_{x_1}^{x_m} [s_2''(x)]^2 \mathrm{d}x + \int_{x_1}^{x_m} [f''(x) - s_2''(x)]^2 \mathrm{d}x \\
&\quad + 2 \int_{x_1}^{x_m} s_2''(x)[f''(x) - s_2''(x)] \mathrm{d}x \,.
\end{aligned}$$

The following is obtained following partial integration

$$\begin{aligned}
\int_{x_1}^{x_m} s_2''(x)[f''(x) - s_2''(x)] \mathrm{d}x &= \sum_{i=1}^{m-1} s_2''(x)[f'(x) - s_2'(x)] \Big|_{x_i}^{x_{i+1}} \\
&\quad - \sum_{i=1}^{m-1} \int_{x_i}^{x_{i+1}} s_2'''(x)[f'(x) - s_2'(x)] \mathrm{d}x = 0 \,,
\end{aligned}$$

because both functions possess the same derivatives at the knots and $s'''(x)$ disappears over $\mathbb{R}$. The second part of the statement is obtained from the definition of ${W_2}^2[x_1, x_m]$ because $\int_{x_1}^{x_m} [f''(x) - s_2''(x)]^2 \mathrm{d}x$ only disappears in the case that $f(x) \equiv s_2(x)$. □

Theorem 7.8. *Each solution to* (S1) *is a quadratic spline.*

Proof. Assumed that another function $f(x)$ solves the problem. Let $s_2(x)$ be the unique quadratic spline with $s_2'(x_i) = f'(x_i)$, $i = 1, 2, \ldots, m$, and $s_2(x_1) = 0$. However, due to Lemma 7.2, this function $s_2(x)$ possesses a smaller total curvature than $f(x)$ such that $f(x)$ cannot solve (S1). □

Due to Theorem 7.8, the optimization task (S1) is equivalent with

$$G(s_2'(x_1), \ldots, s_2'(x_m)) := \mu \sum_{i=1}^{m-1} \frac{1}{\Delta x_i} \left[s_2'(x_{i+1}) - s_2'(x_i)\right]^2 + \sum_{i=1}^{m} k_i (b_i - s_2'(x_i))^2 = \min!$$

The necessary conditions $\partial G / \partial s_2'(x_i) = 0$, $i = 1, 2, \ldots, m$, result in the linear system of equations

$$\begin{pmatrix} \frac{\mu}{\Delta x_1} + k_1 & -\frac{\mu}{\Delta x_1} & 0 & 0 & \cdots & 0 \\ -\frac{\mu}{\Delta x_1} & \lambda_1 + k_2 & -\frac{\mu}{\Delta x_2} & 0 & \cdots & 0 \\ \vdots & \vdots & \vdots & \vdots & \ddots & \vdots \\ 0 & 0 & 0 & -\frac{\mu}{\Delta x_{m-2}} & \lambda_{m-2} + k_{m-1} & -\frac{\mu}{\Delta x_{m-1}} \\ 0 & 0 & 0 & 0 & \frac{-\mu}{\Delta x_{m-1}} & \frac{\mu}{\Delta x_{m-1}} + k_m \end{pmatrix}$$

$$\circ \begin{pmatrix} s_2'(x_1) \\ s_2'(x_2) \\ \vdots \\ s_2'(x_{m-1}) \\ s_2'(x_m) \end{pmatrix} = \begin{pmatrix} k_1 b_1 \\ k_2 b_2 \\ \vdots \\ k_{m-1} b_{m-1} \\ k_m b_m \end{pmatrix}, \quad \text{with } \lambda_i := \frac{\mu}{\Delta x_i} + \frac{\mu}{\Delta x_{i+1}}.$$

The matrix of coefficients is diagonally dominant for $0 \leq \mu$ and therefore invertible. The optimization problem (S1) therefore has a unique solution when an arbitrary $s_2(x_k)$ is given.

Figure 7.10 shows the derivative $s_2'(x)$ of the (S1)-optimal quadratic spline $s_2(x)$ smoothing the points (x_i, b_i):
$(0, -0.45)$, $(1, 1.14)$, $(2, 1.02)$, $(3, 0.39)$, $(4, -0.55)$, $(5, -0.99)$, $(6, 0.42)$, $(7, 1.72)$, $(8, 2.47)$, $(9, 0.58)$, $(10, 0.78)$, $(11, 0.98)$, $(12, 1.08)$, $(13, 3.28)$, $(14, 2.18)$, $(15, 1.12)$, $(16, 0.58)$, $(17, 1.03)$, $(18, 2.22)$, $(20, -0.45)$.
The smoothing parameter is $\mu = 1$ and all weights were set to $k_i = 1$ (see Part III). By specifying that $s_2(x_1) = 0$, the (S1)-optimal quadratic spline $s_2(x)$ in Figure 7.11 results.

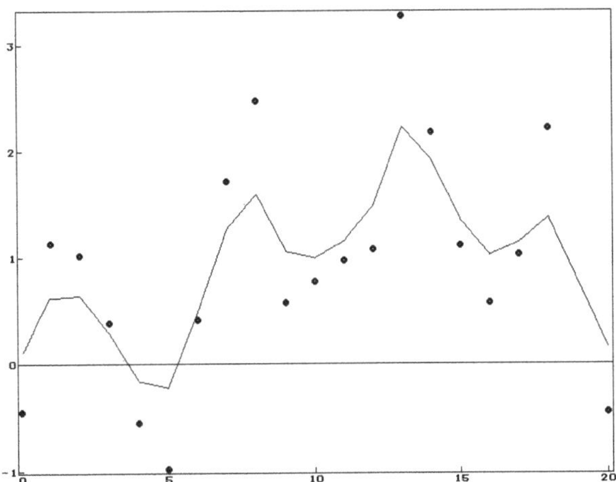

Fig. 7.10: First derivative $s_2'(x)$, see the preceding text for the data.

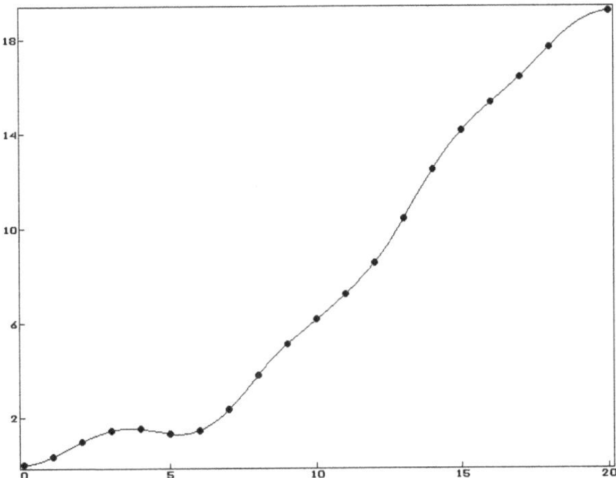

Fig. 7.11: The spline function $s_2(x)$ corresponding to Figure 7.10 where $s_2(0) = 0$. The points are the $(x_i, s_2(x_i))$.

Note 7.1. It is also possible to have a smoothing problem in the form

$$G(s_2) := \mu \int_{x_1}^{x_m} [s_2'(x)]^2 \mathrm{d}x + \sum_{i=1}^{m} k_i(b_i - s_2'(x_i))^2 = \min! \qquad (*)$$

This can be replaced by (see Chapter 7.1)

$$G(s_2) = \mu B^T \circ Z' \circ B + \sum_{i=1}^{m} k_i(b_i - s_2'(x_i))^2 \,.$$

The necessary conditions $\partial G/\partial B_i = 0$ result in the linear system of equations $\mu Z' \circ B + K \circ B = K \circ \mathbf{b}$ or $(\mu Z' + K) \circ B = K \circ \mathbf{b}$.
Here, K is an $(m \times m)$ matrix with weights k_i on the main diagonal and $\mathbf{b} := (b_1, b_2, \ldots, b_m)^T$. The matrix $\mu Z' + K$ is positive definite for all $\mu \geq 0$ and therefore invertible.
However, the problem $(*)$ formulated above does not make sense. The integral and the sum of the squared errors deal with opposing processes. The "smoothing" results in the sense that is represented in the Figure 7.12.

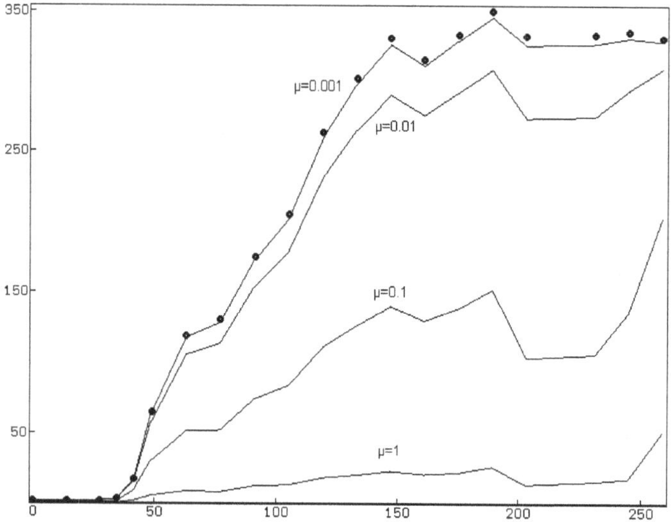

Fig. 7.12: The solutions to $(*)$ for different μ, see text.

Chapter 8

Splines and averaged functions

Averaging is a principle that is often used for the analysis of univariate data. On the one hand it is plausible. Averaging possesses desirable characteristics for commonly used statistical models. In statistics, averaging is the most used estimating function. Incidentally, it was already used in the third century BC.

Things however are not as simple when the data represent the course of functions. What is an "average" course? This problem is explained at the simplest model from the pharmacokinetics. The construction of averaged courses for a set of individual splines is explored. This is done for interpolating cubic splines with regard to a fixed system of knots. It is shown that the average spline can be determined by averaging the individual spline coefficients as well as through the calculation of the interpolating spline function with regard to the average of the function values at the individual knots. This provides the opportunity to determine averaged functional interrelationships by means of splines. The results are illustrated through a computer experiment with a pharmacokinetic model function. The conjecture that "better" averaged splines can be obtained by smoothing the individual splines first is false. The effects that are seen in this case are also illustrated in an example.

A procedure for the construction of so-called normal ranges or reference ranges is subsequently shown. A range around the average spline function is constructed in which a given percent of all individual function values is found. A common system of knots for all individual courses is necessary. This precondition is very drastic. In practice it can only be fulfilled if the observations are planned and executed precisely. But according to Theorem 8.3 it is possible to produce the necessary common system of knots of the individual courses of functions from the data that is on hand. A plan for

the generation of averaged splines as well as for the calculation of desired normal ranges results. As an example, this process is used to describe data about childlike growth.
The calculation of arithmetic averages for minimizing errors in univariate data is a common calculation. When averaging functions, the problem arises that a given function class that represents the applied model is not necessary algebraically closed with regard to this operation. In this case, "average courses" of functions do not make sense in this model. The One-compartment-iv model from pharmacokinetics is used to demonstrate this. The associated mass-time function is $m(t) = D_{ges}e^{-kt}$. The compartment models are observed more closely in the following part of this book.
First the total dose D_{ges} is set to one. Given are the two individual curves $m_1(t) = e^{-0.5t}$ and $m_2(t) = e^{-1.5t}$. The values $m_1(t_i)$ and $m_2(t_i)$ are averaged for each of the chosen 11 knots $t_i = i - 1$, $i = 1, 2, \ldots, 11$. A function $m(t) = e^{-kt}$ is adapted to these averages via the method of least squares. The optimal parameter becomes $k_{\text{mean}} = 0.794$. The averages are not function values of the function $m_{\text{mean}}(t) = e^{-0.794t}$. The pointwise averaging of the individual curves results in a function that does not belong to the observed class of models anymore. It can be analytically proven that for two knots $t_1 \neq t_2$, both of which are not zero, the respective averages do not lie on a function of the form $m(t) = e^{-kt}$.
The averaging of the function parameters results in the value one and leads to the function $m_{\text{parmean}}(t) = e^{-1t}$. No relationship to the averaged function values (see Figure 8.1) can be found here.
The described situation can arise when the model parameters do not linearly fit into the function equation. Seeking averaged courses therefore principally makes sense. The selection of an appropriate model is crucial. For spline functions, the model parameters linearly fit into the function equations. For these model functions, it can be seen that the curve interpolating the averages of the individual function values at the knots and the averaged course of the curves (average of the spline parameters) correspond extremely well. Section 8.1 demonstrates that in this fashion, even for "degenerated" individual courses of functions. The original course is reconstructed well by the averaged spline function. This is especially interesting for measurements with a large scatter.
The following is an often used method for the calculation of an average

course by a member $f(p_1, p_2, \ldots, p_r; x)$ of the observed model functions:

$$F(p_1, p_2, \ldots, p_r) = \sum_{j=1}^{m} \sum_{i=1}^{n} (y_{ij} - f(p_1, p_2, \ldots, p_r; x_j))^2 = \text{min!}$$

n individual functions are described here by the points (x_j, y_{ij}), $j = 1, 2, \ldots, m$, and $i = 1, 2, \ldots, n$. This means that for a common system of knots $x_1 < x_2 < \ldots < x_m$, there exists exactly one measurement $f_i(x_j) = y_{ij}$ for every individual function $f_i(x)$. Furthermore it should be presumed that all individual functions $f(p_1, p_2, \ldots, p_r; x)$ belong to the same function class.

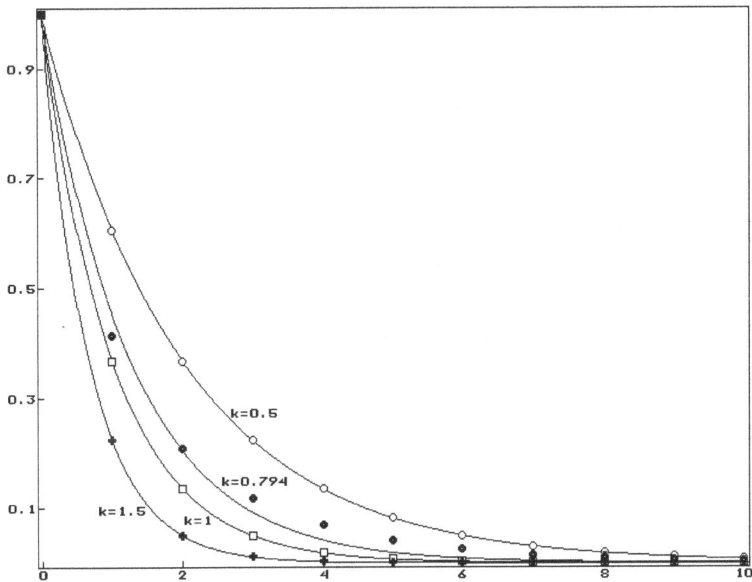

Fig. 8.1: Smoothing the function averages (dots) with $m_{\text{mean}}(t) = e^{-0.794t}$ as compared to averaging parameters ($k = 1$).

All points are equally considered. The parameters p_1 to p_r of the average course are determined such that the sum of the squared errors becomes minimal:

$$
\begin{aligned}
F(p_1, ..., p_r) &= \sum_{j=1}^{m}\sum_{i=1}^{n}(y_{ij} - f(p_1, p_2, \ldots, p_r; x_j))^2 \\
&= \sum_{j=1}^{m}\sum_{i=1}^{n} {y_{ij}}^2 - 2\sum_{j=1}^{m}\sum_{i=1}^{n} y_{ij} f(p_1, p_2, \ldots, p_r; x_j) \\
&\quad + \sum_{j=1}^{m}\sum_{i=1}^{n} f(p_1, p_2, \ldots, p_r; x_j)^2 \\
&= \sum_{j=1}^{m}\sum_{i=1}^{n} {y_{ij}}^2 - 2n\sum_{j=1}^{m}\sum_{i=1}^{n} \frac{y_{ij}}{n} f(p_1, p_2, \ldots, p_r; x_j) \\
&\quad + \sum_{j=1}^{m} n \cdot f(p_1, p_2, \ldots, p_r; x_j)^2 \\
&= \sum_{j=1}^{m}\sum_{i=1}^{n} {y_{ij}}^2 - n\left[2\sum_{j=1}^{m} \overline{y}_j f(p_1, p_2, \ldots, p_r; x_j)\right. \\
&\quad \left. - \sum_{j=1}^{m} f(p_1, p_2, \ldots, p_r; x_j)^2\right].
\end{aligned}
$$

Here $\overline{y}_j$ denotes the average of all measurements in the knot x_j. Otherwise

$$
\begin{aligned}
F^*(p_1, \ldots, p_r) :&= \sum_{j=1}^{m}(\overline{y}_j - f(p_1, p_2, \ldots, p_r; x_j))^2 \\
&= \sum_{j=1}^{m} {\overline{y}_j}^2 - 2\sum_{j=1}^{m} \overline{y}_j f(p_1, p_2, \ldots, p_r; x_j)) \\
&\quad + \sum_{j=1}^{m} f(p_1, p_2, \ldots, p_r; x_j))^2.
\end{aligned}
$$

Since the first partial sums $\sum_{j=1}^{m}\sum_{i=1}^{n} {y_{ij}}^2$ and $\sum_{j=1}^{m} {\overline{y}_j}^2$ are independent of the parameters, they do not effect the determination of p_i, $i = 1, 2, \ldots, r$. With this,

$$
F(p_1, p_2, \ldots, p_r) = \sum_{j=1}^{m}\sum_{i=1}^{n}(y_{ij} - f(p_1, p_2, \ldots, p_r; x_j))^2 = \min!
$$

and

$$
F^*(p_1, p_2, \ldots, p_r) = \sum_{j=1}^{m}(\overline{y}_j - f(p_1, p_2, \ldots, p_r; x_j))^2 = \min!
$$

are equivalent. For the calculation of the optimal parameter set of the averaged courses in this fashion it is insignificant if, under these preconditions, all measurements are used at the same time or if smoothing is carried out with regard to the individual averages $\overline{y}_j$ at the knots. This statement can be generalized insofar that more measurements can be taken into account for every individual course at the knots. However, the number of repetitions for **all** x_j, $j = 1, 2, \ldots, m$, and **all** individual functions must be the same. Otherwise F and F^* are **not** equivalent. It should be made clear that this applies to all model functions $f(p_1, p_2, \ldots, p_r; x)$, independent of the parameters fitting into the function equation linearly or non-linearly. Simulation experiments and their suitable statistical models are applied in the following examples.

8.1 Averaged splines in the case of common knots

The efficiency of the theory of spline functions should be demonstrated when handling standard questions in pharmacokinetics. Averaging splines is a possibility for smoothing measurements. It is shown how a model function in pharmacokinetics is reproduced through averaged splines. Smoothing parameters are taken into account.
First, calculating the area AUC under a model curve is approached with the help of a spline. In pharmacokinetics AUC is used to calculate bioavailability, bioequivalence or clearance. The variation of the calculated AUC-values through measurement errors is then demonstrated on hand of an example. From the theory it is known that a cubic spline function $s(x)$ with regard to the knots x_i, $i = 1, 2, \ldots, m$, is uniquely and completely determined through knowledge of the second derivatives $s''(x_i)$. It is therefore clear that the area under this spline can also be calculated from the points (x_i, y_i) and these values.

Theorem 8.1. *Let $s(x)$ be the uniquely determined interpolating cubic spline function through the points (x_i, y_i), $i = 1, 2, \ldots, m$. Then the following is true*

$$AUC := \int_{x_1}^{x_m} s(x)dx = \frac{1}{2}\sum_{i=1}^{m-1} \Delta x_i(y_i+y_{i+1}) - \frac{1}{24}\sum_{i=1}^{m-1} \Delta {x_i}^3(s''(x_i)+s''(x_{i+1})) \ .$$

Proof. Since $s(x)$ has the form $s(x) = A_i(x - x_i)^3 + B_i(x - x_i)^2 + C_i(x -$

$x_i) + y_i$ over the interval $[x_i, x_{i+1}]$,

$$\int_{x_i}^{x_{i+1}} s(x)dx = \frac{1}{4}A_i\Delta x_i^{\,4} + \frac{1}{3}B_i\Delta x_i^{\,3} + \frac{1}{2}C_i\Delta x_i^{\,2} + y_i\Delta x_i \ .$$

is true. The theorem results if the representations for $A_i = 1/6(s''(x_{i+1}) - s''(x_i))/\Delta x_i$, $B_i = s''(x_i)/2$ and $C_i = \Delta y_i/\Delta x_i - 1/6\Delta x_i(2s''(x_i) + s''(x_{i+1}))$ are inserted into this relationship. □

Deviation from the "true" AUC-value is exclusively defined through the approximation error of $s(x)$ as opposed to the unknown model function $y = f(x)$.
In simulation experiments it needs to be illustrated how the calculation of curve parameters is influenced by errors in measurements.
The starting point is the function $y = m(t) = 200(e^{-0.5t} - e^{-t})$. 21 points were chosen from this curve and exactly stated:

$$\begin{array}{ll}
(\ 0.0,\ \ 0.00000), & (0.5, 34.45402), \\
(\ 1.0, 47.73024), & (1.5, 49.84728), \\
(\ 2.0, 46.50883), & (2.5, 40.88396), \\
(\ 3.0, 34.66862), & (3.5, 28.71531), \\
(\ 4.0, 23.40393), & (4.5, 18.85805), \\
(\ 5.0, 15.06941), & (5.5, 11.96822), \\
(\ 6.0,\ \ 9.46167), & (6.5,\ \ 7.45415), \\
(\ 7.0,\ \ 5.85710), & (7.5,\ \ 4.59293), \\
(\ 8.0,\ \ 3.59604), & (8.5,\ \ 2.81215), \\
(\ 9.0,\ \ 2.19712), & (9.5,\ \ 1.71537), \\
(10.0,\ \ 1.33851). &
\end{array}$$

The exact value of the certain integral of $y = m(t)$ within the boundaries from 0 to 10 amounts to $AUC = 197.31390$ units. The calculation of the area AUS under the uniquely determined interpolating natural cubic spline function $s(t)$ amounts to $AUS = 196.88921$ units. There was no value put here on the best approximation of AUC. An improvement in the value of AUS is possible through the selection of auxiliary conditions, as they are described in Chapter 3 with regard to interpolating cubic splines.
The behavior of AUS should be explored in greater detail in relation to measurement errors.
Each of the given values y_i, $i = 1, 2, \ldots, 21$, shall have an identical normally distributed additive error δ_i with expected value zero and variance σ^2. An

$N(0,1)$ distributed δ_i was produced for all i from 1 to 21. The interpolating natural cubic spline $s_{\hat{y}}(t)$ was then determined for the $\hat{y}_i := y_i + \delta_i$ and the respective AUS value was calculated. The distribution of these AUS values after 10000 runs is represented in Figure 8.2 together with the least squares fitted associated GUASSIAN curve.

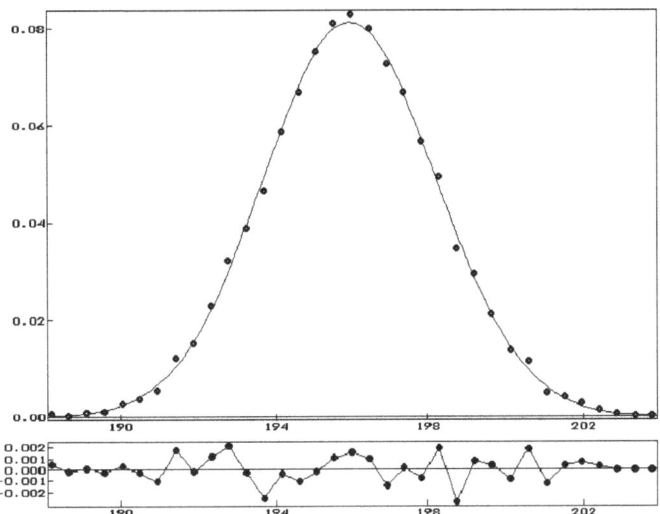

Fig. 8.2: Empirical distribution of 10000 AUS values, see the text for the data. Below: residuals with respect to the best fitted GAUSSIAN curve.

The calculated parameters of this curve are $\mu = 195.96394$ and $\sigma = 2.242513$. The average of the 10000 calculated AUS values is 195.9625. The AUS values are also normally distributed. This is because the second derivatives $s''(t_i)$, $i = 1, 2, \ldots, 21$, are determined by a linear system of equations in which only the knots t_i and the normally distributed $\hat{y}_i$ are used. The sum of normally distributed random variables is also normally distributed. Consequently, according to Theorem 8.1, the assumed normal distribution of the AUS values is justified.

The effects of random errors in measurements are now examined not only with regard to *one* curve parameter, but with regard to the complete spline function. The 21 points of the function $y = m(t) = 200(e^{-0.5t} - e^{-t})$ that are given above are used here again. These are strongly falsified by the normally distributed error with expected value $\mu = 0$ and standard deviation $\sigma = 20$ (!). Figure 8.3 illustrates two examples of 21 incorrect series

of measurements and their respective interpolating natural cubic splines as "snapshots" of the simulation. The points that are drawn in correspond with the 21 initial values $y = m(t)$ given above. The curve represents the interpolating spline through the error prone $\hat{y}_i := y_i + \delta_i$ values. The normal distribution of the errors was chosen for technical reasons.
The averaged spline function that was constructed through the averaging of the spline coefficients already has the form shown in Figure 8.4 after 100 (!) runs. After 5000 runs, the interpolating natural cubic spline function with regard to the original y_i values is modeled closely enough by the average spline function.
How can we calculate the averaged cubic spline function from k available series of observations , each consisting of m points $(x_i, \hat{y}_i^{(n)})$, $i = 1, 2, \ldots, m$, $n = 1, 2, \ldots, k$?

Theorem 8.2. *Let k series of observations of m points $(x_i, \hat{y}_i^{(n)})$, $i = 1, 2, \ldots, m$, $n = 1, 2, \ldots, k$, be given. For the fixed system of knots is presupposed $x_1 < x_2 < \ldots < x_m$. The uniquely determined interpolating cubic spline function $s_n(x)$ is constructed with the auxiliary conditions $s'_n(x_1) = g'_n$ (or $s''_n(x_1) = g''_n$) and $s'_n(x_m) = h'_n$ (or $s''_n(x_m) = h''_n$) for each of the k series of observations $\left(x_i, \hat{y}_i^{(n)}\right)$, $i = 1, 2, \ldots, m$.*
Then the cubic spline function averaged over n that exists through averaging the individual spline coefficients is identical to the interpolating cubic spline function $s_{av}(x)$ with regard to the points $(x_i, \overline{y}_i)$ with $\overline{y}_i := \frac{1}{k}\sum_{n=1}^{k} \hat{y}_i^{(n)}$, $i = 1, 2, \ldots, m$, and the auxiliary conditions

$$s'_{av}(x_1) = \frac{1}{k}\sum_{n=1}^{k} g'_n \ \text{ or } \ s''_{av}(x_1) = \frac{1}{k}\sum_{n=1}^{k} g''_n) \ \textit{and}$$

$$s'_{av}(x_m) = \frac{1}{k}\sum_{n=1}^{k} h'_n \ \text{ or } \ s''_{av}(x_m) = \frac{1}{k}\sum_{n=1}^{k} h''_n) \ .$$

Proof. Since the knots x_i, $i = 1, 2, \ldots, m$, remain fixed it is easy to see that by averaging the spline coefficients $A_i^{(n)}$, $B_i^{(n)}$, $C_i^{(n)}$ and $D_i^{(n)} = \hat{y}_i^{(n)}$, $i = 1, 2, \ldots, m$, $n = 1, 2, \ldots, k$, the result is again a cubic spline function $s_{av}(x)$. It passes through $(x_i, \overline{y}_i)$. Since the spline coefficients are multiples of the derivatives of the knots, $s_{av}(x)$ also fulfills the given auxiliary conditions.
Otherwise, according to Theorem 3.1, the interpolating cubic spline function through $(x_i, \overline{y}_i)$ is uniquely determined with these conditions. From this it follows that both splines are identical. □

If the spline function was decided upon in the selection of the model, then providing averaged function processes (e.g. an averaged growth curve or an averaged calibration curve when calibrating instruments) is possible. The following process is used:

Plan to determine average functional dependencies through splines

1. Chose a system of knots x_i, $i = 1, 2, ..., m$, and **fix** it for all measurements!
2. Determine the $\hat{y}_i^{(n)}$, $i = 1, 2, ..., m$, $n = 1, 2, ..., k$, by measuring the values of each individual curve at the chosen knots!
3. For every knot x_i, $i = 1, 2, ..., m$, build the average $\overline{y}_i = \frac{1}{k} \sum_{n=1}^{k} \hat{y}_i^{(n)}$!
4. Construct the uniquely determined interpolating cubic spline function with respective auxiliary conditions for $(x_i, \overline{y}_i)$, $i = 1, 2, ..., m$!

(8.1)

The effect of the selection of the smoothing parameter on the form of the averaged curve should now be observed. The simplest averaging of the individual interpolating spline functions led to good results in the above simulation. This brings forward the conjecture that the prior smoothing of *every* error prone individual curve followed by averaging the results can further improve the results. This conjecture is incorrect. This is demonstrated in the next simulation example.

Starting point are the 21 values of the function $y = f(t) = 200(e^{-0.5t} - e^{-t})$ given above. A series of measurements $\hat{y}_i := y_i + \delta_i$, $i = 1, 2, \ldots, 21$, results through the addition of normally distributed errors δ_i. This series of measurements $\hat{y}_i := y_i + \delta_i$, $i = 1, 2, \ldots, 21$, should be smoothed by a smoothing natural cubic spline. The algorithm developed in problem (R) was used. All weights were set $k_i = 1$, $i = 1, 2, \ldots, 21$. The natural cubic spline through points $(t_i, \hat{y}_i)$, $i = 1, 2, \ldots, 21$, that possesses the minimal total curvature and accepts the given sum of the squared errors S is calculated.

Due to the accepted error model, the random variable $\sum_{i=1}^{m}(y_i - \hat{y}_i)^2/\sigma^2$ has a χ^2-distribution with m degrees of freedom and an expected value of

m. Dependent on the standard deviation, a smoothing parameter in the magnitude of $S = m\sigma^2 = 21\sigma^2$ should be chosen.

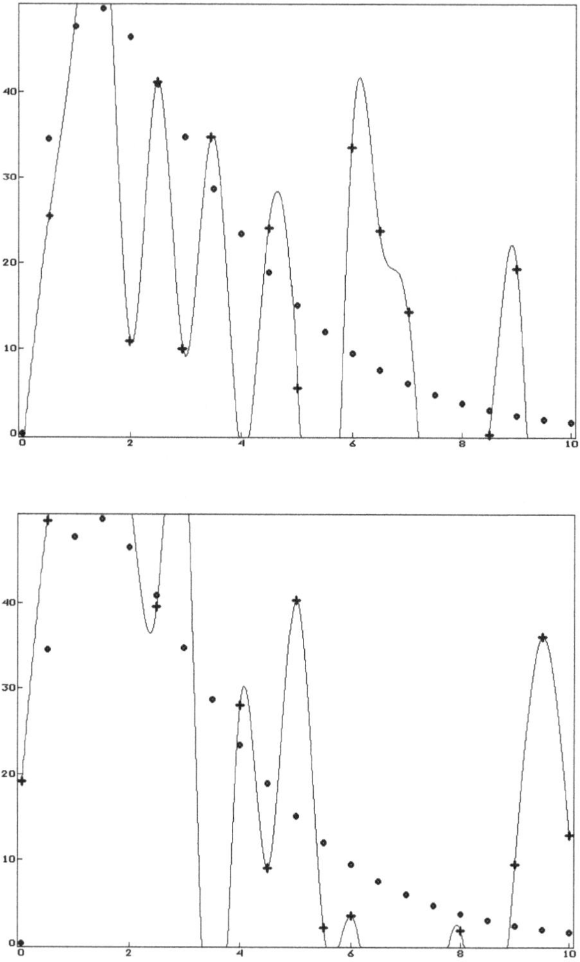

Fig. 8.3: Cubic spline through 21 values $\hat{y}_i := y_i + \delta_i$, see the text for the data.

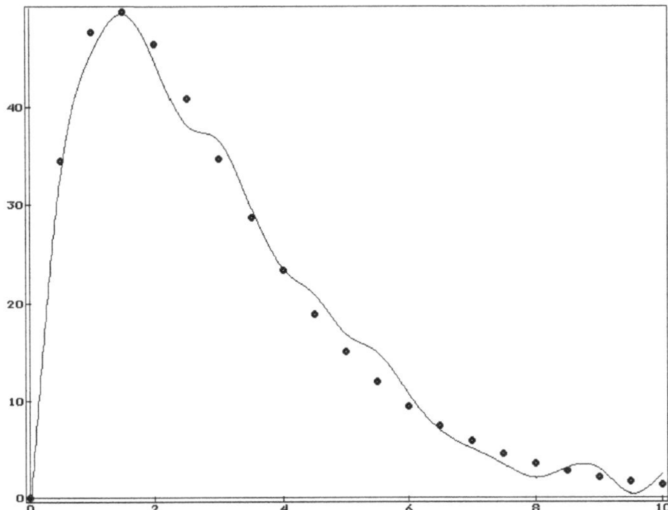

Fig. 8.4: The averaged cubic spline after 100 runs. See the text for the data.

Simulations for different standard deviations $\sigma = 5$, $\sigma = 7.5$, $\sigma = 10$ and $\sigma = 15$ were performed as follows: For given smoothing parameters S, respective smoothing spline functions were calculated for each of the series of measurements $\hat{y}_i := y_i + \delta_i$, $i = 1, 2, \ldots, 21$. The average spline function was built from 10000 of these splines.
The sum of the squared errors SSE_σ with regard to the exact values of y_i is a function of S and σ, $SSE_\sigma = SSE_\sigma(S)$. The curves SSE_σ for each of the fixed σ are illustrated in Figure 8.5.
For small statistical spreads of the applied normal distributions, the sum of the squared errors SSE_σ with regard to the exact y_i of the averaged splines approximately corresponds with the smoothing parameter S. With this, the straight line $SSE = S$ can be used as reference in a respective system of coordinates (see Figure 8.5). The possible values of S and SSE lie on the part of the straight line that is defined by the points $(0, 0)$ and $(S_{\max}, SSE_{\max})$. With this, $SSE_{\max}$ is the sum of the squared errors of the regression line with regard to the exact points y_i. In the observed example the value of $SSE_{\max}$ is 2590.6623 units.
At first glance the results are surprising. With an acceptably large smoothing parameter S, the simulation results with the regression line as an averaged spline function for each given σ. Due to the normal distribution

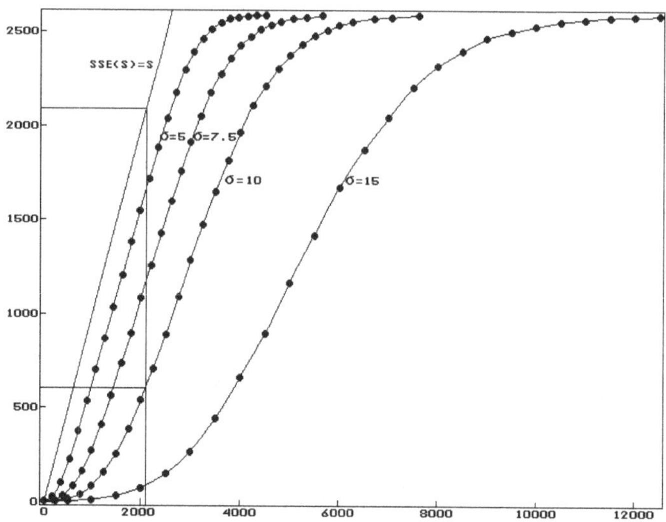

Fig. 8.5: Influence of the smoothing parameter S on the sum of the squared errors of the averaged spline function. See the text for the data.

of error, these straight lines lean toward the regression line with regard to the values of y_i. This can be recognized on hand of the asymptotes of about 2590 units. However, the simulations showed that the averaged spline function at $\sigma \neq 0$ and the suggested smoothing parameter $S = m\sigma^2$ does not result in the interpolating natural cubic spline function through the original points. With this parameter, they also do not correspond with the smoothing spline function through the given exact values of y_i. Figure 8.6 illustrates the relationships between the interpolating spline, the smoothing spline at the original points with $S = 2100$ ($\sigma = 10$), and the averaged spline with the same S developed in the simulation for $s_{av}(x)$. The function $s_{av}(x)$ is the spline function that appears when the original points (x_i, y_i) are smoothed with $S = 605$.
With the stated preconditions, smoothing the individual curves before calculating the average does not make reproducing the original function easier. Non-verifiable shifts are observed.

The already formulated Plan (8.1) for the determination of averaged courses of curves with regard to splines remains intact. Smoothing the individual series of measurements beforehand is not recommended.

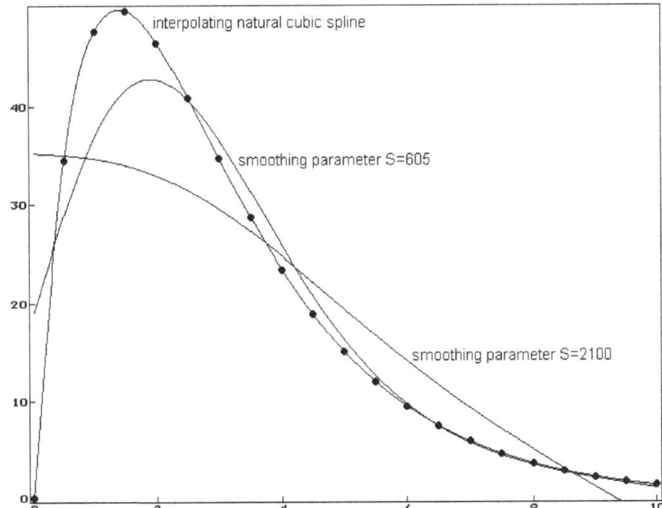

Fig. 8.6: Spline functions with the smoothing parameter $S = 0$ (interpolating natural cubic spline), $S = m\sigma^2 = 2100$ and $S = 605$ (Simulation result). See the text for the data.

8.2 Averaged kinetics and reference ranges

The plan developed in 8.1 for determining an averaged kinetic is now applied to an example. In addition, it is demonstrated how the spline model can be used to calculate reference ranges for courses of values.

The H_2 breathing test is a non-invasive procedure that is used for the diagnosis of various gastroentological diseases. It is used to attain the condition and distribution of intestinal flora. Digestive problems related to different sugars can be recognized and a picture of the paristalsis of the intestine and the transit time of consumed nutrition is obtained. When the test begins the patient consumes a defined amount of a sugar solution. This sugar is broken down bacterially. Hydrogen results which is eliminated by the lungs and can be measured when the patient exhales.

The available data stem from 16 series of measurements of the H_2 breathing test from random chosen patients with alcohol induced cirrhosis of the liver (M. Knoke and U. Thierbach). The patients first washed their mouths with a chlorine hexidin solution to kill any bacteria. They were then given 10 grams of lactulose and at fixed point in time t_i, $i = 1, 2, \ldots, m = 13$, they were given the appropriate concentration $C_n(t_i)$, $n = 1, 2, \ldots, k$, of

H_2. The hydrogen concentrations were observed for a period of 3 hours for each patient. The concentration $C_n(t_i)$, $n = 1, 2, \ldots, k = 16$, of H_2 was measured in the patients' exhaled air at specified times. This means that single values of k individual kinetics were given.
Since a model equation does not exist for the observed H_2 kinetics and it can be assumed that the course is adequately described by the measured value pairs, modeling with splines makes sense. Corresponding to the Plan (8.1), the average of the concentrations $\overline{C}_i = 1/k \sum_{n=1}^{k} C_n(t_i)$, $i = 1, 2, \ldots, m$, was determined for each point in time t_i. Equations of the form $s'_{av}(t_1) = 0$ and $s''_{av}(t_m) = s''_{av}(t_{m-1})$ are suggested as auxiliary conditions for the averaged spline function $s_{av}(x)$ that is to be constructed. This is because no breakdown takes place yet when the measurement begins (breakdown speed is zero) and because concrete assumptions about the breakdown speed or the acceleration at time $t_m = 180$ minutes cannot be made.
The averaged spline for these data were calculated based on the process developed in 8.1 and is illustrated in Figure 8.8. The figure represents the average curve of the H_2 values of all observed individuals. In additional to the clinical interpretation of the individual curves, it is possible to describe a group of patients through this averaged curve. Comparisons with other groups can be based on the averaged curve or of the curve parameters such as derivatives, asymptotes or integrals derived from it.

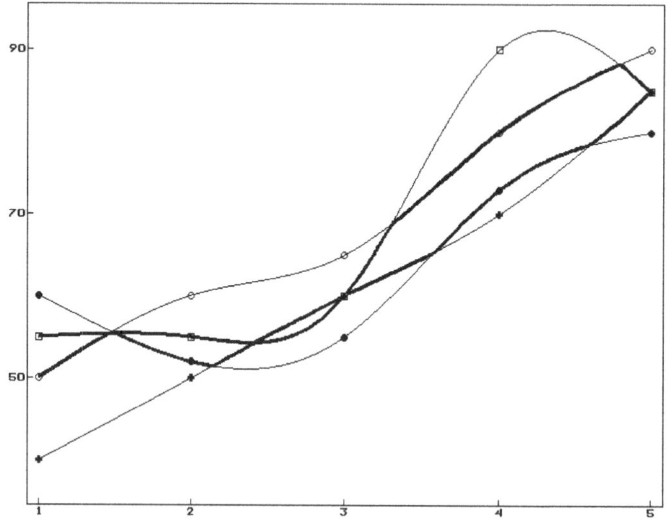

Fig. 8.7: Determining the reference range with individual curves.

Table 8.1: Data of 16 H_2 kinetics, see text.

t [min.]	0	10	20	30	40	50	60	70	80	90	120	150	180
C_1	20	23	15	15	27	23	26	48	43	44	59	84	75
C_2	15	10	15	17	26	24	23	27	31	34	44	40	19
C_3	5	5	4	4	5	4	4	4	4	3	27	69	80
C_4	1	5	10	6	9	8	9	12	10	9	7	7	17
C_5	15	13	12	12	13	12	10	8	12	11	10	17	22
C_6	7	7	8	9	10	14	14	11	16	13	17	74	94
C_7	11	9	10	8	10	11	9	7	7	10	14	40	60
C_8	5	8	6	5	6	4	6	15	19	22	42	45	46
C_9	11	10	8	10	9	9	10	8	10	34	45	47	43
C_{10}	2	2	4	5	3	2	2	2	3	2	17	43	45
C_{11}	8	9	12	7	8	6	6	7	6	9	7	11	58
C_{12}	5	6	3	4	4	7	5	5	4	7	7	9	21
C_{13}	4	6	4	4	4	3	1	4	7	17	23	25	28
C_{14}	13	14	13	13	11	11	10	15	13	13	14	27	39
C_{15}	4	17	15	11	8	7	7	9	11	12	28	38	39
C_{16}	4	7	7	5	5	7	10	8	8	10	9	9	62

What should a reference range be for curves?
In univariate statistics, such a reference region is described as the region that contains a given portion, for example 90%, of the possible values of the observed random variable. Typically symmetrical reference ranges are observed and the extreme values in both possible directions are equally eliminated.
A similar procedure can be used to determine a reference range of functions. A possibility would be to calculate the function values $s_n(t_0)$, $n = 1, 2, \ldots, k$, of the individual splines $s_n(t)$ at a fixed point in time t_0 and mark the extreme values. When t_0 has passed through the observed time interval, all individual curves that have at least one extreme value are eliminated. The entire function is seen as "abnormal" when the observed measurement is "abnormal" at a given point in time. The remaining curves define the reference range. It can not be predicted how many individual curves remain following this procedure. In the most extreme case this set could be empty. In addition, this procedure ignores the fact that extreme courses of curves can become normalized again.

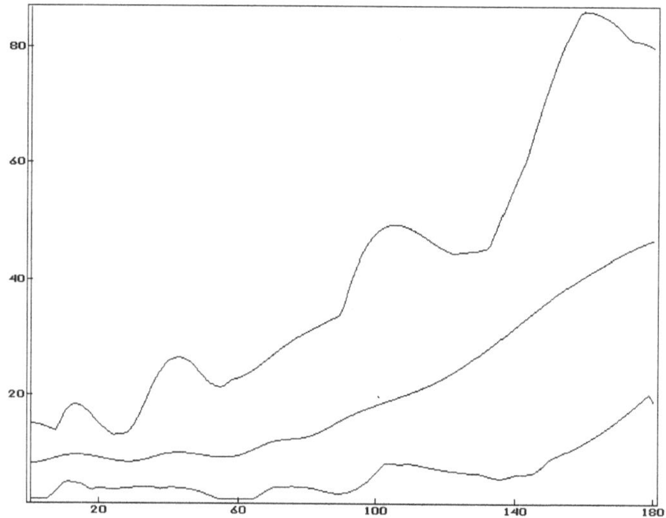

Fig. 8.8: Average H_2 kinetic and the region in which 90% of all function values lie.

It therefore seems better to define the reference range of the course of curves in relation to the dependent variable of the functions. The specified portion of the extreme values from the set of $s_n(t_0)$, $n = 1, 2, \ldots, k$, is eliminated for t_0. The remaining $s_n(t_0)$ define the reference range at time t_0. If t_0 passes through the observed time frame, a reference range for the curves results in which all observed individual values are found, for all point in time, within the percentile identified as normal. The functions forming the boundary of this reference range are continuous, but not necessarily differentiable (see Figure 8.7).

In relation to the individual splines derived from the data, a 90% reference range was calculated and illustrated in Figure 8.8.

The method presented here for the construction of a reference range for curves is not a statistical process! Statistical tolerance estimations assume the knowledge of the distribution model for the observed random variable. Regression models or stochastic processes would have to be discussed here. The observations then have to be realizations of random samples of the respective random variables. This requires observation strategies that are materially problematic.

8.3 Growth curves and averaged splines without common knots

There are many mathematical models for human growth. Attempts to bring concepts of biological growth into a mathematical context began as far back as the 19th century (SCHARF 1981). It has been possible, for a long time now, to make use of rich data. One example is the famous anatomic collection at the Friedrich-Schiller University in Jena. This material was used by KARL PEARSON as a base for his examinations of growth (PEARSON 1894).

Modeling growth with splines has no relation to physiological or other conceptions of growth. It makes it possible to calculate the averaged growth curve and reference ranges with regard to the real data. Specifying the averaged course of the speed of growth does not pose a problem.

The data used in the following were made available by MARTSCHEI (2001). They stem from 100 children in the city of Greifswald, Germany, and contain newborn data and information about weight and height up to the age of 14.

New samples were not taken every single year. Rather, the growth of individual people was observed over a longer period of time. The height $H_n(t_i)$, $n = 1, 2, \ldots, p$, was measured for the nth individual at times t_i,

$i = 1, 2, \ldots, m$.
The raw data first had to be computationally prepared. The date of birth and individual points of observation were used to calculate the actual individual ages. The corresponding height was then assigned. Dividing the time scale into quarters was determined to be sufficient.
Finding a sufficient set of individual data sets that all contained a measurement for the same knots (i.e. for the same quarters) was not possible. The central precondition of Plan (8.1) therefore could not be fulfilled. Additional considerations are necessary in order to calculate the average growth curve through splines.
Applying an additional point of the solution curve to the original set of points does not change the solution to uniquely solvable interpolation tasks in a given function class. This statement is not immediately obvious for spline functions. This is due to the special construction of splines through individual polynomials. The uniquely determined interpolating spline $s(x)$ is constructed for a given set of points. Additional points $(x, s(x))$ are, however, redundant for this construction. This is formulated more exactly in the following theorem.

Theorem 8.3. *Let $s(x)$ be the uniquely determined interpolating natural spline function of degree $(2k-1)$ through the points (x_i, y_i), $i = 1, 2, \ldots, m$, $(k < m)$ let $(x^*, y^* := s(x^*))$ be an arbitrary point on $s(x)$ where $x^* \in [x_1, x_m]$.*
Then the interpolating natural spline function $s^*(x)$ of degree $(2k - 1)$ through the $(m + 1)$ points (x_i, y_i) and (x^*, y^*) is identical to $s(x)$.

Proof. Since the interpolation task can uniquely be solved with the preconditions mentioned in the theorem, the statement follows because $s^*(x_i) = s(x_i) = y_i$, $i = 1, 2, \ldots, m$, and $s^*(x^*) = s(x^*) = y^*$ are true for $s^*(x)$ and $s(x)$. □

The statement of Theorem 8.3 also applies to arbitrary splines under the assumption that $s(x)$ and $s^*(x)$ are constructed with identical boundary conditions. These boundary conditions have to be chosen such that a unique solution of the interpolation task is guaranteed. Examples of such boundary conditions are formulated in Theorem 2.5 and in Theorem 3.1.
Points (x_i, y_i) and the uniquely determinated interpolating cubic spline $s(x)$ with corresponding boundary conditions are given again. An arbitrary proper subinterval $[a, b] \subset [x_1, x_m]$ is chosen now. How can $s(x)$ be reconstructed over $[a, b]$?

Let $x_{j1} < x_{j2} < \ldots < x_{jn}$ denote the n knots in $[a,b]$. We suppose $n > 0$. To the $n+2$ points $(a, s(a))$, $(x_{ji}, s(x_{ji})$, $i = 1, 2, \ldots, n$, and $(b, s(b))$ one constructs the uniquely determined interpolating cubic spline $s_o(x)$ which fulfills the boundary conditions $s_o'(a) = s'(a)$ and $s_o'(b) = s'(b)$ (or $s_o''(a) = s''(a)$ and $s_o''(b) = s''(b)$). Then $s_o(x)$ is identical to $s(x)$ over $[a,b]$.

Note 8.1.

(1) The conclusion of Theorem 8.3 remains correct when, instead of a single (x^*, y^*), a finite number of such value pairs completes the original interpolation task.
(2) The statements of Theorem 8.3 in general are incorrect when $s^*(x)$ is not calculated with respect to **all** original points (x_i, y_i).
(3) The assumption that $s(x)$ and $s^*(x)$ must fulfill the same boundary conditions is absolutely necessary (compare Figure 3.3).

Together with the first note, Theorem 8.3 is useful for the construction of averaged courses of functions and reference ranges. $I_n := [x_{1_n}, x_{m_n}]$ is the interval in which all knots of the nth individual curve, $n = 1, 2, \ldots, p$, lie and $I := \bigcap_{n=1}^{p} I_n$. Furthermore it is assumed that all individual curves can be described by the interpolating spline. A system of knots, as it is called for in Plan (8.1), can then be created over the interval I.
If interest exists regarding the modeling of an averaged functional dependency, the great advantage of Plan 8.1 is that none of the individual curves need to be explicitly calculated. The average curve is determined solely based on the boundary conditions and the averages $\overline{y}_i$. This advantage is not relevant anymore when a common system of knots does not exist. In this case, the concrete calculation of all spline functions is necessary when function values are missing for the new system of knots $x_1' < x_2' < \ldots < x_q'$. The computational results of the growth data are illustrated in Figures 8.9 and 8.10. The average speed of growth corresponds with the first derivative of the averaged growth curve.
An averaging of functions makes sense if splines are used to describe the observed courses. Difficulties do not appear as they are illustrated in Figure 8.1. This is due to the fact that the calculation of the spline coefficients is a linear problem. The calculation of a mean spline is simple if all individual courses have a common system of knots (Plan 8.1). If this situation is not given, a certain amount of additional effort is necessary. This consists in the generation of the missing common knots and the related missing measurements (see Extended Plan in the box). Theorem 8.3 provides the

justification for this procedure. These calculations are located on the interval I. There are observations of all courses only for I. It therefore makes sense that the averaged spline is only observed on I.

Enhanced plan to determine average courses of functions

1. Calculate the uniquely determined interpolating cubic spline function $s_n(x)$, $n = 1, 2, \ldots, p$, with the boundary conditions $s'_n(x_{1_n})$ or $s''_n(x_{1_n})$ and $s'_n(x_{m_n})$ or $s''_n(x_{m_n})$ related to the individual curve through all measured values that are available for that individual!

2. Take as a new system the knots $x'_1 < x'_2 < \ldots < x'_q$ from **all** knots of the p individual curves in interval I! This new system of knots is a subset of the set of knots of all measurements.

3. For each individual curve $s_n(x)$, calculate the vector $s_n(x'_i)$, $n = 1, 2, \ldots, q$, and the derivatives $s'_n(x'_1)$ or $s''_n(x'_1)$ and $s'_n(x'_q)$ or $s''_n(x'_q)$!

4. Construct the averages $\overline{y}_i = \frac{1}{p} \sum\limits_{n=1}^{p} s_n(x'_i)$, $i = 1, 2, \ldots, q$!

5. As an average curve, construct the interpolating cubic spline function through the points $(x'_i, \overline{y}_i)$, $i = 1, 2, \ldots, q$, with the boundary conditions
$s'_{av}(x'_1) = \frac{1}{p} \sum\limits_{n=1}^{p} s_n{}'(x_1{}')$ or $s''_{av}(x'_1) = \frac{1}{p} \sum\limits_{n=1}^{p} s_n{}''(x_1{}')$ and
$s'_{av}(x'_q) = \frac{1}{p} \sum\limits_{n=1}^{p} s_n{}'(x_q{}')$ or $s''_{av}(x'_q) = \frac{1}{p} \sum\limits_{n=1}^{p} s_n{}''(x_q{}')$!

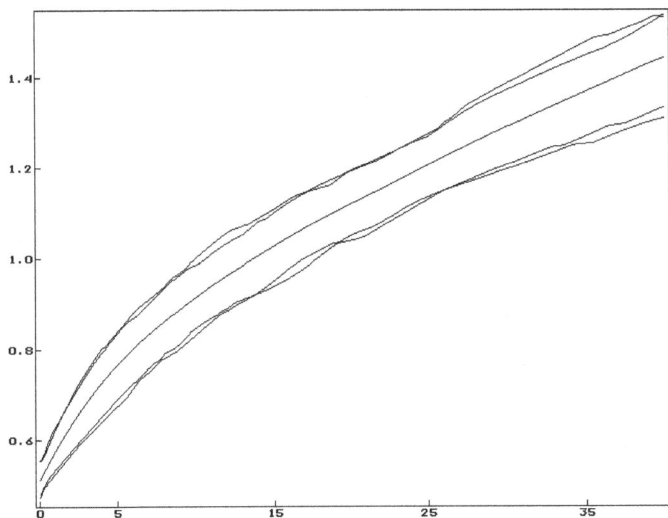

Fig. 8.9: The average growth curve and the 95% and 90% reference range. See the text for the data.

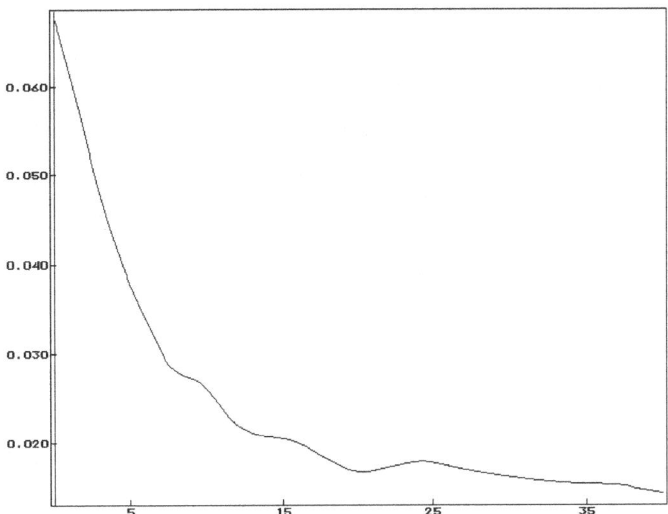

Fig. 8.10: The average speed of growth. See the text for the data.

PART 2
Compartment models

Chapter 9

Concept of a context related mathematical pharmacokinetical model

To aid in physics, mathematics came into contact with medicine. Blood flow models were already studied in the 1850's by HELMHOLTZ, who was a doctor and physicist (KAC 1979). The classic compartment and flow models of pharmacokinetics also follow this tradition. Stochastic methods actually introduced new ways of thought in chemistry and physics in the 19th century. In biological sciences, genetics was the first to take part in such a development. The in those days seemingly forgotten works of MENDEL (1866) bring a theoretical probability model into relation to heredity. None other than R.A.FISHER (1936) made a reproach that the results of MENDEL'S experiments were "too exact". Later inquiries lead to the impression that experimental results were only "good" due to the typical but statistical undependable selection that was used at the time, and because MENDEL'S integrity was undoubtedly not in question (HAGEMANN 1984).

This historical scientific example suggests that the application of mathematical methods particularly implies the analysis and observation of necessary conditions.

Part II of this monograph discusses compartment models, associated probability distributions, statistical parameter estimations and statistical model selection with these aspects in mind. The use of mathematical applications, which are much needed in pharmacokinetics, will be closely pursued.

Pharmacokinetics denotes the description in space and time of intake, distribution, elimination, and metabolism of a drug with respect to an organism. It attains meaning only in the context of pharmacological, biochemical, biophysical, pharmaceutical, and last but not least clinical questions. Fundamental methods are borrowed from the kinetics of chemical reactions. Chemistry offers considerably more experimental possibilities for the study of kinetic models than medicine.

The following modus of observation is typical in pharmacokinetics (**individual kinetics**): The concentration of an applied drug in the blood of a patient is measured at approximately five to ten points in time. Ethical reasons require that the number of measurements should be kept small. The quantitative identification of the course of concentration over time can be derived from the data. This is then included in the processing of the posed problem.
The term pharmacokinetics was created by DOST (1953), who wrote the first monograph on the topic. Since then there is rich literature available, including publications with strictly mathematical approaches. The development of literature in the seventies is referenced by BOXENBAUM (1982). Historical information about pharmacokinetics can be found in DETTLI (1980), WAGNER (1981), RUPP (1980) as well as in monographs (e.g. KMMERLE 1978, GIBALDI/PERRIER 1982).
In the meantime, a variety of questions has been gathering around the concept of pharmacokinetics with regard to a large variety of specialties. The limitation on mathematical aspects already reaches a wide facet from theories to their technically arithmetic handling. Certain questions have repeatedly been dealt with over the years, while others arę not brought to our attention again after they have been solved. Here, therefore attempts were made to show origins of solutions together with their concrete citation. The goal of the following treatise is to recommend an adequate method for the evaluation of individual kinetics with the aim to:

- link to the approved classic models of pharmacokinetics,
- clearly differentiate between the relations of variants of these models and those of data evaluation attempts,
- make visible, which model variants can be identified from the models on hand, and which can be interpreted according to assumptions,
- give a method for the calculation of the parameter and study its qualities, and
- objectively make the choice of a suitable pharmacokinetical model with the data on hand.

The propositions derived in the following essentially refer to two-compartment models. Under different points of view, it is explained that this reduction in consideration of the available observations of an individual kinetics is meaningful.
Mathematical models for pharmacokinetics will now be introduced. Deterministic descriptions of pharmacokinetical events form the beginning here.

The Two-compartment model for iv bolus administration is observed in detail. The somewhat extensive solution to the corresponding linear initial value problem does not present a problem. However, it is not fully found in literature about pharmacokinetics.
Further deterministic modelings are introduced: theoretical system descriptions, differential equations with delay, and nonlinear differential equations. Calculability and identifiability are concepts which fundamentally contribute to the support of the interpretation of results of a pharmacokinetical experiment. That said, an overview can be given for two-compartment models for iv bolus administration. The corresponding propositions found in the literature are of local character, or refer to mass courses of time, and therefore to non-comparable situations.
Linearization of model equations, which are even found in common pharmacokinetic textbooks, are often contained in recommended computer programs for the evaluation of data. It remains almost unnoticed that this influences parameter calculations in pharmacokinetics. Such differences are clarified in the example in Chapter 15.
Putting together deterministic and stochastic methods is essential in order to accomplish the goals formulated above. Compartment models and associated residence time distributions are handled in Chapter 13. Qualities of these, and the possibility of censoring, i.e. the distribution is defined with respect to the observed time interval, are referred to. It was not made clear in the literature that the concentration-time function observed in one compartment does not necessarily reflects the residence time of pharmacon molecules in an organism.
Stochastic differential equations, stochastic processes, and regression models are mentioned as additional stochastic attempts. These procedures are valuable in that they can possibly help to interpret pharmacokinetic events. The problem of attaining quantitative propositions about data evaluation procedures may be a difficult mathematical problem. However, it is not possible to check the legitimacy of the given assumptions of an error model for a regression attempt in the scope of individual kinetics. Pharmacological experiments are not reproducible for an individual under the same conditions. For this reason, literature about method of least squares (MLS) calculations of model parameters of interest, found under the heading "regression", are to be understood as approximations provided that they concern individual kinetics and not population kinetics.
The method of parameter calculation in Chapter 15 is explained in a different statistical context: A residence time distribution stands in relation

to a compartment model, and its parameters correspond with those of a pharmacokinetic model. The observation of the time course of the concentration provides a concrete sample. The parameters are estimated based on the varied minimum-χ^2-method . If required, an incomplete observation can be taken into account through censoring. The calculability and identifiability of a pharmacokinetical model have to be examined with regard to the stochastic model underlying the parameter calculation.

The given parameter estimate does not make sense for all two-compartment models for iv bolus administration. The models that can be estimated are mentioned in Proposition 13.22. The advantage of the varied minimum-χ^2-method over the general application of the MLS-calculation however is that the estimators can be given characteristics (Chapter 13.3). Existence, uniqueness, invariance property, asymptotic unbiasedness, as well as knowledge about the asymptotic distribution of the varied minimum-χ^2-estimator are included. The fundamental theorem of FISHER, PEARSON, NEYMAN, and CRAMER (CRAMER 1946) plays a major role with regard to the proof of these propositions. The given method of proof can possibly also be applied to the estimable two-compartment model for iv bolus administration, which depends on the residence time distribution of three parameters. However, the expressions that arise from this become too confusing.

From the observation of the concentration-time course one gets residence time data in grouped form. The standard statistic for evaluation of such data is the χ^2-statistic. Two problems can be processed with that, parameter estimation and model selection as a statistical test. As is well known, using the data twice in subsequent statistical procedures is problematic. It is an advantage of the χ^2-test that merely a loss of degrees of freedom arises here. Let n be the cell number and m the number of parameters to be estimated. The asymptotic distribution of the χ^2-test statistic is the χ^2-distribution with $(n - m - 1)$ degrees of freedom if the parameters were estimated according to the varied minimum-χ^2-method. This is also right for the maximum-likelihood (ML) estimated parameters if the estimation is based on the group frequencies of grouped data. The χ^2-teststatistic does not have any limiting χ^2-distribution, when the ML-estimation of the parameters stem from the original sample data (CHERNOFF and LEHMANN 1954).

The minimum-χ^2-estimation is a special kind of the minimum distance estimation method and has a very welcome robustness property: Accepted the real but unknown watched residence time distribution $G(t)$ is not member of the distribution family developed from the compartment model. That

means the proposed model is incorrect. Then a minimum distance estimator is a consistent estimator for the specific parameter, which selects the best approximation of $G(t)$ in the parametric distribution family (PARR 1981). So the robustness ensures a certain compensation of the model error. This must be taken into consideration since the data also contain information about disturbance variables as well as measuring errors.
The problem of model selection in pharmacokinetics can, through varied minimum-χ^2-estimates of the parameters based on the last observed results, be reduced to the use of the classic PEARSON test.
An expansion of the described treatment of parameter estimates and the selection of models for models with more than two compartments does not make sense for several reasons:

(1) One can no longer explicitly indicate the roots of the characteristic polynomial of the system of differential equations. A system for model variants is therefore missing.
(2) Proofs concerning the asymptotic distribution of estimators cannot be used for with this method.
(3) The number and the information content of the available observations (Where does a second compartment identify itself? Can measurements be taken there at all?) suggest the limitation on simple attempts.

However, it is worthwhile to try to apply this procedure to nonlinear examples.
The planning of statistical experiments, like the hallmark of given tests of goodness of fit, only seem empirically possible in the context of computer experiments. Such examinations are not the topic of the book on hand.
Recently, it has been observed that there is amplified interest in population based interpreted pharmacokinetics. A first bibliography about this can be found in YUH et al. (1994). Instead of individual kinetics, concentration time courses characterizing a population are of interest.
With regard to a statistical view of a population, difficulties arise for compartment models because real nonlinear model functions do not form algebraically complete sets. Part I of the book on hand proposes another possibility for the treatment of these problems by using spline interpolations, spline approximations and spline averaging. The minimum-χ^2-method offers another approach for the averaging of kinetics (see Chapter 15.3).

Chapter 10

Compartment models

To a pharmacologist, the term pharmacokinetics implies methods to describe the distribution in space and time of an applied drug in an organism. Furthermore, the pharmacological experiment, the evaluation, the identification of the qualities of drugs, the identification of the reaction of an individual (individual kinetics) or a population (population kinetics) to the administration of an active agent, as well as the use of the results, for example for an individually calculated pharmacotherapy, are included.
A pharmacokinetic experiment insists that the spatial and temporal distributions of an applied drug quantity should be measured in an organism (test subject, patient, laboratory animal). This is possible through the use of radioactively marked substances (tracer kinetics).
Unmarked pharmaka merely permit the observation of temporal courses of concentration, which are most often carried out in relatively easily accessible bodily liquids such as in the blood.
The assessment of experimental results occurs with regard to a mathematical model of the agent distribution in the body. The studies of the qualities of such mathematical models as well as methods used for the planning and evaluation of experiments should be understood in a narrower sense under the term pharmacokinetics.
The starting point for mathematical modeling is the pharmacokinetic experiment: An individual is given the dose DOS of a drug. At fixed points in time t_i after the beginning of the test, the concentrations $cm(t_i) = cm_i$ of the drug are determined, for example, in the blood. The number r of measurements is kept small in order to avoid burdening the test subjects. The number lies approximately in the magnitude of $r = 10$.
Because of the in-vivo conditions, the dependence of the temperature on the process is left out. The spatial distribution cannot be observed here. The

measurements are placed in connection with a concentration time function $c(t)$, which is derived from a mathematical model.
In comparison with tracer kinetics, which make use of partial differential equations and are very closely related to the methods of kinetics of chemical reactions (JACQUEZ 1988 and KAJIYA/KODAMA/ABE 1985 give a concien-cious summary), simpler mathematical models can be expected here. This expectation is confirmed in the scope of a master thesis (BOLDT 1987) where a space-time model is studied in combination with concentration measurements over time.
One of the goals of the work on hand is to recommend mathematical models that are adequate for the available observations of individual kinetics.
Compartment is described as (McINTOSH/McINTOSH 1980) "quantities of a substance having uniform and distinguishable kinetics of transformation or transport" (ATKINS 1969), "each of which is homogeneous and well mixed and ... interacts by exchanging material" (JACQUEZ 1988). For example, compartments can be:

- conditions to be distinguished physiologically,
- regions that can be described anatomically,
- differentiable behavior of a chemical substance under subtly differentiated conditions or
- chemically altered parts of a substance.

The temporal functions of interest refer to the mass or concentration of the observed agent, are defined on $\mathbb{R}^+$ or a subset of $\mathbb{R}^+$, and accept only positive real values.
The application of compartment models is not restricted to pharmacology. Such models are also of interest for other branches of medicine, social sciences, economics, behavioral biology, and ecology (e.g. BROWN 1980, JÄGER 1989).
With regard to the function $c(t)$ which is to be calculated, arbitrary freedom is available to reproduce the measured values. However, models are of interest which can adequately interpret possibilities for the meanings of parameters. The classic one- and two-compartment models will be introduced next. One must on the one hand refer to these traditional linear models in the interest of the comparability of findings. On the other hand, for the majority of applications, these also prove to be completely sufficient.

10.1 One-compartment model

A certain dose DOS of a drug is introduced into the bloodstream intravenously, and spreads out into the body through processes which will not be described in greater detail here. The function $c(t)$, which describes the concentration of the drug in the blood, is defined by the homogeneous linear differential equation with initial values

$$\frac{d}{dt} = -k_{el}c(t), \quad k_{el} > 0, \; c(0) = c^0 \ . \tag{10.1}$$

Definition 10.1. *The initial value problem* (10.1) *is called a One-compartment model for iv bolus administration or* ***One-compartment-iv model*** *for short.*

Lemma 10.1. *For the One-compartment-iv model,*

$$c(t) = c^0 \mathrm{e}^{-k_{el}t}$$

is the solution to the initial value problem (10.1).

The parameter c^0 is called **initial concentration** or **initial value** and k_{el} is the **elimination constant**. These parameters can be calculated from measurements and are easily interpreted. The (fictitious!) **distribution volume** $V_d = DOS/c^0$, **elimination half-life** $t_{1/2} = \ln 2/k_{el}$ as well as the **clearance** $CL = k_{el}V_d$ are calculated from these. The use of these terms is typical in medical literature.
If an extravascular application is carried out, for example, an i.m. injection, then the invasion may be described by an inhomogeneous linear differential equation with initial values

$$\frac{d}{dt}c_i(t) = -k_i c_i(t), \quad k_i > 0, \quad c_i(0) = c_i^0,$$

with the solution

$$c_i(t) = c_i^0 \mathrm{e}^{-k_i t}.$$

k_i is the **invasion constant**. Superposition is presupposed and the following initial value problem is formulated for the concentration of the drug in the blood $c(t)$:

$$\frac{d}{dt}c(t) = -k_{el}c(t) + k_i c_i^0 \mathrm{e}^{-k_i t}, \quad k_i > 0, \quad k_{el} > 0, \quad c(0) = 0. \tag{10.2}$$

Definition 10.2. *The initial value problem* (10.2) *is called a One-compartment model with first order input or* ***One-compartment-ev model*** *for short.*

Lemma 10.2. *For the One-compartment-ev model, the initial value problem* (10.2) *has the following solution:*

Case 1 $k_{el} = k_i =: k$

$$c(t) = c_i^o kt\mathrm{e}^{-kt} \ ;$$

Case 2 $k_{el} \neq k_i$

$$c(t) = \frac{c_i k_i}{k_{el} - k_i}[\mathrm{e}^{-k_i t} - \mathrm{e}^{-k_{el} t}] \ ,$$

in short form:

$$c(t) = A[\mathrm{e}^{at} - \mathrm{e}^{bt}] \ .$$

Proof. This ordinary linear differential equation of 1st order with constant coefficients is solved via the method of the variation of constants. □

The parameters of the adaptable function $c(t)$ are to be calculated from measurements cm_i, $i = 1, \ldots, r$, which are on hand.

Definition 10.3. *In agreement with their typical use in literature, the parameters k, k_i, k_{el}, c^o, c_i^o used above will be defined as **model parameters** and the parameters a, b, A will be defined as **system parameters**. The corresponding agreement also applies to more-compartments models.*

Remark 10.1.

(1) The existence and uniqueness of the solutions to the given differential equations are founded in the theory of ordinary differential equations.
(2) The given differential equations are linear. Linear compartment models always describe pharmacokinetical models whose accompanying defining differential equation is linear. It is typical to simply call them compartment models.
(3) The curve adaptation of the measurements is given by a, b and A. Does a one-to-one correspondence exist between the system parameters and the model parameters? The problem with regard to the calculability of model parameters is brought to awareness here. It is discussed in greater detail in Chapter 12.
(4) On which of the cases mentioned in Lemma 10.2 does the experimenter base the evaluation of the measured concentrations? The importance of this decision for the interpretation of the observations is obvious.

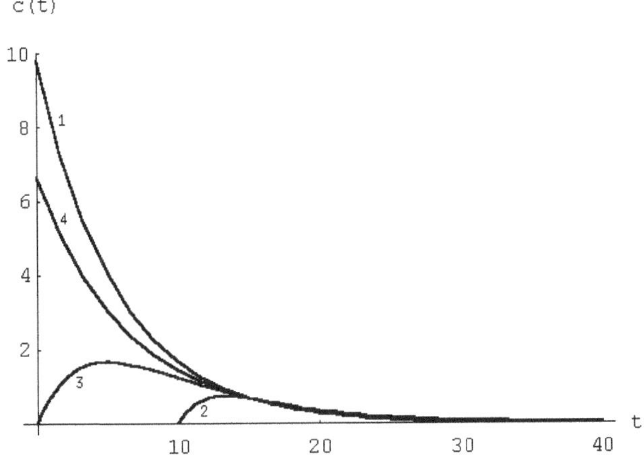

Fig. 10.1: Four types of concentration-time courses, see text.

Figure 10.1 shows that with only a few measurements, the subjective choice of a model is problematic. Four types of functions are shown:

1. $c(t) = Ae^{-at}$
2. $c(t) = A(e^{-a(t-t_0)} - e^{b(t-t_0)})$
3. $c(t) = A(e^{-at} - e^{-bt})$
4. $c(t) = Ae^{-at} - Be^{-bt}$.

The functions of varying types practically are not differentiable in $[14, \infty]$. They are different because of different readings, or by observations of the informative part of the definition ranges. The parameters are:

Type	A	B	t_0	a	b
1	9.80	-	-	0.1759	-
2	3.00	-	10	0.2000	0.4000
3	21.60	-	-	0.1778	0.2196
4	6.25	−0.3935	-	0.1609	0.0686

Although it has already been dealt with by DOST (1953), numerous monographs about pharmacokinetics (for example GIBALDI/PERRIER 1982, JACQUEZ 1988, VAN ROSSUM 1977) do not deal with Case 1 of Lemma 10.2. The separate identification of invasion and elimination processes is discussed in great detail by SCHELER (1980) and GLADTKE/V. HATTINGBERG

(1977). A function

$$c_{lag}(t) = \frac{c_i^o k_i}{k_{el} - k_i} \left[\mathrm{e}^{-k_i(t-t_{lag})} - \mathrm{e}^{-k_{el}(t-t_{lag})} \right]$$

serves to describe a process when an administered drug appears in the blood with certain delay t_{lag} (lag-time, the typical symbol used in literature). In comparison with Case 2, Lemma 10.2, this function additionally contains the parameter t_{lag}.

10.2 Two-compartment models

Concentration changes are again proportionally related to concentration. The two-compartment model is then represented by a system of linear differential equations:

$$\frac{\mathrm{d}}{\mathrm{d}t} c(t) = Kc(t) + I(t) \tag{10.3}$$

with

$$\begin{aligned} K &= \begin{pmatrix} -(k_{10} + k_{12}) & k_{21} \\ k_{12} & -(k_{20} + k_{21}) \end{pmatrix}, \\ c(t) &= [c_1(t), c_2(t)]^T, \quad \text{and} \\ I(t) &= [I_1(t), I_2(t)]^T. \end{aligned}$$

$k_{ij} \geq 0$ is presupposed for all constants that appear and obviously correspond to the model. The notation k_{ij} indicates the transport of a substance from compartment i to compartment j. If a zero appears as the index, it means that an interchange with the environment of the compartment system has taken place. $I_1(t)$ and $I_2(t)$ describe the invasion of the drug. Illustration 10.2 shows a scheme of the general two-compartment model. The linear system of two ordinary differential equations of 1st order with constant coefficients (10.3) can be transformed into a linear ordinary differential equation of 2nd order with constant coefficients because of the equivalence between such systems of differential equations and differential equations of higher order: The second of the equations in (10.3) is reformulated into $c_1(t)$,

$$c_1(t) = [c_2'(t) + (k_{20} + k_{21})c_2(t) - I_2(t)]/k_{12}, \tag{10.4}$$

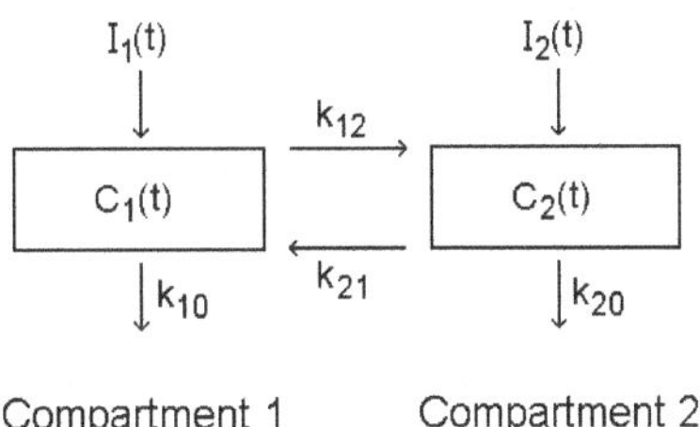

Fig. 10.2: A scheme of the two-compartment model.

and is substituted into the first of the equations. The solution is:

$$c_2'' + (k_{10} + k_{12} + k_{20} + k_{21})c_2' + [(k_{10} + k_{12})(k_{20} + k_{21}) - k_{12}k_{21}]c_2 = I_2' + (k_{10} + k_{12})I_2 + k_{12}I_1 \ . \quad (10.5)$$

If $c_2(t)$ is the solution to (10.5), $c_1(t)$ yields from (10.4). Existence and uniqueness of the solutions to (10.3) are guaranteed for given initial conditions by the theory of linear differential equations with constant coefficients.

Lemma 10.3. *Let $K \neq 0$. Then the characteristic polynomial*

$$\begin{aligned} P_K(\lambda) &= \det|K - \lambda E| \\ &= \lambda^2 + (k_{10} + k_{12} + k_{20} + k_{21})\lambda + k_{10}k_{20} + k_{10}k_{21} + k_{20}k_{12} \end{aligned}$$

for (10.3), *respectively* (10.5), *has real negative roots. Here E describes the unit of the corresponding matrix ring.*

Proof. The roots of $P_K(\lambda)$ are $\lambda_{1/2} = -(k_{10}+k_{12}+k_{20}+k_{21})/2 \pm [(k_{10}+k_{12}+k_{20}+k_{21})^2 - 4(k_{10}k_{20}+k_{10}k_{21}+k_{20}k_{12})]^{1/2}/2 = -(k_{10}+k_{12}+k_{20}+k_{21})/2 \pm [(k_{10}+k_{12}-k_{20}-k_{21})^2 + 4k_{21}k_{12}]^{1/2}/2$. They are always real because $k_{ij} > 0$. The negativity in the case of a double root, as well as in the case of different roots, is justified after some elementary reformulation because $k_{ij} > 0$. □

If $I_1(t) = I_2(t) = 0$, then (10.3) is called a homogeneous system. Due to Lemma 10.3, a qualitative identification of the solutions $c_h(t)$ to the homogeneous system (10.3) is possible:

Corollary 10.1. *The solutions to the homogeneous system of differential equations* (10.3) *are non-oscillating and non-negative real functions converging to zero for t toward infinity.*

Definition 10.4. *The homogeneous initial value problem defined by the homogeneous system* (10.3), *and the initial conditions* $c_{1h}(t) = c^o$, *and* $c_{2h} = 0$ *is called a* ***Two-compartment model for iv bolus administration*** *or* ***Two-compartment-iv model*** *for short. It will be numerated again by* (10.3).

The interpretation is clear: Through the intravenous administration of a drug, an initial concentration of $c_{1h}(0) = c^o$ arises in compartment 1 (identified with blood circulation), while $c_{2h}(0) = 0$ is presumed.
Since a complete solution to the homogeneous initial value problem (10.3) was not found in the literature about pharmacokinetics, but the model should still be studied in detail, it is now given.

Lemma 10.4. *Let* $\det K \neq 0$ *and* λ_1 *as well as* λ_2 *be differing roots of the characteristic polynomial of* (10.3). *The general solution to the homogeneous initial value problem* (10.3) *for three different cases is then given as follows:*

Case 1: $k_{12} \neq 0$

$$\begin{aligned}
c_{1h}(t) &= \frac{c^0}{\lambda_1 - \lambda_2}\left[(\lambda_1 + k_{20} + k_{21})\mathrm{e}^{\lambda_1 t}\right. \\
&\qquad \left. -(\lambda_2 + k_{20} + k_{21})\mathrm{e}^{\lambda_2 t}\right] \ , \\
c_{1h}(t) &= A\mathrm{e}^{\lambda_1 t} + B\mathrm{e}^{\lambda_2 t} \ , \\
c_{2h}(t) &= \frac{k_{12}c^0}{\lambda_1 - \lambda_2}\left[\mathrm{e}^{\lambda_1 t} - \mathrm{e}^{\lambda_2 t}\right] \ , \\
c_{2h}(t) &= C\mathrm{e}^{\lambda_1 t} + D\mathrm{e}^{\lambda_2 t} \ ;
\end{aligned}$$

Case 2: $k_{21} \neq 0$

$$\begin{aligned}
c_{1h}(t) &= \frac{c^0}{\lambda_2 - \lambda_1}\left[(\lambda_2 + k_{10} + k_{12})\mathrm{e}^{\lambda_1 t}\right. \\
&\qquad \left. -(\lambda_1 + k_{10} + k_{12})\mathrm{e}^{\lambda_2 t}\right] \ , \\
c_{1h}(t) &= A\mathrm{e}^{\lambda_1 t} + B\mathrm{e}^{\lambda_2 t} \ , \\
c_{2h}(t) &= \frac{c^0(\lambda_1 + k_{10} + k_{12})(\lambda_2 + k_{10} + k_{12})}{k_{21}(\lambda_1 - \lambda_2)}\left[\mathrm{e}^{\lambda_2 t} - \mathrm{e}^{\lambda_1 t}\right], \\
c_{2h}(t) &= C\mathrm{e}^{\lambda_1 t} + D\mathrm{e}^{\lambda_2 t} \ ;
\end{aligned}$$

Case 3: $k_{12} = k_{21} = 0$ *and* $k_{10} \neq k_{20}$

$$\begin{aligned}
c_{1h}(t) &= c^o\mathrm{e}^{-k_{10}t} \ , \\
c_{2h}(t) &= 0 \ .
\end{aligned}$$

A, B, C, D, λ_1 and $\lambda_2 \in \mathbb{R}$ serve as system parameters.

Proof. Since only real roots appear, $Fe^{\lambda_1 t}$ and $Ge^{\lambda_2 t}$, $F, G \in \mathbb{R}$, generate the solutions. In Case 1, c_{2h} is obtained from (10.5) and c_{1h} from (10.4). In Case 2, the equation of second order from (10.3) is solvable in c_1. Case 3 allows direct integration. The initial conditions in the respective cases clearly define the constants such that $F = k_{12}c^o/(\lambda_1 - \lambda_2)$, $G = k_{12}c^o/(\lambda_2 - \lambda_1)$; $F = c^o(\lambda_2 + k_{10} + k_{12})/(\lambda_2 - \lambda_1)$, $G = c^o(\lambda_1 + k_{10} + k_{12})/(\lambda_1 - \lambda_2)$ and $F = c^o$, $G = 0$. □

Lemma 10.5. *Let* $\det K \neq 0$ *and* λ *be a double root of the characteristic polynomial in* (10.3). *The general solution to the homogeneous initial value problem* (10.3) *for three different cases is then given as follows:*

Case 1: $k_{10} + k_{12} = k_{20} + k_{21}$ *and* $k_{21} = 0$, *so* $k_{10} + k_{12} = k_{20}$

$$c_{1h}(t) = c^o e^{-k_{20}t}, c_{2h}(t) = k_{12}c^o t e^{-k_{20}t} ,$$
$$c_{2h}(t) = Ate^{-k_{20}t} ;$$

Case 2: $k_{10} + k_{12} = k_{20} + k_{21}$ *and* $k_{12} = 0$, *so* $k_{20} + k_{21} = k_{10}$

$$c_{1h}(t) = c^o e^{-k_{10}t}, c_{2h}(t) = 0;$$

Case 3: $k_{10} + k_{12} = k_{20} + k_{21}$ *and* $k_{12} = k_{21} = 0$, *so* $k_{10} = k_{20}$

$$c_{1h}(t) = c^o e^{-k_{10}t}, c_{2h}(t) = 0.$$

$A \in \mathbb{R}$ serves as a system parameter.

Proof. The radicand of the roots of the characteristic polynomial disappears exactly when one of the three coefficient constellations given in Lemma 10.5 sets in. The general solution given for the homogeneous initial value problem are again available in accordance with the theory of linear ordinary differential equations, or by direct integration.
Case 1 yields $c_{1h}(t) = Ge^{-k_{20}t}/k_{12}$, $c_{2h}(t) = (F + Gt)e^{-k_{20}t}$; the second yields $c_{1h}(t) = (F + Gt)e^{-k_{10}t}$, $c_{2h}(t) = Ge^{-k_{10}t}/k_{21}$ and the third yields $c_{1h}(t) = Fe^{-k_{10}t}$, $c_{2h}(t) = Ge^{-k_{10}t}$. The initial conditions yield $F = 0$ and $G = k_{12}c^o$ in case 1. Otherwise $F = c^o$ and $G = 0$. □

The One-compartment-iv model proves to be a special case of the Two-compartment-iv model: Use Lemma 10.5, set $k_{el} = k_{10} + k_{12}$ in Case 1 and set $k_{el} = k_{10}$ in the two remaining cases.

Lemma 10.6. *Let* $\det K = 0$. *The general solution to the homogeneous initial value problem* (10.3) *for two different cases is then given as follows:*

Case 1: $k_{10}+k_{12}+k_{20}+k_{21}\neq 0$

1.1. $k_{12}\neq 0$

$$c_{1h}(t)=\frac{c^o}{k_{10}+k_{12}+k_{20}+k_{21}}\left[\{k_{10}+k_{12}\}\times \mathrm{e}^{-\{k_{10}+k_{12}+k_{20}+k_{21}\}t}+\{k_{20}+k_{21}\}\right],$$

$$c_{1h}(t)=A\mathrm{e}^{\lambda_1 t}+B\ ,$$

$$c_{2h}(t)=\frac{c^o k_{12}}{k_{10}+k_{12}+k_{20}+k_{21}}\left[1-\mathrm{e}^{-\{k_{10}+k_{12}+k_{20}+k_{21}\}t}\right],$$

$$c_{2h}(t)=C(1-\mathrm{e}^{\lambda_1 t})\ ;$$

1.2. $k_{21}\neq 0$

$$c_{1h}(t)=\frac{c^0}{k_{10}+k_{12}+k_{20}+k_{21}}\left[\{k_{10}+k_{12}\}\times \mathrm{e}^{-\{k_{10}+k_{12}+k_{20}+k_{21}\}t}+\{k_{20}+k_{21}\}\right],$$

$$c_{1h}(t)=A\mathrm{e}^{\lambda_1 t}+B\ ,$$

$$c_{2h}(t)=\frac{c^o(k_{10}+k_{12})(k_{20}+k_{21})}{k_{10}+k_{12}+k_{20}+k_{21}}\times\left[1-\mathrm{e}^{-\{k_{10}+k_{12}+k_{20}+k_{21}\}t}\right],$$

$$c_{2h}(t)=C(1-\mathrm{e}^{\lambda_1 t})\ ;$$

1.3.1. $k_{12}=k_{21}=0$ *with* $k_{10}\neq 0$ *and* $k_{20}=0$

$$c_{1h}(t)=c^o\mathrm{e}^{-k_{10}t}, c_{2h}(t)=0;$$

1.3.2. $k_{12}=k_{21}=0$ *with* $k_{10}=0$ *and* $k_{20}\neq 0$

$$c_{1h}(t)=c^o, c_{2h}(t)=0;$$

Case 2: $k_{10}+k_{12}+k_{20}+k_{21}=0$

This is the trivial case where $K=0$. *It remains unnoticed.*

A, B, C and $\lambda_1\in\mathbb{R}$ serve as system parameters.

Proof. Under the mentioned assumptions

$$P_K(\lambda)=\lambda^2+(k_{10}+k_{12}+k_{20}+k_{21})\lambda$$

is valid. Case 1 concerns roots which are different from each other, and case 2 concerns double roots. The solutions are found as usual. The functions $F\mathrm{e}^{-(k_{10}+k_{12}+k_{20}+k_{21})t}$ and $G\mathrm{e}^{0t}=G$ form a fundamental set for the solutions. The following yields from the initial conditions for the constants

F and G (real numbers)

in Case 1.1.:

$$F = \frac{-k_{12}c^0}{k_{10} + k_{12} + k_{20} + k_{21}} \text{ and } G = -F\ ;$$

in Case 1.2.:

$$F = \frac{c^0(k_{10} + k_{12})}{k_{10} + k_{12} + k_{20} + k_{21}} \text{ and } G = c^0 - F\ ;$$

in Case 1.3.:

$$F = c^o \text{ and } G = 0\ .$$

□

We have $\det K = 0$ exactly when $k_{10}k_{20} + k_{10}k_{21} + k_{12}k_{20} = 0$ is true. Due to the fact that the k_{ij} are non-negative, it is necessary that certain coefficients are zero.
Eight cases have to be distinguished, two of them are identical:

1.	k_{10}, k_{12}	2.	k_{10}, k_{20}
3.	k_{10}, k_{12}, k_{21}	4.	k_{10}, k_{12}, k_{20}
5.	k_{10}, k_{20}, k_{21}	6.	k_{20}, k_{21}
7.	k_{21}, k_{12}, k_{20}	8.	k_{20}, k_{10}

are respectively zero.
The One-compartment-iv model appears as a special case in Case 7. The homogeneous initial value problem (10.3) is thus observed in the entirety.
If **extravascular administration** is modeled, then inhomogeneities $I_1(t)$ and $I_2(t)$ given in (10.3) have to be put into concrete terms. The inconceivable variety of possible attempts is again restricted by the knowledge available about biochemical, physical etc. qualities and the resorption behavior of the observed drugs.
As with the One-compartment-iv model, the invasion process of the drug is normally presumed by a function $c_{01}(t) = c_{01}^o e^{-k_{01}t}$. This is equivalent to $I_1(t) = k_{01}c_{01}^o e^{-k_{01}t}$ because (10.3) is a differential equation. $k_{01} > 0$ is once again called an **invasion constant**; c_{01}^0 is the parameter corresponding with the invasion process and its initial condition $c_{01}(t) = c_{01}^o$.

Definition 10.5. *The initial value problem defined by the system of differential equations* (10.3) *with inhomogeneities $I_1(t) = k_{01}c_{01}^o e^{-k_{01}t}$ and $I_2(t) = 0$ as well as the initial conditions $c_1(t) = 0$ and $c_2(t) = 0$ is called a* ***Two-compartment model with first order input*** *or* ***Two-compartment-ev model*** *for short and is subsequently also described as an inhomogeneous initial value problem* (10.3).

Next, for all cases where attention should be paid, special solutions $c_{1i}(t)$ and $c_{2i}(t)$ are indicated for the inhomogeneous initial value problem (10.3).

Lemma 10.7. *Let* $\det K \neq 0$ *and* $-k_{01}$ *not be a root of the characteristic polynomial in* (10.3). *Then*

$$c_{1i}(t) = (k_{20} + k_{21} - k_{01})k_{01}c_{01}^0 \mathrm{e}^{-k_{01}t}/P(-k_{01}),$$
$$c_{2i}(t) = k_{01}k_{12}c_{01}^0 \mathrm{e}^{-k_{01}t}/P(-k_{01})$$

is a particular solution to the inhomogeneous initial value problem (10.3).

Proof. The differential equation (10.5) is solved under the assumption that $k_{12} \neq 0$. The inhomogeneity is a quasi polynomial, such that a particular solution exists in the form of $c_{2i}(t) = F\mathrm{e}^{-k_{01}t}$, $F \in \mathbb{R}$. $F = k_{12}k_{01}c_{01}^o/P_K(-k_{01})$ is obtained through substitution in (10.5). $c_{1i}(t)$ is obtained from (10.4). For $k_{21} \neq 0$, a differential equation of 2nd order $c_1(t)$ can be derived from (10.3). The same particular solution is obtained. The case $k_{12} = k_{21} = 0$ makes direct integration possible. The corresponding solution is a special case. □

Lemma 10.8. *Let* $\det K \neq 0$ *and* $-k_{01}$ *be a single root of the characteristic polynomial in* (10.3). *A particular solution to the inhomogeneous initial value problem* (10.3) *for three different cases is then given as follows:*

Case 1: $k_{12} \neq 0$

$$c_{1i}(t) = \frac{k_{01}c_{01}^0\left[1 + (k_{20} + k_{21} - k_{01})\,t\right]}{k_{10} + k_{12} + k_{20} + k_{21} - 2k_{01}} e^{-k_{01}t},$$
$$c_{2i}(t) = \frac{k_{01}k_{12}c_{01}^0 t}{k_{10} + k_{12} + k_{20} + k_{21} - 2k_{01}} e^{-k_{01}t};$$

Case 2: $k_{21} \neq 0$

$$c_{1i}(t) = \frac{(k_{20} + k_{21} - k_{01})\,k_{01}c_{01}^0 t}{k_{10} + k_{12} + k_{20} + k_{21} - 2k_{01}} e^{-k_{01}t},$$

$$c_{2i}(t) = \frac{\left[(k_{20} + k_{21} - k_{01})\,t - 1\right]\left[k_{10} + k_{12} - k_{01}\right]k_{01}c_{01}^0}{k_{10} + k_{12} + k_{20} + k_{21} - 2k_{01}} e^{-k_{01}t};$$

Case 3: $k_{12} = k_{21} = 0$ *and* $k_{10} \neq k_{20}$

$$c_{1i}(t) = \frac{\left[1 + (k_{20} - k_{01})\,t\right]k_{01}c_{01}^0}{(k_{10} + k_{20} - 2k_{01})} e^{-k_{01}t},$$
$$c_{2i}(t) = 0\,.$$

Proof. For the attempt $c_{2i}(t) = Fte^{-k_{01}t}$, the constant $F = k_{01}k_{12}c_{01}^0/(k_{10} + k_{12} + k_{20} + k_{21} - 2k_{01})$ is obtained as explained in the proof of Lemma 10.7. In the second case with attempt $c_{1i}(t) = Fte^{-k_{01}t}$, the constant $F = (k_{20} + k_{21} - k_{01})k_{01}c_{01}^0/(k_{10} + k_{12} + k_{20} + k_{21} - 2k_{01})$ is obtained in the same way. The third solution follows from the first. □

Remark 10.2. $k_{10} + k_{12} + k_{20} + k_{21} - 2k_{01} \neq 0$ is true because $-k_{01}$ is a single root of $P_K(\lambda)$.

Lemma 10.9. *Let* $\det K \neq 0$ *and* $-k_{01}$ *be a double root of the characteristic polynomial in* (10.3). *A particular solution to the inhomogeneous initial value problem* (10.3) *for three different cases is then given as follows:*

Case 1: $k_{10} + k_{12} = k_{20} + k_{21}, k_{12} \neq 0, k_{21} = 0$

$$c_{1i}(t) = k_{20}c_{01}^0 te^{-k_{20}t} ,$$
$$c_{2i}(t) = k_{20}k_{12}c_{01}^0 t^2 e^{-k_{20}t}/2 ;$$

Case 2: $k_{10} + k_{12} = k_{20} + k_{21}, k_{12} = 0, k_{21} \neq 0$

$$c_{1i}(t) = 0 ,$$
$$c_{2i}(t) = -k_{10}c_{01}^0 e^{-k_{10}t}/k_{21} ;$$

Case 3: $k_{10} + k_{12} = k_{20} + k_{21}, k_{12} = k_{21} = 0, k_{10} = k_{20} \neq 0$

$$c_{1i}(t) = k_{10}c_{01}^0 te^{-k_{10}t} ,$$
$$c_{2i}(t) = 0 .$$

Proof. In the first case the inhomogeneous differential equation (10.5) is solved by means of $c_{2i}(t) = Ft^2e^{-k_{01}t}$. Comparison of coefficients results in $F = k_{01}k_{12}c_{01}^0/2$. The relations $k_{01} = k_{20} = k_{10} + k_{12}$ yield from the assumptions with regard to k_{ij}. The solution function $c_{1i}(t)$ from (10.4) simplifies itself to that of the given form. In the second case, (10.3) is transformed into a differential equation of second order in c_1. The attempt $c_{1i}(t) = Ft^2e^{-k_{01}t}$ leads to $F = (k_{20}+k_{21}-k_{01})k_{01}c_{01}^0/2$, where the relation $F = 0$ yields due to $k_{01} = k_{10} = k_{20} + k_{21}$. Thereby $c_{1i}(t) = 0$, and the second solution function can be obtained from an equation in (10.3). Direct integration is possible in the third case. This particular solution is the special case for $k_{12} = 0$ from the first case. □

Lemma 10.10. *Let* $\det K = 0$ *and* $-k_{01}$ *not be a root of the characteristic* *polynomial in* (10.3)*. Then*

$$c_{1i}(t) = \frac{(k_{20} + k_{21} - k_{01})\, c_{01}^{0}}{k_{01} - k_{10} - k_{12} - k_{20} - k_{21}} e^{-k_{01}t} ,$$

$$c_{2i}(t) = \frac{k_{12} c_{01}^{0}}{k_{01} - k_{10} - k_{12} - k_{20} - k_{21}} e^{-k_{01}t}$$

is a particular solution to the inhomogeneous initial value problem (10.3).

Proof. This proof is similar to the proof of Lemma 10.7. All of the different cases here also lead to the same solution. □

Lemma 10.11. *Let* $\det K = 0$ *and* $-k_{10} \neq 0$ *be a single root of the characteristic polynomial in* (10.3).
A particular solution to the inhomogeneous initial value problem (10.3) *for three different cases is then given as follows:*

Case 1: $k_{12} \neq 0$

$$c_{1i}(t) = \frac{k_{01} c_{01}^{0} \left[1 + (k_{20} + k_{21} - k_{01})\, t\right]}{k_{10} + k_{12} + k_{20} + k_{21} - 2k_{01}} e^{-k_{01}t} ,$$

$$c_{2i}(t) = \frac{k_{01} k_{12} c_{01}^{0} t}{k_{10} + k_{12} + k_{20} + k_{21} - 2k_{01}} e^{-k_{01}t} ;$$

Case 2: $k_{21} \neq 0$

$$c_{1i}(t) = \frac{(k_{20} + k_{21} - k_{01})\, k_{01} c_{01}^{0} t}{k_{10} + k_{12} + k_{20} + k_{21} - 2k_{01}} e^{-k_{01}t} ,$$

$$c_{2i}(t) = \frac{\left[(k_{20} + k_{21} - k_{01})\, t - 1\right] \left[k_{10} + k_{12} - k_{01}\right] k_{01} c_{01}^{0}}{k_{10} + k_{12} + k_{20} + k_{21} - 2k_{01}} e^{-k_{01}t} ;$$

Case 3: $k_{12} = k_{21} = 0$ *and* $k_{10}k_{20} = 0$

3.1. $k_{10} \neq 0$ *and* $k_{20} \neq 0$

$$c_{1i}(t) = (1 + k_{01}t)\, c_{01}^{0} e^{-k_{01}t} ,$$

$$c_{2i}(t) = 0 ;$$

3.2. $k_{10} = 0$ *and* $k_{20} \neq 0$

$$c_{1i}(t) = -c_{01}^{0} e^{-k_{01}t} ,$$

$$c_{2i}(t) = 0 .$$

Proof. The same distinctions in cases are obtained and the same solutions are arrived at as in Lemma 10.8. Due to the assumptions, the 3rd case still has to be specified. □

Lemma 10.12. *Let* $\det K = 0$ *and* $-k_{01}$ *be a double root of the characteristic polynomial in* (10.3). *Then*

$$c_{1i}(t) = 0\,,$$
$$c_{2i}(t) = 0$$

is the unique solution to the inhomogeneous initial value problem (10.3).

Proof. $k_{01} = 0$ necessarily follows under the mentioned assumptions. □

The solution to the inhomogeneous initial value problem (10.3) is obtainable by the fact that the constants of its general solution are determined by the initial values. This general solution is the sum of the general solution of the associated homogeneous differential equation system and a particular solution to the inhomogeneous initial value problem (10.3).
Although the solution to the inhomogeneous initial value problem (10.3) is not examined further here, its general form will be listed together with its corresponding prerequisites in the three following lemmata. More cases than the listed ones do not need to be distinguished. There will be no detailed explanation for this statement. It is based on the solution methods of the mentioned differential equations, the respective special form of the roots of $P_K(\lambda)$ and avoiding a division by zero. Attention should be brought to the fact that the solution to the homogeneous initial value problem (10.3) is not allowed to be confused with the general solution to the homogeneous system (10.3).

Lemma 10.13. *The homogeneous system of differential equations* (10.3) *has the following general solution* $c_{1ha}(t), c_{2ha}(t)$*:*
Proposition 1
Let $\det K \neq 0$ *and* $\lambda_1 \neq \lambda_2$ *be the roots of the characteristic polynomial in* (10.3).

Case 1: $k_{12} \neq 0$

$$c_{1ha}(t) = Ae^{\lambda_1 t} + Be^{\lambda_2 t}\,,$$
$$c_{2ha}(t) = Ce^{\lambda_1 t} + De^{\lambda_2 t}\,;$$

Case 2: $k_{21} \neq 0$

$$c_{1ha}(t) = Ae^{\lambda_1 t} + Be^{\lambda_2 t}\,,$$
$$c_{2ha}(t) = Ce^{\lambda_1 t} + De^{\lambda_2 t}\,;$$

Case 3: $k_{12} = k_{12} = 0$ *and* $k_{10} \neq k_{20}$

$$c_{1ha}(t) = Ae^{-k_{10}t} ,$$
$$c_{2ha}(t) = Be^{-k_{20}t} .$$

Proposition 2
Let $\det K = 0$ *and* λ *be the double root of the characteristic polynomial in* (10.3).

Case 1: $k_{10} + k_{12} = k_{20} + k_{21}, k_{12} \neq 0$ *and* $k_{21} = 0$

$$c_{1ha}(t) = Ae^{-k_{20}t} ,$$
$$c_{2ha}(t) = (B + Ct)\, e^{-k_{20}t} ;$$

Case 2: $k_{10} + k_{12} = k_{20} + k_{21}, k_{21} \neq 0$ *and* $k_{12} = 0$

$$c_{1ha}(t) = (A + Bt)\, e^{-k_{10}t} ,$$
$$c_{2ha}(t) = Ce^{-k_{10}t} ;$$

Case 3: $k_{10} + k_{12} = k_{20} + k_{21}$ *and* $k_{12} = k_{21} = 0$

$$c_{1ha}(t) = Ae^{-k_{10}t} ,$$
$$c_{2ha}(t) = Be^{-k_{10}t} .$$

Proposition 3
Let $\det K \neq 0$ *and* λ *be the double root of the characteristic polynomial in* (10.3) *with* $\lambda \neq 0$.

Case 1: $k_{12} \neq 0$ *or* $k_{21} \neq 0$

$$c_{1ha}(t) = Ae^{\lambda t} + B ,$$
$$c_{2ha}(t) = C\left[1 - e^{\lambda t}\right] ;$$

Case 2: $k_{12} = k_{21} = 0$ *and* $k_{10} \neq 0$

$$c_{1ha}(t) = Ae^{-k_{10}t} ,$$
$$c_{2ha}(t) = B ;$$

Case 3: $k_{12} = k_{21} = 0$ *and* $k_{20} \neq 0$

$$c_{1ha}(t) = A ,$$
$$c_{2ha}(t) = Be^{-k_{20}t} .$$

Lemma 10.14. *Let* $\det K \neq 0$. *Then the following general solution* $c_1(t), c_2(t)$ *is associated with the inhomogeneous initial value problem* (10.3)*:*
Proposition 1
The roots λ_1 *and* λ_2 *of the characteristic polynomial are different from each other and* $P_K(-k_{01}) \neq 0$.

Case 1: $k_{12} \neq 0$

$$c_1(t) = Ae^{\lambda_1 t} + Be^{\lambda_2 t} + Ce^{-k_{01}t} ,$$
$$c_2(t) = De^{\lambda_1 t} + Ee^{\lambda_2 t} + Fe^{-k_{01}t} ;$$

Case 2: $k_{21} \neq 0$

$$c_1(t) = Ae^{\lambda_1 t} + Be^{\lambda_2 t} + Ce^{-k_{01}t} ,$$
$$c_2(t) = De^{\lambda_1 t} + Ee^{\lambda_2 t} + Fe^{-k_{01}t} ;$$

Case 3: $k_{12} = k_{21} = 0, k_{10} \neq 0, k_{20} \neq 0, k_{10} \neq k_{20}$

$$c_1(t) = Ae^{-k_{10}t} + Be^{-k_{01}t} ,$$
$$c_2(t) = Ce^{-k_{20}t} + De^{-k_{01}t} .$$

Proposition 2
The roots of the characteristic polynomial coincide and $P_K(-k_{01}) \neq 0$.

Case 1: $k_{10} + k_{12} = k_{20} + k_{21}, k_{12} \neq 0, k_{21} = 0$

$$c_1(t) = Ae^{-k_{20}t} + Be^{-k_{01}t} ,$$
$$c_2(t) = (C + Dt)\, e^{-k_{20}t} + Ee^{-k_{01}t} ;$$

Case 2: $k_{10} + k_{12} = k_{20} + k_{21}, k_{12} = 0, k_{21} \neq 0$

$$c_1(t) = (A + Bt)\, e^{-k_{10}t} + Ce^{-k_{01}t} ,$$
$$c_2(t) = De^{-k_{10}t} + Ee^{-k_{01}t} ;$$

Case 3: $k_{10} + k_{12} = k_{20} + k_{21}, k_{12} = k_{21} = 0$

$$c_1(t) = Ae^{-k_{10}t} + Be^{-k_{01}t} ,$$
$$c_2(t) = Ce^{-k_{10}t} + De^{-k_{01}t} .$$

Proposition 3
Let $\lambda_1 \neq \lambda_2$ *and* $-k_{01}$ *be a single root of the characteristic polynomial.*

Case 1: $k_{12} \neq 0$

$$c_1(t) = Ae^{\lambda_1 t} + Be^{\lambda_2 t} + (C + Dt)\, e^{-k_{01}t} ,$$
$$c_2(t) = Ee^{\lambda_1 t} + Fe^{\lambda_2 t} + Gte^{-k_{01}t} ;$$

Case 2: $k_{21} \neq 0$

$$c_1(t) = Ae^{\lambda_1 t} + Be^{\lambda_2 t} + Cte^{-k_{01}t} ,$$
$$c_2(t) = De^{\lambda_1 t} + Ee^{\lambda_2 t} + (F + Gt)\, e^{-k_{01}t} ;$$

Case 3: $k_{12} = k_{21} = 0, k_{10} \neq 0, k_{10} \neq k_{20}, k_{10} \neq 0, k_{20} \neq 0$

$$c_1(t) = Ae^{-k_{10}t} + (B + Ct)\, e^{-k_{01}t} \,,$$
$$c_2(t) = De^{-k_{20}t} \,.$$

Proposition 4
Let $\lambda_1 = \lambda_2$ and $-k_{01}$ be the double root of the characteristic polynomial.

Case 1: $k_{10} + k_{12} = k_{20} + k_{21}, k_{12} \neq 0, k_{21} = 0$

$$c_1(t) = (A + Bt)\, e^{-k_{20}t} \,,$$
$$c_2(t) = \left(C + Dt + Et^2\right) e^{-k_{20}t} \,;$$

Case 2: $k_{10} + k_{12} = k_{20} + k_{21}, k_{12} = 0, k_{21} \neq 0$

$$c_1(t) = (A + Bt)\, e^{-k_{10}t} \,,$$
$$c_2(t) = Ce^{-k_{10}t} \,;$$

Case 3: $k_{10} + k_{12} = k_{20} + k_{21}, k_{12} = k_{21} = 0, k_{10} = k_{20} \neq 0$

$$c_1(t) = (A + Bt)\, e^{-k_{10}t} \,,$$
$$c_2(t) = Ce^{-k_{10}t} \,.$$

Lemma 10.15. *Let* $\det K = 0$. *Then the following general solution is associated with the inhomogeneous initial value problem* (10.3)*:*
Proposition 1
Let $\lambda_1 \neq \lambda_2$ and $P_K(-k_{01}) \neq 0$.

Case 1: $k_{12} \neq 0$ *or* $k_{21} \neq 0$

$$c_1(t) = Ae^{\lambda_1 t} + Be^{-k_{01}t} + C \,,$$
$$c_2(t) = De^{\lambda_1 t} + Ee^{-k_{01}t} + F \,;$$

Case 2: $k_{12} = k_{21} = 0, k_{10} \neq 0, k_{20} = 0$

$$c_1(t) = Ae^{-k_{10}t} + Be^{-k_{01}t} \,,$$
$$c_2(t) = C + De^{-k_{01}t} \,;$$

Case 3: $k_{12} = k_{21} = 0, k_{10} = 0, k_{20} \neq 0$

$$c_1(t) = A + Be^{-k_{01}t} \,,$$
$$c_2(t) = Ce^{-k_{20}t} + De^{-k_{01}t} \,.$$

Proposition 2
Let $\lambda_1 = \lambda_2$ and $P_K(-k_{01}) \neq 0$.

$$c_1(t) = Ae^{-k_{01}t} \,,$$
$$c_2(t) = Be^{-k_{01}t} \,.$$

Proposition 3
Let $\lambda_1 = \lambda_2$ and $-k_{01} \neq 0$ be a single root of the characteristic polynomial.

Case 1: $k_{12} \neq 0$

$$c_1(t) = Ae^{\lambda_1 t} + (B + Ct)\, e^{-k_{01}t} + D\,,$$
$$c_2(t) = Ee^{\lambda_1 t} + Fte^{-k_{01}t} + G\,;$$

Case 2: $k_{21} \neq 0$

$$c_1(t) = Ae^{\lambda_1 t} + Bte^{-k_{01}t} + C\,,$$
$$c_2(t) = De^{\lambda_1 t} + (E + Ft)\, e^{-k_{01}t} + G\,;$$

Case 3: $k_{12} = k_{21} = 0, k_{10}k_{20} = 0, k_{10} \neq 0, k_{20} = 0$

$$c_1(t) = Ae^{-k_{10}t} + (B + Ct)\, e^{-k_{01}t}\,,$$
$$c_2(t) = D\,;$$

Case 4: $k_{12} = k_{21} = 0, k_{10}k_{20} = 0, k_{10} = 0, k_{20} \neq 0$

$$c_1(t) = Ae^{-k_{01}t} + B\,,$$
$$c_2(t) = Ce^{-k_{20}t}\,.$$

Proposition 4
Let $\lambda_1 = \lambda_2$ and $-k_{01}$ be the double root of the characteristic polynomial. Then

$$c_1(t) = 0\,,$$
$$c_2(t) = 0$$

is the only solution.

The solutions to the inhomogeneous initial value problem (10.3) are not listed here. Their determination is unproblematic.
The calculation of the homogeneous and of the inhomogeneous initial value problem (10.3) is carried out in a systematic way which is pre-marked by the theory of linear differential equations. Figure 10.2 clarifies which special two-compartment model corresponds to the given conditions of the solution. The opposite procedure is typical in the practice of pharmacokinetics: The solution to the corresponding differential equation is searched for and a specified two-compartment model and its parameters are determined from the measurements. It is necessary to know the relationship between a used model function and the related pharmacokinetical models for the correct interpretation of results.

Lemma 10.16. *The time courses of concentrations of the one-compartment model appear in several special cases of the two-compartment model.*

Proof. The function $c(t)$ indicated in Lemma 10.1 (it is observed in the application compartment) appears in three cases in accordance with Lemma 10.5:

1. for $k_{21} = 0, k_{12} \neq 0$ and $k_{10} + k_{12} = k_{el}$;
2. for $k_{12} = 0$ and $k_{10} = k_{el}$;
3. for $k_{12} = k_{21} = 0$ and $k_{10} = k_{el}$.

In addition, $k_{12} = k_{20} = k_{21} = 0$ yields from Lemma 10.8.
Two cases have to be distinguished with regard to the inhomogeneous equation. If $k_{el} = k_i = k$ is assumed, then the function $c(t)$ given in Lemma 10.2 is also the solution to the homogeneous initial value problem defined by (10.3) and the starting conditions $c_{1h}(0) = 0, c_{2h}(0)$. This corresponds with Case 2 in Lemma 10.5, where $k_{20} = 0$ is assumed. It is however also identical to the function $c_1(t)$, which yields from Lemma 10.14, Proposition 4, Case 1 where $k_{21} = 0, k_{10} + k_{12} = k$ and the starting conditions are $c_1(0) = 0$ as well as $c_2(0) = 0$.
If $k_{el} \neq k_i$ is assumed, then the function $c(t)$ provided in Lemma 10.2 is also the solution to the homogeneous initial value problem defined by (10.3), $k_{12} = k_{20} = 0$ and the initial conditions $c_{1h}(0) = 0$, and $c_{2h}(0) = 0$. It is also appropriately solved in Lemma 10.15, Proposition 1 with the use of the initial conditions $c_1(0) = 0$ and $c_2(0) = 0$. □

It cannot be generalized that the inhomogeneous case of the one-compartment model comes from the homogeneous case of the two-compartment model. Here the inhomogeneities again are exponential functions with respect to the processes of first order.

Acceptance of a one-compartment model under reference to measured concentrations in the blood is not an argument against other models. If necessary, other observations have to be taken into account, for example the process of elimination, in order to be able to decide in favor of more detailed valid mathematical descriptions. The consequences with regard to possible interpretations of results are obvious. SCHELER (1980) dedicates a section of his book to the discussion of the one-compartment model with parallel elimination paths.
With regard to a clear description, the difference between an incompletely observed or part of a two(or more)-compartment model and a one-compartment model per se will be able to be distinguished in Definition 10.6. The

concept of a compartment described earlier is therefore composed more clearly:

Definition 10.6. *In pharmacokinetics, a quantity of a substance in an organism which is characterized by specific kinetics of transformation or transport processes, is described as a compartment. Compartments can interact by exchanging substances. Substance absorption (invasion) and substance excretion (elimination) can each only be carried out in one way per compartment.*

Several variants of substance eliminations are therefore modelled as couplings with other compartments. So a one-compartment pharmacokinetic model with parallel elimination paths has to be understood as a special two-compartment model.

10.3 More-compartment models

The correct application of the two-compartment model requires a systematic and complete development of the model equation. The system of the solutions even for the mathematically uncomplicated inhomogeneous initial value problem formulated in Definition 10.5 gets very confusing under the point of view of the application.

This really also applies to more-compartment models which are once again understood here as systems of linear differential equations with constant coefficients. Its mathematical treatment is predominated in pharmacokinetics with the method of LAPLACE transformation. On page 50, or page 64, GIBALDI/PERRIER (1982) mention that the concentration-time-functions of interest are obtained as sums of exponential terms. Summands of type $A_i t e^{a_i t}$ which appear in the complete solution to the homogeneous initial value problem (10.3) remain unnoticed. These remarks also apply to other authors, for example WOLF/HEINZEL/KOSS/BOZLER (1977). The solution to the homogeneous initial value problem given in FELDMANN/SCHNEIDER (1976) is based on the reformulation of an eigenvalue problem. BELLMAN (1983) solves the same question with the method of the matrix exponential. JACQUEZ (1988) discusses specially structured more-compartment systems (catenary systems, mammilary systems). Attention should also be brought to ANDERSON (1983) and KAJIYA/KODAMA/ABE (1985). The term "multicompartment systems" in K. M. S. HUMAK (1983) is used in correlation with regression attempts.

If for the matrix $K = (k_{ij})$ of a model, $k_{ij} = 0$ for $i \geq j + 2$ is true, K is called a HESSENBERG matrix, then the solutions to the associated system of differential equations appear as convolutions of exponential functions (HEARON/LONDON 1972).
The difficulty in the application of more-compartment models is not due to the solution to the differential system of equations. It is due to the discrepancy between the number as well as the location of the practically feasible measurements and the observations necessary for the "adequately exact" parameter calculation. The decision that is needed for a type of function contained in the solution set is connected with this: Do measurements sway around a monotonous function or do they indicate subdued oscillations which are part of the solution set of more-compartment systems (FELDMANN/SCHNEIDER 1976, page 249)? There are various interpretations of the results of a pharmacokinetic experiment!
With reference to Corollary 10.1 of Lemma 10.3, no periodical functions need to be taken into consideration in a two-compartment model. MATIS/PATTEN (1979) give a relatively simple example of a five-compartment model, which posesses complex roots of the characteristic polynomial. The three-compartment model with this property, which can be found in JACQUEZ (1972, page 67), is much less complicated.
One must justify in respective applications whether complicated models are meaningful. The attempt of an eleven-compartment system to describe the metabolism of iodine (GRASS/HABERMEHL 1980) raises more questions than can be conclusively answered.

Chapter 11

Other deterministic models

In this monograph, the term compartment model is used for linear differential equations. A bridge can be made from a system of linear differential equations with constant coefficients to equivalent differential equations of higher order. Let modeling begin with the formulation

$$\mathcal{D}c\left(t\right) = I\left(t\right), \tag{11.1}$$

where $\mathcal{D}$ is a differential operator, $c(t)$ is a concentration-time-function and $I(t)$ is a so-called input function. It will be called a system theoretical description of the pharmacokinetical process of interest. To determine $c(t)$, the VOLTERRA integral equation

$$c\left(t\right) = \mathcal{D}^{-1}I\left(t\right) = \int\limits_{0}^{t} K\left(t,u\right)I\left(u\right)\mathrm{d}u$$

needs to be solved. In the case of a linear differential operator, the kernel function (time invariance presupposed) is of the form $K(t-u)$, consequently

$$c\left(t\right) = \int\limits_{0}^{t} K\left(t-u\right)I\left(u\right)\mathrm{d}u\ . \tag{11.2}$$

This will be demonstrated for an example at the Two-compartment-ev model. The differential equation (10.5) is obtained with respect to (10.3). It is abbreviated and written in the form:

$$c''\left(t\right) + k_1 c'\left(t\right) + k_1 c\left(t\right) = f\left(t\right)\ .$$

The initial conditions are $c(0) = c'(0) = c''(0) = 0$. The corresponding indexing of compartments is ignored and the inhomogenities are not further specified. With respect to the known rules of the formation of the LAPLACE transformation of the derivative of a function (the required prerequisites are to be fulfilled) one obtains

$$P(s)\,C(s) = F(s)\ ,$$

$$C(s) = \int_0^\infty \mathrm{e}^{-st} c(t)\,\mathrm{d}t\ ,$$

$$F(s) = \int_0^\infty \mathrm{e}^{-st} f(t)\,\mathrm{d}t\ ,$$

with regard to the initial conditions from the last differential equation. In this context, $P(s)$ is the characteristic polynomial from (10.3) and s denotes the variable of the LAPLACE transformation. Let $C(s) = F(s)/P(s)$ be true, whereby $1/P(s)$ is the LAPLACE transformation of the function

$$g_1(t) = \frac{\mathrm{e}^{\lambda_1 t} - \mathrm{e}^{\lambda_2 t}}{\lambda_1 - \lambda_2} \quad \text{for} \quad \lambda_1 \neq \lambda_2\ ,$$

or

$$g_2(t) = t\mathrm{e}^{\lambda_t} \quad \text{for} \quad \lambda_1 = \lambda_2 = \lambda\ .$$

The λ_i represent the roots of the characteristic polynomial in (10.3). With regard to the convolution theorem for the LAPLACE transformation (see BRONSTEIN/SEMENDJAJEW/MUSIOL/MUEHLIG 1997, page 664 for example),

$$c(t) = \int_0^t f(t-u)\,g_i(u)\,\mathrm{d}u;\ i = 1, 2\ ,$$

yields for the two different cases. Since the convolution product is commutative,

$$c(t) = \int_0^t g_i(t-u)\,f(u)\,\mathrm{d}u;\ i = 1, 2\ ,$$

is true. A function g_i therefore corresponds with the kernel function in (11.2) for the Two-compartment-ev model.

The convolution integral also allows for a probability theoretical interpretation provided that the functions $K(t-u)$ and $I(u)$ satisfy certain conditions: $c(t)$ can be understood as the density of the sum of two independent random variables with densities $K(.)$ and $I(.)$.

Linear systems have been substantially examined in mathematics. They also play a big role in control theory and signal theory in which subject specific terminology is partly used.
The theory of linear systems was used early in pharmacokinetics (TEORELL 1937). JACQUEZ (1972, Chapter 6) points to this in his book. CUTLER (1978) clearly presents the topic and gives a literature summary. The observation that boundaries are set with regard to the application in pharmacokinetics, made in VAN ROSSUM/VAN GINNEKEN (1980, pages 61 cf.), is worth mentioning. An example of this is given and reference is made to corresponding literature. Further works are dedicated to special applications, calculation problems, and so on (e.g. IGA et al 1986, VENG-PEDERSEN/GILLESPIE 1986).
There is varying motivation behind the use of the formulation (11.1). Problems with regard to parameter identifiability can be a reason to move from the analysis of compartment models to so to speak summary descriptions of the events by means of corresponding models. The term "model independent pharmacokinetics" is used in this case. The identification of the kinetic behavior of a drug does not occur with the transfer constants k_{ij} of compartment models any more. It comes from the moments of the concentration-time-function (for example, YAMAOKA/NAKAGAWA/UNO 1978, WEISS 1984) in (11.2) and is actually a characteristic in the context of probability theory. MATIS/WEHRLY/METZLER (1983) give a representation of the associated mathematical field and particularily the required prerequisites for such a procedure. CHANTER (1985) draws attention to an incorrect interpretation. This was the starting point of a clarifying discussion (LANDAW/KATZ 1985, GILLESPIE/VENG-PEDERSEN 1985, BENET 1985).
Many authors say that the improved interpretability of a "model independent" parameter is an advantage. GIBALDI/PERRIER (1982) devote one chapter in their book to the system theoretical description of the metabolism of drugs in connection with physiological events. KEILSON/KESTER/WATERHOUSE (1978) bring the kinetics of a drug in relation to blood circulation. They make use of a MARKOV process though. CUTLER(1978, 1979, 1981) also takes physiological models of blood circulation into account. WEISS/FOERSTER (1979) put their model of the type (11.1) in relation to pulmonary circulation. HIMMELSTEIN/LUTZ (1979) summarize non-linear attempts as well as partial differential equations in their summary article about the physiologically based formations of models.

11.1 Compartment models with delay

The starting point of the following considerations is the one-compartment model (10.1). On the other hand, it should mathematically be described that the speed of a change in concentration is determined by an earlier condition. An after effect should therefore be considered. The problem leads to so-called differential equations with delay (MYSCHKIS 1955) and should not be mistaken for the shift that was worked into model (10.2) through a parameter t_{lag}.
The solutions to such differential equations are no longer found to be necessary in closed form even for the case of constant coefficients. The methods that are used for parameter calculation require a large number of measurements. Compartment models which referred to differential equations with delay were presented by KANYAR/ELLER/GYOERI (1981) and GYOERI/ELLER (1981). The attempt by KRISZTIN (1984) is more general and is connected with integro-differential equations.
The One-compartment-iv model with regard to delay time τ will now be discussed and used as a simple example. With reference to (10.1) the differential equation with delay is

$$\frac{\mathrm{d}}{\mathrm{d}t}c(t) = -k_{el}c(t-\tau) \tag{11.3}$$

for real $t \geq 0, \tau \geq 0$ and $k_{el} > 0$.
$c(t-\tau)$ is defined on $[-\tau, \infty)$. The clear attainability of a unique solution to (11.3) puts the handicap of a starting condition in the form of a continuous function $g(t)$ on $[-\tau, 0]$. Let $g(t) \equiv G$. The solution to (11.3) is obtained as follows:
$g(t) \equiv G$ is given on $[-\tau, 0]$. Therefore the relation $c'(t) = -k_{el}c(t-\tau) = -k_{el}G$ exists on $[0, \tau]$. Integration yields $c(t) = -k_{el}Gt + H$. Relation $H = G$ yields from the demanded continuity in 0. So

$$c(t) = (-k_{el}t + 1)\,G\,.$$

The differential equation

$$c'(t) = -k_{el}c(t-\tau) = -k_{el}\left[-k_{el}(t-\tau)+1\right]G$$

is solved on $[\tau, 2\tau]$. The demanded continuity of the solution delivers the integration constant, etc. By mathematical induction it can be shown that

$$c(t) = G\left[1 + \sum_{j=1}^{m}\frac{(-k_{el})^j}{j}\left(t - \{j-1\}\tau\right)^j\right]$$

with $(m-1)\tau \le t \le m\tau$ is the solution to (11.3). Its dependence on τ is labeled with the notation $c(t) = c_\tau(t)$. Let $k_{el} = 1$ and $g(t) = 1$ on $(-\infty, 0)$. The behavior of the solution $c_\tau(t)$ can be characterized as follows:

Proposition 11.1. *(*MYSCHKIS *1955, page 67)* $\tau_0 \in [0,1)$ *exists such that the function* $c_\tau(t)$ *in* $[0,\infty)$ *has roots for all* $\tau > \tau_0$. *If* N_τ *represents the smallest of these roots, then* $c_\tau(t)$ *decreases monotonically in* $[0, N_\tau + \tau]$. *If* $0 \le \tau \le \tau_0$, *then* $c_\tau(t) > 0$ *is true, and* $c_\tau(t)$ *decreases monotonically in* $[0,\infty)$. $N_\tau = \infty$ *is then set.*
Further, $N_{\tau 1} \le N_{\tau 2}$ *for* $\tau_1 > \tau_2 \ge 0$ *as well as* $c_{\tau 1}(t) = c_{\tau 2}(t)$ *for* $0 \le t < N_{\tau 1}$ *and* $\tau_1 > \tau_2 \ge 0$ *are true.*

Figure 11.1 may explain the content of Proposition 11.1. It clarifies that with increasing τ, the solutions $c_\tau(t)$ to (11.3) become less suitable for the description of pharmacokinetical processes; negative values of concentration do not make sense. It is proven by MYSCHKIS (1955, chapter 17), that for a sufficient large τ, every solution to (11.3) has a set of roots which is not bounded from above. On the other hand, if τ approaches zero, (11.3) becomes the typical one-compartment-iv model (10.1).

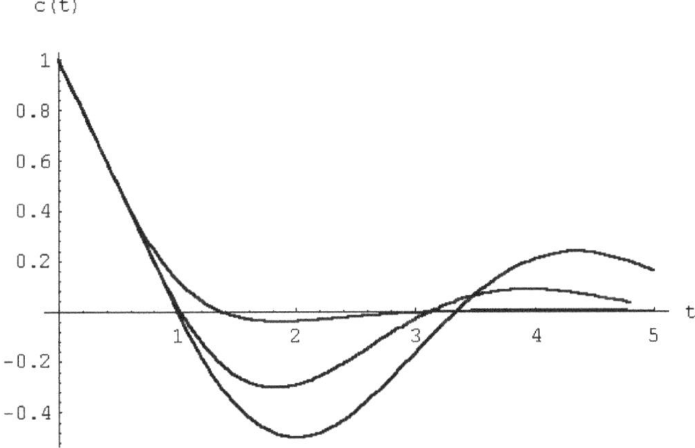

Fig. 11.1: Solution to the differential equation $c'(t) = -kc(t-\tau)$ for $k = 1$, $g(t) = 1$ and $\tau = 0.5$, $\tau = 0.8$, $\tau = 1$. The oscillation increases with τ.

The given example (11.3) shows the connection of the order of magnitude of τ and the mathematical qualities of the solution to (11.3). This can however contradict physiological, biochemical, etc knowledge about the length of a possible after effect. So besides the adoptions to the measurements, the adequacy of compartment models with delay must also be judged.

11.2 Nonlinear kinetics

The ideas developed by BROWN (1902) with regard to the course of the expiration of a substratum enzyme reaction are illustrated by

$$E + S \underset{k_2}{\overset{k_1}{\rightleftharpoons}} ES \overset{k_3}{\rightarrow} E + P \,.$$

With regard to these ideas, the following nonlinear system of differential equations of the enzyme kinetics was formulated by HENRI (1903) as well as by MICHAELIS/MENTEN (1913):

$$\begin{aligned} c'_S(t) &= -k_1 c_E(t)\, c_S(t) + k_2 c_{ES}(t) \\ c'_E(t) &= -k_1 c_E(t)\, c_s(t) + k_2 c_{ES}(t) + k_3 c_{ES}(t) \\ c'_{ES}(t) &= k_1 c_E(t)\, c_S(t) - k_2 c_{ES}(t) - k_3 c_{ES}(t) \\ c'_P(t) &= k_3 c_{ES}(t) \,. \end{aligned} \tag{11.4}$$

It describes the concentration-time courses of enzyme E, substratum S, enzyme substratum complex ES, and reaction product P. A closed solution to (11.4) generally cannot be given. By observing the mass balance

$$\frac{\mathrm{d}}{\mathrm{d}t}\left[c_E(t) + c_{ES}(t)\right] = 0$$

as well as under the supposition

$$c'_S(t) = 0 \quad \text{(quasi-steady-state)},$$

MICHAELIS and MENTEN (HENRI as well earlier) obtained the differential equation

$$c'_S(t) = \frac{-v_{\max}}{K_M + c_S(t)} c_S(t) \,. \tag{11.5}$$

The idea of deriving a differential equation for the time course $c_S(t)$ of the substratum concentration by the less limiting supposition

$$c'_{ES}(t) = 0 \quad \text{(steady-state)}$$

from (11.4) stems from BODENSTEIN (1913). An equation of type (11.5) is obtained again but K_M has to be interpreted differently. WALTER (1977) is referred to for this as well as broader considerations about approximative solutions to systems of differential equations (11.4).

SCHREIBER (2006) examined numeric methods of the solution to (11.4). In an example she demonstrates that the calculation of the parameters k_i of (11.4) from measured product concentration values generally is not unique. Equations are no longer linear if attempts of type (11.5) are used in conjunction with systems of differential equations connected with compartment models. For example, SCHELER (1980) demonstrates such model formations in connection with capacity bound elimination processes. The work of WAGNER (1974) offers an excellent summary of the meaningfulness of nonlinear pharmacokinetic models. Special drugs as well as unusual features of physiological processes which required the application of such descriptions are named here. Attention should furthermore be brought to the consolidation of theoretical descriptions in the context of enzyme kinetics which is validly conveyed by KERNEVEZ (1980). A generalization of (11.5) is the so called HILL equation

$$c'(t) = \frac{-v_{\max}}{K_M^n + c^n(t)}\, c^n(t)\,, \qquad 1 \le n \in \mathbb{N}\,.$$

MCINTOSH/MCINTOSH (1980) and LASCH (1987) discuss the practical handling of enzyme kinetics models. The comparative considerations of the errors that arise through the application of the usual graphic solution methods need to be emphasized here. Linearizations of (11.5), for example in LINEWEAVER/BURK (1934), for the purpose of parameter calculation are obsolete.

Growth as a time-dependent process is of interest for different natural sciences, especially in medicine. The efforts to mathematize the ideas of biological growth go back nearly 100 years (SCHARF 1981) farther than the beginnings of pharmacokinetics. PESCHEL/MENDE (1986, page 28) list differential equations for growth. They are derived from the hyperlogistical differential equation

$$x'(t) = E\,[x(t) - F]^k\,[G - \{x(t) - H\}^w]^n\ .$$

Besides classic compartment theory and the expansions indicated above, no examples could be found in pharmacokinetics. However, mathematical models of growth offer newer interpretation possibilities for pharmacokinetics. An example is observed:

A special case of the V.BERTALANFFY growth model (1941)

$$y' = Fy^n - Gy^m, \qquad F, G \in \mathbb{R}\,,$$

is the equation

$$y' = Fy - Gy^2$$

given by VERHULST (1838). This is somewhat transcribed with the symbols used for the concentration-time-function:

$$c'(t) = [F - Gc(t)]\, c(t) \ . \tag{11.6}$$

(10.1) is obtained for $G = 0$ and $F < 0$. The parameter G additionally offers the possibility to take a concentration dependent change in elimination into account. (11.6) is a BERNOULLI differential equation. The solution can be written in the form:

$$c(t) = \frac{FK}{GK + e^{-Ft}} \ ,$$

where

$$K = \frac{c(0)}{F - Gc(0)}$$

contains the initial condition $c(0)$. If delay should be taken into account, (11.6) can experience a modification:

$$c'(t) = [F - Gc(t - \tau)]c(t) \ ,$$

τ describes the delay. A function with initial conditions in the form of a function continuous on $[-\tau, 0]$ has to be formulated again for this differential equation with delay. The work of SCHUETTE (1979) contains propositions about the existence and uniqueness of the solutions of this initial value problem, as well as propositions concerning the stability and the monotonic behavior of distinguished cases.

The difficulties with regard to the application of differential equations with delay were pointed at above.

Chapter 12

Calculability and identifiability

As indicated at the end of Definition 10.3, the structural indentifiability of linear compartment models are of interest in the following sense:
A linear more-compartment model is predefined. The inhomogeneities, or input functions, or the initial conditions imply in which way and into which compartment the drug is applied. The time courses of concentrations of this drug are observable in certain compartments. Curve fitting to the measurements delivers a set of system parameters.
How many of the model parameters can be determined from the available system parameters? Are the model parameters uniquely defined by the system parameters? Does this assignment have continuity qualities that represent the so to speak counterpart of the continuous dependence of the solutions to the differential equation from the parameters?
Besides a list of literature regarding this topic, KUSUOKA/MAEDA/-KODAMA (1985) give a definition of local identifiability. They state that the identification problem formulated with regard to this is not generally solved for linear more-compartment models. Their results are fundamentally based on the application of a theorem of the classic analysis concerning the inverse function of the map between system and model parameters.
Certain classes of compartment models are observed. An example of a non-identifiable three-compartment model is represented in Figure 12.1.
VAJDA(1981) defines and studies structure equivalence of compartment models. The reciprocity theorem, or the proposition about implicit functions also come into play here. DELFORGE (1981) provides necessary and sufficient conditions for local identifiability in compartment models. KAIZU/HIRATA/KURATA (1983) published counterexamples of this. ANDERSON (1983) occupies himself with the problematic area of the identifiability of compartment models for mass balance equations. Due to the

initial conditions $c(0) = c^0 = DOS/V_d$, an additional parameter has to be introduced at the transition to the corresponding differential equations for concentrations. The illustrations from ANDERSON (1983, pages 103-104) and FELDMANN/SCHNEIDER(1976, page 250) concerning the identifiability of the two-compartment model are therefore only applicable with regard to time courses of masses.
A section of the book by BOURNE (1995) is left to the explanations of identifiability problems.

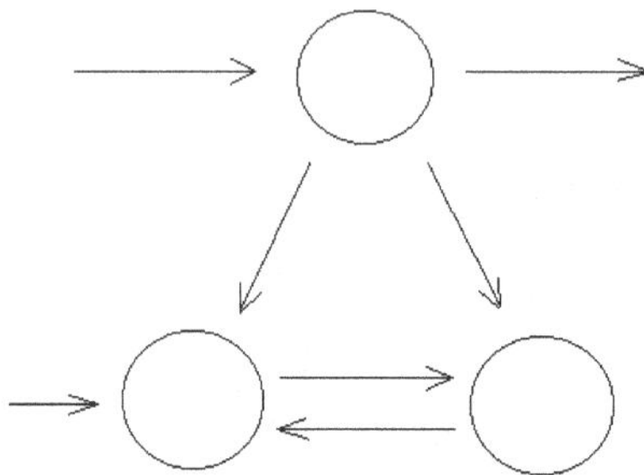

Fig. 12.1: Non-identifiable three-compartment model.

A complete overview will now be given for homogeneous initial value problems (10.3) concerning the correspondences between observable time courses of concentrations and the calculability of model parameters. Some terms are provided first.

Definition 12.1. *The determination whether the observation of drug concentration in the application compartment, in the alternative compartment or in both compartments is carried out is called* ***observation mode****. When nothing else is presumed, the application should be carried out in only one compartment. This is marked with the number* 1.

Definition 12.2. *A compartment model is called* ***permissible*** *when there is at least one $k_{ij} \neq 0$, and when for all i from $k_{ij} \neq 0$ it yields that: There is an m with $k_{mi} \neq 0$.*

Permissible models do not contain any compartments which have exits but no inputs. $c_i(t) = 0$ results if for compartment i the conditions $k_{ij} \neq 0$ are true for at least one j and $k_{mi} = 0$ for all m less than or equal to the

number of the compartments of the model. This compartment does not communicate with the rest.
An abbreviated description of the $2^4 = 16$ Two-compartment-iv models listed in Table 12.1 makes the following formulations easier.
It is agreed on that

$$\delta_{k_{ij}} = \begin{cases} 1 & \text{for } k_{ij} \neq 0 \\ 0 & \text{otherwise} \end{cases} .$$

Definition 12.3. *A two-compartment model* (10.3) *is described by the vector* $(k_{10}, k_{12}, k_{20}, k_{21})$ *of the constants of the model or by the natural number who's dual coding is* $(\delta_{k10}, \delta_{k12}, \delta_{k20}, \delta_{k21})$.

Proposition 12.1. *For the application in compartment* 1, *there are* 9 *permissible Two-compartment-iv models* (10.3).

Proof. $k_{12} \neq 0$ is true for 8 models. They are permissible and marked by the natural numbers 4, 5, 6, 7, 12, 13, 14, 15 . To that comes the proper one-compartment model which is coded by 8. □

With the use of the given coding scheme, 259 permissible models can be identified in the case of three compartments. The situation here is that each compartment communicates with every other compartment, that every compartment eliminates, and that the application of the drug is carried out in exactly one compartment. One proper One-compartment-iv model and two proper Two-compartment-iv models are special cases.
With regard to the description of pharmacokinetical processes, only simple examples should be used.

Definition 12.4. *A permissible compartment model is called* ***calculable regarding an observation mode*** *when all model parameters as well as the initial values can uniquely be determined from the system parameters.*

Proposition 12.2. *By observation of the time courses of concentration in compartment* 1, *there are exactly* 5 *calculable Two-compartment-iv models.*

Proof. Let $c^0 \neq 0$. This initial concentration has to be taken into account in addition to the model parameters. The 9 permissible models are examined:
Model 4 is calculable. $c_{1h}(t) = c^0 e^{-k_{12}t} = Ae^{at}$ is true because $\det K = 0$ and Lemma 10.6, Case 1.1.

Model 5 is calculable.

$$c_{1h}(t) = \frac{c^0 k_{12}}{k_{12}+k_{21}} \left[e^{-\{k_{12}+k_{21}\}t} + k_{21} \right]$$

is true because $\det K = 0$ and because of Lemma 10.6, Case 1.1. If $\hat{c}_{1h}(t) = Ae^{at} + B$ is fitted to the measurements, then $c^0 = -aA$, $k_{12} = -aA/(A+B)$ and $k_{21} = -aB/(A+B)$ arise.
Model 6 is not calculable. $c_{1h}(t) = c^0 e^{-k_{12}t}$ is true for unequal roots of the characteristic polynomial due to Lemma 10.4, case 1 and $\det K \neq 0$. If $\hat{c}(t) = Ae^{at}$ is fitted to the measurements, then c^0, k_{12} and k_{20} cannot be calculated uniquely from a and A. The assumption $\lambda_1 = \lambda_2$, which means $k_{12} = k_{20}$, leads to $c_{1h}(t) = c^0 e^{-k_{20}t}$ according to Lemma 10.5. There is a one-to-one correspondence between model and system parameters here.
Model 7 is calculable. One has $\det K \neq 0$ and $\lambda_1 \neq \lambda_2$. Double roots would require that certain additional k_{ij} be zero. With regard to Lemma 10.4, Case 2,

$$c_{1h}(t) = \frac{c^0}{\lambda_2 - \lambda_1} \left[(\lambda_2 + k_{12}) e^{\lambda_1 t} - (\lambda_1 + k_{12}) e^{\lambda_2 t} \right]$$

is true. The system parameters from $\hat{c}(t) = Ae^{at} + Be^{bt}$, with use of the relations $\lambda_1 \lambda_2 = k_{20} k_{12}$ and $\lambda_1 + \lambda_2 = -(k_{12} + k_{20} + k_{21})$, allow for the unique calculation of the model parameters:

$$\begin{aligned} k_{12} &= \frac{-(aA + bB)}{A+B}, \\ k_{20} &= \frac{-ab(A+B)}{aA+bB}, \\ k_{21} &= \frac{-(a-b)^2 AB}{(A+B)(aA+bB)} \quad \text{and} \\ c^0 &= A + B . \end{aligned}$$

Model 8 is calculable. It is the proper One-compartment-iv model.
Model 12 is not calculable. $c_{1h}(t) = c^0 e^{-(k_{10}+k_{12})t}$ is true due to Lemma 10.6, Case 1.1. and $\det K = 0$. The 2 system parameters calculated from the observations do not allow for the unique determination of c^0, k_{10} and k_{12}.

Model 13 is calculable. It can be handled like model 7. The equations are

$$\begin{aligned} k_{10} &= \frac{-ab\,(A+B)}{bA+aB}\,, \\ k_{12} &= \frac{-\,(a-b)^2\,AB}{(A+B)\,(bA+aB)}\,, \\ k_{21} &= \frac{-\,(bA+aB)}{A+B} \quad \text{and} \\ c^0 &= A+B\,. \end{aligned}$$

Model 14 is not calculable. It is handled like model 7. One has $c_{1h}(t) = c^0 e^{-(k_{10}+k_{12})t}$ and c^0, k_{10} and k_{12} cannot be uniquely calculated from the 2 system parameters.
Model 15 is not calculable. Four model parameters and additional c^0 contrast with the maximum obtainable number of 4 system parameters.
It can be realized without effort that the denominators of the calculations given for models 5, 7, and 13 are not equal to zero. □

Definition 12.5. *A permissible model is called* ***identifiable*** *when exactly one calculable model exists with regard to a given observation mode and predefined system parameters.*

Proposition 12.3. *With regard to the observation in compartment* 1*, there is exactly one identifiable Two-compartment-iv model* (10.3)*. This is the model* $(0, k_{12}, 0, k_{21})$ *which does not take elimination into account.*

Proof. With regard to the preceding proposition it is determined that:

(1) System parameters a, A, b, B are predefined. Then $(0, k_{12}, k_{20}, k_{21})$ as well as $(k_{10}, k_{12}, 0, k_{21})$ are associated calculable models.
(2) System parameters a and A are predefined. Then $(0, k_{12}, 0, 0)$ and $(k_{10}, 0, 0, 0)$ are associated calculable models.
(3) System parameters a, A, B are predefined. Then $(0, k_{12}, 0, k_{21})$ is the only associated calculable model: In accordance with the complete solution of the homogeneous initial value problem (10.3), $c_{1h}(t) = Ae^{at} + B$ is true due to Lemma 10.6, Cases 1.1 and 1.2. It is also derived from Lemma 10.4 when one of the roots of the characteristic polynomial becomes zero. The latter is equivalent to $\det K = 0$. Therefore, only $c_{1h}(t)$ has to be taken into account with regard to Lemma 10.6. Because $c^0 \neq 0$, only the three models listed at the end of Lemma 10.6 (at least two of the $k_{ij} = 0$ have to be zero) can be considered as calculable. Since $k_{10} + k_{12} = A \neq 0$ and $k_{20} + k_{21} = B \neq 0$ are needed (see Lemma

10.6), $(0, k_{12}, 0, k_{21})$ is left as the only calculable Two-compartment-iv model that is associated with the system parameters a, A, B. □

The One-compartment-iv model (10.1) and the One-compartment-ev model (10.2) are identifiable. This does not need to be justified any further.

Proposition 12.4. *There are exactly two calculable Two-compartment-iv models with reference to observation in compartment* 2. *They are also identifiable. These models are* $(0, k_{12}, 0, 0)$ *and* $(0, k_{12}, k_{20}, 0)$.

The proof follows according to that of the two previous propositions and can be avoided here.
With respect to the system parameters a, A, b, B, the models $(0, k_{12}, k_{20}, k_{21})$ and $(k_{10}, k_{12}, 0, k_{21})$ aren't identifiable in case of observation in compartment 1.
The investigation of monotonicity of $c_{1h}(t)$ also does not provide any more clues.

Proposition 12.5. *With regard to observation in compartment* 1, *the Two-compartment-iv models* $(0, k_{12}, k_{20}, k_{21})$ *and* $(k_{10}, k_{12}, 0, k_{21})$ *are not differentiable with regard to the monotonic behavior of the time courses of concentration* $c_{1h}(t)$.

Proof. $c'_{1h}(t) < 0$ is shown respectively. For the second model from Lemma 10.4, Case 1, one has

$$c'_{1h}(t) = \frac{c^0}{\lambda_1 - \lambda_2}\left[\lambda_1(\lambda_1 + k_{21})e^{\lambda_1 t} - \lambda_2(\lambda_2 + k_{21})e^{\lambda_2 t}\right] .$$

Let $\lambda_1 < \lambda_2 < 0$. Then $c'_{1h}(t) < 0$ is true exactly when $\lambda_1(\lambda_1 + k_{21})e^{\lambda_1 t} > \lambda_2(\lambda_2 + k_{21})e^{\lambda_2 t}$. From the representation of λ_i given in the proof of Lemma 10.3 follow $\lambda_1 + k_{21} < 0$ and $\lambda_2 + k_{21} > 0$, consequently, $c'_{1h}(t) < 0$. For the first model, the proposition arises from the analogous considerations; $\lambda_1 + k_{12}$ and $\lambda_2 + k_{12}$ are shown in different signs. □

The above propositions relay that the information gained about the Two-compartment-iv model by observations in compartment 1 and compartment 2 varies. This also applies to the number of respectively calculable parameters.
BENET (1972) observes so-called mammilary models; i.e. every compartment exclusively communicates with a central compartment which is the only eliminating one, and in which the applied substance is observed. Without proof, he gives $(2n - 1)$ as the maximum number of calculable model

parameters k_{ij}, n is the number of compartments. This does not contradict the above propositions.

The explanations however require the correction in MEIER/RETTIG/HESS (1981, pages 426 and 427) with regard to the number of calculable model parameters when observing one or more compartments.

The result from CHAU (1977) is remarkable. He notes that for non-calculable Two-compartment-iv models (10.3), the uncertain model parameters can be restricted by inequalities. The thresholds in these inequalities are given with the system parameters.

The propositions concerning local calculability and uniqueness are given by CHERRAULT/GUILLEZ (1981) under the assumptions that the roots λ_j of the characteristic polynomial of the observed n-compartment model are different from each other and every function $c_i(t), i = 1, \ldots, n$, is the sum of terms of the form $A_{ij}e^{\lambda_j t}$. It is clear that these prerequisites are restrictions of the generality of the propositions.

Generalizations of Proposition 13.2 to Proposition 13.5 cannot be expected for n-compartment models. The distinction of cases with regard to k_{ij} already become confusing for $n = 3$. The roots of the characteristic polynomial are not easy to handle. These are no longer explicit functions of the k_{ij} for $n > 3$.

An iv pharmacokinetic experiment as well as an ev pharmacokinetic experiment are required for the determination of parameters for Two-compartment-ev models (GIBSON/TAYLOR/COLBURN 1987). This topic will not be looked at in depth here.

The results of this section are summarized in Table 12.1.

Table 12.1: Two-compartment-iv models: Overview of permissibility, calculability and identifiability.

Model	k_{10}	k_{12}	k_{20}	k_{21}	permissible	calculable with observ. in	
						Comp. 1	Comp. 2
0	0	0	0	0	no		
1	0	0	0	1	no		
2	0	0	1	0	no		
3	0	0	1	1	no		
4	0	1	0	0	yes	yes	yes, ident
5	0	1	0	1	yes	yes, ident	no
6	0	1	1	0	yes	no	yes, ident
7	0	1	1	1	yes	yes	no
8	1	0	0	0	yes	yes	no
9	1	0	0	1	no		
10	1	0	1	0	no		
11	1	0	1	1	no		
12	1	1	0	0	yes	no	no
13	1	1	0	1	yes	yes	no
14	1	1	1	0	yes	no	no
15	1	1	1	1	yes	no	no

Chapter 13

Compartment models and associated residence time distributions

The stochastic modeling of kinetic processes is inaugurated through two different considerations:

- the interactions between a drug and an organism are considered to be random events,
- the measurements are considered to be erroneous.

The possibility of bringing the classic compartment models of pharmacokinetics in connection with a stochastic mathematical model is examined. This leads to residence time distributions. Its qualities are represented in Chapter 13.1. The truncation problem is discussed in Chapter 13.2. Parameter estimation for the residence time distributions is found in Chapter 14.3. Residence time distributions and their connection to compartment models were studied under different points of view (see JACQUEZ 1988, VENG-PEDERSEN 1991, BIEBLER 1989, CHENG/GILLESPIE/JUSKO 1994, YU/WEHRLY 2004 for example). A complete description of the associated distributions is given for the Two-compartment model for iv bolus administration in this section.
Statistical design of experiments and investigation of robustness are not dealt with here. Examples of the processing of such questions are given by SCHALL/LUUS (1992), or TOD/ROCCHISANI (1996). The latter deals with D-optimal experimental plans.

13.1 Unbounded residence times

Every non-negative continuous real function whose integral over $\mathbb{R}$ equals 1 defines a probability distribution. The solutions of linear differential equations treated in Chapter 10 in connection with compartment models are real non-negative continuous functions defined on $[0, \infty)$.

Definition 13.1. *Let $c(t) = c(t, a, b, A, B)$ be a function that corresponds with one of the observed compartment models, is dependent on system parameters a, b, A, B, and integratable on $[0, \infty)$. Then*

$$f_c(t) = \frac{c(t)}{AUC} \quad \text{with } AUC = \int_0^\infty c(t)\,\mathrm{d}t < \infty$$

*denotes the **standardized concentration-time-function** of $c(t)$.*

The fact that $f_c(t)$ is continuous for the given $c(t, a, b, A, B)$ does not require any more consideration. Not every function $c(t)$ associated with a compartment model can be assigned a standardized function $f_c(t)$: The function $c(t) = A\mathrm{e}^{\lambda_1 t} + B$ from Lemma 10.6 is only locally integratable.
A random variable has to be defined in a suitable way in order to be able to probabilistically interpret the standardized concentration-time-functions of compartment models. The propositions refer to the measurable spaces $(\mathbb{R}^+, \mathcal{B})$, $\mathcal{B}$ the Borel algebra, or $(\mathbb{R}^{n+}, \mathcal{B}^n)$.

Definition 13.2. *The duration of presence, synonymously: residence time, of a drug molecule in an organism is regarded to be a random variable X. With respect to a compartment model, let $m_e(t)$ denote the drug quantity of applied dose DOS eliminated from the organism up to time t. The probability distribution of X, the residence time distribution, is defined as*

$$F_X(t) = Prob\,(X < t) = \frac{m_e\,(t)}{DOS}\,.$$

The density is denoted by $f_X(t)$.

Agreement

Probability distribution and density are defined on all $\mathbb{R}$. Without it being expressed, the courses of concentration over time $c(t)$ which are brought into relation are therefore thought to be extended from the positive axis to the whole spectrum of real numbers with the value zero. The same technical simplifications concern the derived probability distributions, the density derived from $c(t)$, as well as other functions.
How much Definition 13.2 makes sense and how much it can be the basis of the evaluation of a pharmacokinetic experiment has to be examined for the respective model.

Proposition 13.1. *For a One-compartment-iv model* (10.1), *the residence time of a drug molecule in the organism is exponentially distributed. $f_X(t) = f_c(t)$ is true.*

Proof. $c(t) = c^0 \mathrm{e}^{-k_{10}t}$ with $c^0 = DOS/V_d$ is true with regard to the model mentioned in Lemma 10.1. Multiplication by the distribution volume V_d yields the amount of substance $m(t) = V_d c(t)$. The balance equation $DOS = m_e(t) + m(t)$ gives $m_e(t) = DOS - m(t)$, $F_X(t) = 1 - \mathrm{e}^{-k_{10}t}$, and $f_X(t) = f_c(t) = k_{10}\mathrm{e}^{-k_{10}t}$ is easy to check. □

This compartment model then refers to a dose independent residence time distribution, whereby the density function is immediately available by standardization of the time course of concentration.

Proposition 13.2. *The following is true under the assumption $k_i \neq k_{el}$ for the One-compartment-ev model* (10.2)*:*

(1) *The random variable X has the distribution function:*

$$F_X(t) = 1 + \frac{k_{el}k_i}{k_{el} - k_i}\left(\frac{e^{-k_{el}t}}{k_{el}} - \frac{e^{-k_it}}{k_i}\right).$$

(2) *$f_X(t) = f_c(t)$ is true for the density function.*
(3) *$f_X(t)$ is the linear combination of two densities of exponential distributions.*

Proof. According to the pharmacokinetic model, $m_e'(t) = k_{el}m(t)$, $m(t) = V_d c(t) = DOS\ k_i[\mathrm{e}^{-k_it} - \mathrm{e}^{-k_{el}t}]/(k_{el} - k_i)$, with $V_d = DOS/c_i^0$. Integration leads to

$$m_e(t) = DOS + DOS\frac{k_{el}k_i}{k_{el} - k_i}\left(\frac{e^{-k_{el}t}}{k_{el}} - \frac{e^{-k_it}}{k_i}\right),$$

whereby the integration constant DOS yields from the initial condition $m_e(0) = 0$. The given function $F_X(t)$ is obtained with regard to Definition 13.2. It is continuous again as a sum of continuous functions and is monotonically non-decreasing because

$$F_X'(t) = f_X(t) = \frac{k_{el}k_i}{k_{el} - k_i}\left(e^{-k_it} - e^{-k_{el}t}\right) > 0$$

except in the trivial case. In addition, since $F_X(0) = 0$ and $F_X(\infty) = 1$, $F_X(t)$ is a distribution function. While observing $AUC = c_i^0/k_{el}$, $f_X(t) = f_c(t)$ becomes visible. This density is a linear combination; $f_X(t) = Kk_i\mathrm{e}^{-k_it} + (1-K)k_{el}\mathrm{e}^{-k_{el}t}$, $K = k_{el}/(k_{el} - k_i)$. □

Proposition 13.3. *The following is true under the assumption $k_i = k_{el} = k$ for the One-compartment-ev model* (10.2)*:*

(1) *The random variable X is Gamma distributed and has the distribution function*

$$F_X(t) = 1 - (kt + 1)\, e^{-kt} \; .$$

(2) *The following is true for the density function:*

$$f_X(t) = f_c(t) = k^2 t e^{-kt} \; .$$

Proof. The proof follows the same considerations as the ones of the previous proposition. □

The three previous propositions mean that the stochastic model is meaningfully in relation to one-compartment models (10.1) and (10.2). It should be stressed that the distribution functions are independent from the applied drug quantity DOS and that the densities correspond with the observable time courses of concentration $c(t)$. This is fundamentally based on the qualities of the pharmacokinetic model:
Instead of the relation $c'(t) = -k_{el}c(t)$, consider a so-called elimination process of 0th order $c'(t) = -k_{el}$, $c(0) = c^0$. The solution $c(t)$ only applies to the interval $[0, c^0/k_{el}]$ where it isn't negative. $m_c(t) = V_d c(t) = DOS - k_{el}V_d t$, $m_e(t) = k_{el}V_d t$, $F_X(t) = V_d k_{el} t/DOS$ and $f_X(t) = V_d k_{el}/DOS$ yield from $c(t) = c^0 - k_{el}t$. Function values $F_X(t) = 1$ and $f_X(t) = 0$ are set for $t > c^0/k_{el}$. The distribution is dependent on DOS. In addition, $f_c(t) \neq f_X(t)$ is true because $f_c(t) = 2(1 - k_{el}t/c^0)k_{el}/c^0$.
The attempt $c'(t) = -k_{el}c^2(t)$, $c(0) = c^0$ leads to $c(t) = c^0/(c^0 k_{el} t + 1)$, $F_X(t) = 1 - 1/(DOSk_{el}t + 1)$ and $f_X(t) = DOSk_{el}/(DOSk_{el}t + 1)^2$. But $f_c(t)$ does not exist because $c(t)$ is only locally integratable.
These examples prove that the proposition "The time course of drug concentration in plasma can usually be regarded as a statistical distribution curve." given in GIBALDI/PERRIER (1982, page 410) is incorrect. VENG-PEDERSEN (1989) also is not precise enough in his explanations in Section 3. 'Statistical Moments'. The function $c(t)/AUC$ does not necessarily has the property (2) given there.
The associated residence time distributions are derived for the Two-compartment-iv model (10.3). Administration is carried out in compartment 1. Of the nine permissible models, seven are still of interest since $(0, k_{12}, 0, 0)$ and $(0, k_{12}, 0, k_{21})$ describe processes without elimination. The residence time cannot be looked at as a random variable anymore for these models. Case $(k_{10}, 0, 0, 0)$ was already dealt with. It is intuitively clear that $F_X(t)$ can't be a probability distribution for $(k_{10}, k_{12}, 0, 0)$. The

applied dose is only partly eliminated from the organism. The rest remains in compartment 2. One can calculate that $F_X(t)$ is truly limited by $k_{10}/(k_{10}+k_{12}) < 1$ with regard to Lemma 10.6, Case 1.1. The remaining five cases have to be dealt with in greater detail.

Proposition 13.4. $(0, k_{12}, k_{20}, 0)$ *is a Two-compartment-iv model* (10.3). *Then the following is true under the assumption* $k_{12} \neq k_{20}$ *for administration in compartment* 1*:*

(1) *The random variable* X *has the distribution function*
$$F_X(t) = 1 + \frac{k_{12}k_{20}}{k_{20}-k_{12}}\left(\frac{e^{-k_{20}t}}{k_{20}} - \frac{e^{-k_{12}t}}{k_{12}}\right).$$
(2) $f_X(t) = f_{c_2}(t)$ *is true for the density function.*
(3) $f_X(t)$ *is a linear combination of the densities of two exponential distributions.*

Proof. Lemma 10.4, case 1 yields
$$c_2(t) = \frac{c^0 k_{12}}{k_{20}-k_{12}}\left(e^{-k_{12}t} - e^{-k_{20}t}\right)$$
because $\det K \neq 0$ and $k_{12} \neq k_{20}$. The function that describes the elimination amount of the substance is defined by $m_e'(t) = k_{20}V_d c_2(t)$. V is the distribution volume with regard to $c^0 = DOS/V_d$. With that, $f_X(t) = F_X'(t) = m_e'(t)/DOS = k_{12}k_{20}[\mathrm{e}^{-k_{12}t} - \mathrm{e}^{-k_{20}t}]/(k_{20}-k_{12}) = c_2(t)/AUC = f_{c_2}(t)$ with $AUC = c^0/k_{20}$. Integration with regard to the starting condition $F_X(0) = 0$ yields the function $F_X(t)$ which is mentioned in the proposition. The fact that it is a distribution function yields from $F_X(0) = 0, F_X(\infty) = 1, F_X(t)$ is continuous and $F_X(t)$ is monotonically non-decreasing because $f_X(t) > 0$. Finally, $f_X(t) = Kk_{12}\mathrm{e}^{-k_{12}t} + (1-K)k_{20}\mathrm{e}^{-k_{20}t}$ for $K = k_{20}/(k_{20}-k_{12})$ is true. □

Observation in compartment 2 does not mean only the calculability and identifiabiliy of the model $(0, k_{12}, k_{20}, 0)$ (compare Table 12.1), but also knowledge of the residence time distribution. No such information can be obtained from observations in compartment 1 alone.

Proposition 13.5. $(0, k_{12}, k_{20}, 0)$ *is a Two-compartment-iv model* (10.3). *Then the following is true under the assumption* $k_{12} = k_{20} = k$ *at administration in compartment* 1*:*

(1) *The random variable* X *is Gamma distributed and has the distribution function*
$$F_X(t) = 1 - (1 + kt)\,e^{-kt}.$$

(2) $f_X(t) = f_{c_2}(t) = k^2 te^{-kt}$ *is true for the density.*

Proof. Under the given assumptions, $\det K \neq 0$ and k is a double root of the characteristic polynomial given in (10.3). The assertions can be derived from $c_2(t) = c^0 kte^{-kt}$ with regard to Lemma 10.6. □

The analogies of Propositions 13.2 and 13.4 as well as 13.3 and 13.5 are conspicuous. The Two-compartment-iv model that is discussed can also be interpreted as a One-compartment-ev model.

Proposition 13.6. $(0, k_{12}, k_{20}, k_{21})$ *is a Two-compartment-iv model* (10.3). *Then the following is true for administration in compartment* 1*:*

(1) *The random variable* X *has the distribution function*

$$F_X(t) = 1 - \frac{\lambda_2 e^{\lambda_1 t} - \lambda_1 e^{\lambda_2 t}}{\lambda_2 - \lambda_1}.$$

(2) $f_X(t) = f_{c_2}(t)$ *is true for the density.*

(3) $f_X(t)$ *is a linear combination of the densities of two exponential distributions.*

Proof. $\det K \neq 0$ is true for the model. Double roots cannot appear. Just like it can be seen in the proof for Lemma 10.3, the consequence of this would be $k_{12} = 0$ or $k_{21} = 0$. $c_2(t) = c^0 k_{12}[e^{\lambda_1 t} - e^{\lambda_2 t}]/(\lambda_1 - \lambda_2)$ is true due to Lemma 10.4, Case 1. $m_e(t)$ is defined through $m_e'(t) = k_{20} V c_2(t)$. With that, $f_X(t) = F_X'(t) = \lambda_1\lambda_2[e^{\lambda_1 t} - e^{\lambda_2 t}]/(\lambda_1 - \lambda_2) = c_2(t)/AUC = f_{c_2}(t)$ where $AUC = c^0 k_{12}/(\lambda_1\lambda_2)$. The given function $F_X(t)$ can be obtained by integration. $F_X(0) = 0$ is the initial condition. $F_X(0) = 0$ and $F_X(\infty) = 1$ are true. $F_X(t)$ is continuous and $F_X(t)$ is monotonically non-decreasing because $f_X(t) > 0$. $F_X(t)$ is therefore a distribution function. Furthermore, the identity $f_X(t) = K(-\lambda_1)e^{\lambda_1 t} + (1-K)(-\lambda_2)e^{\lambda_2 t}$ exists for $K = \lambda_2/(\lambda_2 - \lambda_1)$. Take into account that $\lambda_i < 0$ (Lemma 10.3). □

Proposition 13.7. $(k_{10}, k_{12}, 0, k_{21})$ *is a Two-compartment-iv model* (10.3). *Then the following is true for administration in compartment* 1*:*

(1) *The random variable* X *has the distribution function*

$$F_X(t) = 1 - \frac{k_{10}}{\lambda_2 - \lambda_1}\left[\left(1 + \frac{k_{21}}{\lambda_1}\right)e^{\lambda_1 t} - \left(1 + \frac{k_{21}}{\lambda_2}\right)e^{\lambda_2 t}\right].$$

(2) $f_X(t) = f_{c_1}(t)$ *is true for the density.*

(3) $f_X(t)$ *is a linear combination of the densities of two exponential distributions.*

Proof. $\det K \neq 0$ is true for the model. Double roots of the characteristic polynomial do not appear because $k_{12} \neq 0$ and $k_{21} \neq 0$ (compare with the proof from Lemma 10.3). $c_1(t) = c^0[(\lambda_1+k_{21})\mathrm{e}^{\lambda_1 t}-(\lambda_2+k_{21})\mathrm{e}^{\lambda_2 t}]/(\lambda_1-\lambda_2)$ is given by Lemma 10.4, Case 1. $m_e(t)$ is defined by $m_e'(t) = k_{10}Vc_1(t)$. From this one gets

$$f_X(t) = \frac{k_{10}}{\lambda_1 - \lambda_2}\left[(\lambda_1 + k_{21})\,e^{\lambda_1 t} - (\lambda_2 + k_{21})\,e^{\lambda_2 t}\right] .$$

Integration and the initial condition $F_X(0) = 0$ yield the function $F_X(t)$. $F_X(0) = 0$ and $F_X(\infty) = 1$ are true. $F_X(t)$ is continuous and $F_X(t)$ is monotonically non-decreasing because $f_X(t) > 0$. $F_X(t)$ is therefore a distribution function. Furthermore, $f_X(t) = c_1(t)/AUC$ with $AUC = c^0/k_{10}$ as well as $f_X(t) = K(-\lambda_1)\mathrm{e}^{\lambda_1 t} + (1-K)(-\lambda_2)\mathrm{e}^{\lambda_2 t}$ with $K = (k_{10} + \lambda_2)/(\lambda_2 - \lambda_1)$. Take into account that $\lambda_i < 0$ and $\lambda_1\lambda_2 = k_{10}k_{21}$. □

Proposition 13.8. *$(k_{10}, k_{12}, k_{20}, 0)$ is a Two-compartment-iv model* (10.3). *Then the following is true under the assumption $\lambda_1 \neq \lambda_2$ for administration in compartment* 1*:*

(1) *The random variable X has the distribution function*
$$F_X(t) = 1 - \frac{1}{\lambda_2 - \lambda_1}\left[(\lambda_2 + k_{10})\,e^{\lambda_1 t} - (\lambda_1 + k_{10})\,e^{\lambda_2 t}\right] .$$
(2) *$f_X(t) = Lf_{c_1}(t)+(1-L)f_{c_2}(t)$ with $L = -k_{10}/\lambda_1$ is true for the density.*
(3) *$f_X(t)$ is a linear combination of the densities of two exponential distributions.*

Proof. $\det K \neq 0$ and $\lambda_1 = -(k_{10} + k_{12})$, $\lambda_2 = -k_{20}$ are true under the given assumptions. Appropriately with respect to Lemma 10.4, Case 1, $c_1(t) = c^0\mathrm{e}^{\lambda_1 t}$ and $c_2(t) = c^0 k_{12}[\mathrm{e}^{\lambda_1 t} - \mathrm{e}^{\lambda_2 t}]/(\lambda_1 - \lambda_2)$. The following is obtained from the attempt $m_e'(t) = k_{10}V_d c_1(t) + k_{20}V_d c_2(t)$:

$$f_X(t) = \frac{1}{\lambda_1 - \lambda_2}\left[\lambda_1(\lambda_2 + k_{10})\,e^{\lambda_1 t} - \lambda_2(\lambda_1 + k_{10})\,e^{\lambda_2 t}\right] .$$

$\lambda_1\lambda_2 = k_{10}k_{20} + k_{20}k_{12}$ is used here. Integration of $f_X(t)$ and the initial condition $F_X(0) = 0$ yields the given function $F_X(t)$. It is a distribution function. $f_{c_1}(t) = -\lambda_1\mathrm{e}^{\lambda_1 t}$ and $f_{c_2}(t) = \lambda_1\lambda_2[\mathrm{e}^{\lambda_1 t} - \mathrm{e}^{\lambda_2 t}]/(\lambda_1 - \lambda_2)$ because of $AUC1 = -c^0/\lambda_1$ and $AUC2 = c^0 k_{12}/(\lambda_1\lambda_2)$. Statement 2 can be checked with this. After all, $f_X(t) = K(-\lambda_1)\mathrm{e}^{\lambda_1 t} + (1-K)(-\lambda_2)\mathrm{e}^{\lambda_2 t}$ is true for $K = (k_{10} + \lambda_2)/(\lambda_2 - \lambda_1)$. □

Proposition 13.9. *$(k_{10}, k_{12}, k_{20}, 0)$ is a Two-compartment model* (10.3). *Then the following is true under the assumption $\lambda_1 = \lambda_2$ for administration in compartment* 1*:*

(1) *The random variable X has the distribution function*

$$F_X(t) = 1 - (k_{12}t + 1)e^{-k_{20}t}\,.$$

(2) $f_X(t) = Lf_{c_1}(t) + (1 - L)f_{c_2}(t)$ *is true for the density,* $L = k_{10}/(k_{10} + k_{12})$.

(3) $f_x(t)$ *is a convex linear combination of the densities of a exponential distribution and a Gamma distribution.*

Proof. $\det K \neq 0$ and $k_{20} = k_{10} + k_{12}$ are true under the given assumptions. $c_1(t) = c^0 e^{-k_{20}t}$ and $c_2(t) = c^0 k_{12} t e^{-k_{20}t}$ with regard to Lemma 10.5, Case 1. $f_X(t) = (k_{12}k_{20}t + k_{10})e^{-k_{20}t}$ yields from the attempt $m_e'(t) = k_{10}V_d c_1(t) + k_{20}V_d c_2(t)$. Integration and observation of $F_X(0) = 0$ yields the above function $F_X(t)$. It does not need to be shown in greater detail here that it is a distribution function. The integrals $AUC1 = c^0/k_{20}^2$ and $AUC2 = c^0 k_{12}/k_{20}^2$ lead to $f_{c_1}(t) = k_{20}e^{-k_{20}t}$ as well as $f_{c_2}(t) = k_{20}^2 t e^{-k_{20}t}$. Comparisons of coefficients between f_X and f_{c_1} as well as f_{c_2} yield the relation formulated in 2. The linear combination is convex because $0 < L = k_{10}/(k_{10} + k_{12}) < 1$. □

Proposition 13.10. $(k_{10}, k_{12}, k_{20}, k_{21})$ *is a Two-compartment-iv model. Then the following is true for administration in compartment* 1*:*

(1) *The random variable $\underline{X}$ has the distribution function*

$$F_X(t) = 1 - \frac{1}{\lambda_2 - \lambda_1}\left[(\lambda_2 + k_{10})\,e^{\lambda_1 t} - (\lambda_1 + k_{10})\,e^{\lambda_2 t}\right]\,.$$

(2) $f_X(t) = Lf_{c_1}(t) + (1 - L)f_{c_2}(t)$, $L = k_{10}(k_{20} + k_{21})/(\lambda_1\lambda_2)$ *is true for the density.*

(3) $f_X(t)$ *is a linear combination of the densities of two exponential distributions.*

Proof. $\det K \neq 0$ as well as $\lambda_1 \neq \lambda_2$ are true under the given assumptions. If a double root appears, $k_{12} = 0$ or $k_{21} = 0$ would be necessary (see the proof for Lemma 10.3). Consequently, $c_1(t)$ and $c_2(t)$ are obtainable with regard to Lemma 10.4, Case 1. $m_e(t)$ is given by the attempt $m_e'(t) = k_{10}V_d c_1(t) + k_{20}V_d c_2(t)$. This yields

$$f_X(t) = \frac{1}{\lambda_1 - \lambda_2}\left[(\lambda_2 + k_{10})\,\lambda_1 e^{\lambda_1 t} - (\lambda_1 + k_{10})\,\lambda_2 e^{\lambda_2 t}\right]\,.$$

$F_X(t)$ arises in the above form by integration. $F_X(0) = 0$ is the initial condition. It does not need to be demonstrated here that $F_X(t)$ is a distribution function. The integrals $AUC1 = c^0(k_{20} + k_{21})/(\lambda_1\lambda_2)$ and $AUC2 = c^0 k_{12}/(\lambda_1\lambda_2)$ are involved in the determination of

$$f_{c_1}(t) = \frac{\lambda_1\lambda_2}{\lambda_1 - \lambda_2}\left[(\lambda_1 + k_{20} + k_{21})\, e^{\lambda_1 t} - (\lambda_2 + k_{20} + k_{21})\, e^{\lambda_2 t}\right]$$

as well as

$$f_{c_2}(t) = \frac{\lambda_1\lambda_2}{\lambda_1 - \lambda_2}\left(e^{\lambda_1 t} - e^{\lambda_2 t}\right) .$$

Statement 2 can be checked with this. After all, $f_X(t) = K(-\lambda_1)\mathrm{e}^{\lambda_1 t} + (1-K)(-\lambda_2)\mathrm{e}^{\lambda_2 t}$ is true for $K = -(k_{10} + \lambda_2)/(\lambda_1 - \lambda_2)$. □

13.2 Properties of distributions of unbounded residence times

Associated residence time distributions were derived for the Two-compartment-iv models. Not every one of these models corresponds with such a distribution. A stochastic description of a pharmacokinetic process proves to be well interpretable. It is interesting in itself that linear combinations of the densities of exponential distributions, or the densities of special Gamma distributions, can be understood as a density of a random variable that can be clearly described. As explained in Proposition 13.3, time courses of concentration do not necessarily define residence time distributions. Pharmacokinetics literature about this topic is found to be vague. On the other hand, a family of probability distributions is the starting point of the mathematical modeling of pharmacokinetic processes. For example, WEISS (1983) studies Gamma distributed residence times and PIOTROVSKII (1987) deals with WEIBULL distributed residence times. Adequate conditions should be formulated in order to bring such ideas into connection with observable time courses of concentrations.

Proposition 13.11. *Let the residence time X of a molecule in an organism be a continuously distributed random variable with density $f_X(t)$. Suppose elimination of the drug is only carried out in the observation compartment, mass and concentration are connected by the equation $m(t) = V_d c(t)$, V_d is a distribution volume, and elimination can be described by $m_e'(t) = k_{el} m(t)$. Here k_{el} is an elimination constant. Then the following is true:*
For every density $f_X(t)$, a time course of concentration $c(t)$ exists such that $f_X(t) = f_c(t)$.

Proof. With regard to Definition 13.2, $f_X(t) = m_e'(t)/DOS$ is true. $c(t) = DOSf_X(t)/(V_d k_{el})$ then immediately follows from the assumption that $m_e'(t) = k_{el}m(t)$. □

The proposition from PIOTROVSKII (1987, formula 7) is rendered more precisely by Proposition 13.11 with regard to the agreement of $f_c(t)$ and $f_X(t)$ as well as their required assumptions. Linearity of the compartment model, or the corresponding differential equation, is therefore not required. The pharmacokinetic interpretation of WEIBULL- and Gamma distributed residence times therefore requires a look at the coherences between the stochastic and a suitable pharmacokinetic model of thought. As shown in Chapter 13.1, only densities $f_X(t) = k^2 t\mathrm{e}^{-kt}$ are to be observed with regard to the classes of WEIBULL and Gamma probability distributions in the context of the Two-compartment-iv model. They appear in the case of double roots of the characteristic polynomial and indicate special cases of the model.

What can be said about the monotonic behavior of the probability densities examined here?

Proposition 13.12. *The residence time densities associated with the Two-compartment-iv models* $(0, k_{12}, k_{20}, 0)$ *and* $(0, k_{12}, k_{20}, k_{21})$ *take on exactly one maximum on* $[0, \infty)$.

Proof. The densities are given in the proofs of Propositions 13.4, 13.5 and 13.6. The derivatives have to be examined. □

Proposition 13.13. *The residence time density associated with the Two-compartment-iv model* $(k_{10}, k_{12}, 0, k_{21})$ *is a strictly monotone decreasing function.*

Proof. $f_X(t) = f_{c1}(t)$ with regard to Proposition 13.7. The assertion was proven in Proposition 12.5. □

Proposition 13.14. *The residence time densities associated with the Two-compartment-iv models* $(k_{10}, k_{12}, k_{20}, 0)$ *and* $(k_{10}, k_{12}, k_{20}, k_{21})$ *show varying monotonic behavior dependent on* k_{ij}.

Proof. The corresponding examples are only given for the simpler of the models. If $k_{10} = 4, k_{12} = 2$ and $k_{20} = 1$, then $f_X(t) = 18\mathrm{e}^{-6t}/5 + 2\mathrm{e}^{-t}/5$ decreases monotonically on $[0, \infty)$. $f_X(t) = 9\mathrm{e}^{-3t} - 8\mathrm{e}^{-4t}$ takes on a maximum in $t^\star = \ln(32/27) > 0$ for $k_{10} = 1, k_{12} = 2$ and $k_{20} = 4$. □

The convex linear combination of monotonically decreasing functions is monotonically decreasing again.

Proposition 13.15. *Let $f(t), g(t)$ and $h(t)$ be strictly monotonically decreasing real functions and $f(t) = Kg(t) + (1 - K)h(t)$. Then $0 \leq K \leq 1$ isn't necessarily valid.*

Proof. $g(t) = L\mathrm{e}^{-Lt}$ and $h(t) = M\mathrm{e}^{-Mt}$ are used as examples. $f(t)$ monotonically strictly decreases for $K = 2, L = 3$ and $M = 4$. □

In contrast to formula (9) in PIOTROVSKII (1987), these explanations show that linear combinations of exponential terms related to densities of residence time distributions do not necessarily have to be convex.
The idea to use the standardized solutions of differential equations as densities and to classify probability distributions in this way possibly goes back to K. PEARSON (1894). In the quoted work the differential equation

$$f'(t) = \frac{K + t}{L_0 + L_1 t + L_2 t^2} f(t) = R(t) f(t)$$

is formulated. Its standardized solutions form the so-called PEARSON system of probability density functions. It is remarkable that the corresponding probability distributions, foreseen by exceptions, are uniquely determined by their first four moments. The PEARSON system lends itself to the application of the computational experimental examination of qualities of statistical methods because conditions can be systematized and alternative hypotheses can be formulated parametrically. In this point of view it is asked if the residence time distributions that are in relation to the differential equation (10.3) and are derived in Chapter 13.1 are of the PEARSON type.
The majority of probabiliy distributions that arise in practical statistics can be traced to the PEARSON system. ELDERTON/JOHNSON (1969) provide an systematic overview concerning the PEARSON distributions.
Next it is determined that the exponential distributions and the Gamma distributions are special cases of the so-called type III of the PEARSON system (ELDERTON/JOHNSON 1969).

Proposition 13.16. *Let $f(t) = KL\mathrm{e}^{-Lt} + (1 - K)M\mathrm{e}^{-Mt}; K, 0 < L$ and $0 < M$ be real numbers. $f(t)$ is a density of a* PEARSON *distribution if and only if $L = M$.*

Proof. $f(t)$ is a exponential distribution for $L = M$. If $f(t)$ is a PEARSON distribution, then $f'(t) = R(t)f(t)$. The coefficient comparison for $f'(t) = K(-L^2)\mathrm{e}^{-Lt} + (1-K)(-M^2)\mathrm{e}^{-Mt} = R(t)f(t)$ yields two equations from which $L = M$ follows. □

Proposition 13.17. *Let* $f(t) = K\mathrm{e}^{-Kt} + (L-K)Lt\mathrm{e}^{-Lt}; 0 < K$ *and* $0 < L$ *be real numbers.* $f(t)$ *is the density of a* PEARSON *distribution if and only if* $K = L$.

Proof. $f(t)$ is the density of a exponential distribution for $K = L$. Let $f'(t) = R(t)f(t)$. The coefficient comparison for $f'(t) = [L^2 - 2KL - (L-K)L^2t]\mathrm{e}^{-Lt}$ yields two equations from which the assertion follows. □

Proposition 13.18. *A residence time distribution associated with a permissible Two-compartment-iv model* (10.3) *belongs to the* PEARSON *system if and only if the model can be interpreted as a One-compartment-iv model* (10.1) *or as a One-compartment-ev model* (10.2) *with* $k_i = k_{el}$.

Proof. The one-compartment models mentioned lead to exponential or Gamma distributions. The residence time distributions of the permissible two-compartment models, cf. Propositions 13.1 to 13.10, do not belong to the PEARSON system. This can be seen in the application of Propositions 13.16 and 13.17. Take into account $k_{20} = k_{10} + k_{12}$ and $k_{ij} > 0$ in the case of Proposition 13.9 in connection with Proposition 13.17. □

There are residence time distributions associated with Two-compartment-iv models (10.3) that do not belong to the PEARSON system but are already uniquely characterized by no more than four parameters. The first four moments of the distributions can work here. In pharmacokinetics literature, especially in connection with the evaluation of results of experiments, so-called 'model independent parameters' are given. The 'mean residence time' MRT (the usual abbreviation here) should be looked at more closely. It is defined as

$$MRT = \frac{\int_0^\infty t\,c(t)\,\mathrm{d}t}{\int_0^\infty c(t)\,\mathrm{d}t}\,.$$

Both from the description as mean residence time and from the definition, it can be seen that this parameter refers to the theory of probability. It is the expected value of a random variable that is even denoted as residence time. Therefore it respectively has to be proven that the observation of a time course of concentration also implies the observation of

Table 13.1: The expected values of the residence time distributions associated with Two-compartment-iv models.

Model	Distribution	Expected Value (MRT)
(1,0,0,0)	Proposition 13.1	$1/k_{10}$
(0,1,1,0)	Proposition 13.4	$1/k_{12} + 1/k_{20}$
	Proposition 13.5	$2/k$
(0,1,1,1)	Proposition 13.6	$-(1/\lambda_1 + 1/\lambda_2)$,
(1,1,0,1)	Proposition 13.7	$-[1/\lambda_1 + 1/\lambda_2 + k_{10}/(\lambda_1\lambda_2)]$
(1,1,1,0)	Proposition 13.8	$1/(k_{10} + k_{12}) + 1/k_{20} - k_{10}/(k_{10} + k_{12})k_{20}$
	Proposition 13.9	$2/(k_{10} + k_{12}) - k_{10}/(k_{10} + k_{12})^2$
(1,1,1,1)	Proposition 13.10	$-[1/\lambda_1 + 1/\lambda_2 + k_{10}/(\lambda_1\lambda_2)]$

this random variable. The definition of the mean residence time of a drug molecule in the body given by Yamaoka/NAKAGAWA/UNO(1978) is insufficient. That the sense of the parameter MRT gives reason for misunderstandings is shown in the discussion around the article by CHANTER(1985), by BENET (1985), LANDAW/KATZ (1985), GILLESPIE/VENG-PEDERSEN (1985), as well as in the opening words of the publisher of this journal. At this point, attention must be drawn to the article by VENG-PEDERSEN (1989). MATIS/WEHRLY/METZLER (1983) as well as EISENFELD (1981) also deal with mean residence times in connection with pharmacokinetic questions, the latter with respect to Markov processes though. Residence time distributions and parameters derived from them are also specially observed for circulation and flow models (LANSKY 1996, SMITH et al. 1997). Residence times however, without further reductions being necessary, are calculated from the constants of linear compartment models (VARON et al. 1995).

Statistical estimations due to the moment method for the identification of probability distributions are not unproblematic from the view of the statistical estimation theory. Although they are arithmetically practicable and under quite general conditions asymptotically unbiased and asymptotic normally distributed, the variance of such estimators is too large even for extensive samples. It is found that the estimate of the second moment is already very inaccurate for simple examples.

13.3 Truncation

Residence time X is a random variable with values in $\mathbb{R}^+$. It is possible that the empirical density roughly approaches $f_X(t)$ since the observations of a pharmacokinetic experiment are carried out on a finite interval (t_1, t_r). This is expected when $P(X < t_1) + P(X > t_r)$ has a noteworthy magnitude. The transition to a truncated distribution can be meaningful if the experimental conditions do not allow for an extension of the observation interval. Truncation indicates the transition to another random variable, and to another distribution. The density f_{X_g} of the distribution truncated at points t_1 and t_r is given by

$$f_{X_g}(t) = \frac{f_X(t)}{\int\limits_{t_1}^{t_k} f_X(t)\,\mathrm{d}t}$$

and exists in the case that the integral in the denominator is not equal to zero. Here X_g denotes a random variable different from X. The domain of X is the whole positive real axis, the domain of X_g is the interval $[t_1;\, t_2]$. Take now for example the exponential distribution with density $f_X(t) = \lambda \mathrm{e}^{-\lambda t}$ and truncation on $[0;\, T]$. For the expectations hold $E[X] = 1 = 1/\lambda$ and $E[X_g] = [1 + T\lambda - \mathrm{e}^{T\lambda}]/[\lambda - \lambda \mathrm{e}^{T\lambda}]$, respectively. The hazard function of X is constant $1/\lambda$. The hazard function $H_{X_g}(t) = \lambda \mathrm{e}^{-\lambda t}/[\mathrm{e}^{-\lambda t} - \mathrm{e}^{-\lambda T}]$ goes to infinity for t running to T.
The parameters of the truncated distribution generally cannot be equated with those of the original distribution. The relations between them are demonstrated by RASCH (1978) for the expected values of normal distributions. Truncation will briefly be introduced for residence time distributions in connection with Two-compartment-iv models.

Definition 13.3. *Let $c(t) = c(t, a, b, A, B)$ be a function that corresponds with one of the observed compartment models, is dependent on the param-*

eters a, b, A, B and can be integrated on (t_1, t_r). Then

$$f_{cg}(t) = \frac{c(t)}{\int\limits_{t_1}^{t_r} c(t)\,\mathrm{d}t}$$

denotes the truncated in t_1 and t_r, standardized concentration-time-function that is associated with $c(t)$.

The following proposition gives a bit of orientation in the case that truncation has to be included in stochastic modeling because of a given observation situation.

Proposition 13.19. *Let f_X be the density of a residence time distribution, f_c, f_{c_1}, f_{c_2} standardized concentration-time functions and $f_{Xg}, f_{cg}, f_{c_1 g}, f_{c_2 g}$ the corresponding truncated in t_1 and t_r functions. The following is then true:*

(1) $f_X(t) = f_c(t)$ yields $f_{Xg} = f_{cg}$.
(2) $f_{Xg}(t) = f_{cg}(t)$ generally doesn't yield $f_X(t) = f_c(t)$.
(3) $f_X(t) = Kf_{c_1}(t) + (1-K)f_{c_2}(t)$, $K \in \mathbb{R}$ doesn't yield $f_{Xg}(t) = Kf_{c_1 g}(t) + (1-K)f_{c_2 g}(t)$.

The following proposition is used later.

Proposition 13.20. *The truncated in the points t_1 and t_r density of the residence time distribution that is associated with the One-compartment-iv model has the form*

$$f_{Xg}(t) = \frac{k_{10}e^{-k_{10}t}}{e^{-k_{10}t_1} - e^{-k_{10}t_r}}\ .$$

The simple proofs need not to be indicated here.

Chapter 14

Other stochastic models

14.1 Stochastic differential equations

$y(t) = \mathcal{D}x(t)$ denotes a linear differential equation. Randomness can then varyingly be taken into account in the attempt of a stochastic differential equation: Inhomogeneity is understood as a random variable, the coefficients of the differential equation, or the operator, are understood as random variables, and the initial conditions are also understood as random variables. There are extensive representations about theories of stochastic differential equations. JACQUEZ (1988) dedicates a chapter in his book to the representation of two principle ways in which compartment models can be connected with stochastic attempts (stochastic compartment models). In one case, the inhomogeneities are seen as random variables, and in the second case, the model parameters of a compartment model are seen as random variables. Special propositions are not indicated.
CAMPELLO et al. (1978) discuss parameter estimation for stochastic differential equations in relation with compartment models.
An application of a multi-compartment system with stochastic inhomogeneities for the description of the accumulation of radioactive Polonium in a human organism is given by KAJIYA/KODAMA/ABE (1985, pages 156 cf.).
The One-compartment-iv model (10.1) has been generalized to a stochastic differential equation by DITLEVSEN/DEGAETANO (2005).

14.2 Stochastic processes

The ability to describe a stochastic process of the distribution of a set of drug molecules in a system of compartments is explained by IOSIFESCU/TAUTU (1973, pages 231 cf.) with regard to an example from

WIGGIN (1960). The expected value (as a function of time) of the set of drug molecules in a compartment is given as a convolution integral. This is another probability theoretical interpretation of the convolution integral (cf. Chapter 10.3). ADOMIAN (1983) deals with this problem in greater detail. He briefly mentions stochastic compartment models and refers to BELLMAN (1983) with regard to this topic.
Birth-and-death processes or bifurcation processes seem most suitable for the modeling of events observed in pharmacokinetics. It can be seen that the expected value function of this stochastic process which comes from the One-compartment-iv model (10.1) has the form $E_X(t) = A\mathrm{e}^{-at}$. This yields from a pure death process (the death rates are independent from time and the deaths are stochastically independent events). The correspondences of the parameters to be interpreted are obvious. Diffusion models would have to be used if one is not satisified with a discrete phase space (this concept is not used uniformly in the theory of the MARKOV processes though). In connection with this, a result from KELLER/KERSTING/ROESLER (1988) where processes can be approximated through birth-and-death processes, seems worth mentioning. The so-called SKOROHOD topology is referred to. More dimensional MARKOV processes can possibly be referred to for the description of exchange events (analogous with more-compartment models) (LAHRES 1964, pages 88 cf.). FADDY (1985) deals with non-linear stochastic attempts. MATIS/WEHRLY (1981) and the literature mentioned there refer to stochastic compartment models. Furthermore, MATIS/CARTER (1972), HERRMANN(1978), EISENFELD (1979), AGRAFIOTIS (1985), PARTHASARATHY/MAYLISWAMI (1981), MATIS/WEHRLY/METZLER (1983), MATIS/WEHRLY (1972, 1979, 1998) as well as YU/WEHRLY (2004) are mentioned. Chapter 10.3 refers to circulation and flux models (see WEISS (1990) for example). They are listed here once more, since they offer both deterministic and stochastic possibilities for descriptions. LANSKY (1996) and SMITH et al. (1997) describe circulation processes with the help of stochastic models and bring residence time distributions in relation to them.
Adversely, the proper application of these attempts is influenced by problems in parameter estimation:
On one hand, maximum-likelihood estimations for homogeneous MARKOV chains (qualities of the estimators should be known in all cases) are only available under assumptions that can hardly be proven here. On the other hand, the realization of a mathematical sample of a MARKOV chain (it describes the repeated independent observations of the process at a finite

amount of points in time) is not possible due to the conditions of pharmacokinetic experiments. Stochastic processes also do not prove to be practical in the context of pharmacokinetics due to another reason. The strong and in a pharmacokinetic experiment non-verifiable assumptions which must be made according to an adequately efficient theory are important. An example:
It is known that the exponential distributions are the only lifetime distributions which have not degenerated and for which the probability of survival and the conditional probability of survival (the observed object already lived for a certain time t_0) coincide. The supposition of an exponentially distributed life time is therefore necessary for the existence of the MARKOV quality of a birth-and-death process which has been asserted. At least estimations and tests are available under further conditions in this case. The question whether measurements on hand can be understood as samples in the sense of the estimation theory of MARKOV processes still needs to be answered.
This example shows that a corresponding birth-and-death process can hardly be calculated via easily understood mathematical qualities for a somewhat more complicated pharmacokinetic model (cf. Chapter 11.2). The degenerated 'memoryless' lifetime distribution is now discussed. Its density function is:

$$f_{t^*}(t) = \begin{cases} 1 \text{ for } t = t^* \\ 0 \text{ otherwise.} \end{cases}$$

The observed event is deterministic. All objects have an equal lifetime. The term density function is only correct here when $f_{t^*}(t)$, the accompanying distribution function $F_{t^*}(t)$, and the formation of the derivitaves are understood in the sense of the distribution theory in the mathematical analysis. The DIRAC-function is often used as a simple example in system theoretical and probability theoretical oriented pharmacokinetics without mentioning these theoretical foundations.

Plausibility and the good interpretation of parameters are some of the advantages of the modelling of phamacokinetic events by stochastic processes.

14.3 Regression attempts

Most publications that refer to stochastic aspects for descriptions of pharmacokinetic processes start out with regression attempts. The language that is used requires precision. General regression (the regression function is given by the conditional expectations), and special regression (the regression function is distinguished by minimal residual variance in a predefined function class) are different. The former is the proper regression attempt in the sense of stochastics. A two-dimensional random variable has to be named. It has to be proven that the observed concentrations over time are the conditional expectations. The second method can then be understood as an approximation in the numeric sense. It is often connected with a probabilistic model. The GAUSS-MARKOV-Theorem cannot be used anymore for the nonlinear functions typical in pharmacokinetics.

If the type of functional dependence should be examined, the linearity test of FISHER already requires that (formulated for the topic that is dealt with here) the number of concentration measurements is larger than the number of distinct measuring times. Only few concentration measurements (at most one each time) can be taken from an individual in a pharmacokinetic experiment. In a statistical sense, a repetition of the measurements is not possible due to biological reasons. In principle, the assumptions with regard to the error model of a regression attempt cannot be checked. For this reason, recommendations to handle individual kinetics as regression problems are not seen as a good idea. For example, SHEINER (1984, 1985, 1986) has a different opinion.

The objections do not hold when a regression model of pharmacokinetic interest should be judged for a sample of individuals in a population. BEAL (1984) explains such methods for population pharmacokinetics and YUH et al. (1994) give an early literature overview.

Chapter 15

Calculation methods related to compartment models

An essential task is the calculation of model parameters being interested of an individual kinetics or a population kinetics with regard to observation data. The adapted fitting principle has to correspond with the character of the mathematical model. Approximations correspond with deterministic attempts and statistical estimations with stochastic attempts. The characterization of methods of calculation is an essential prerequisite for the proper interpretation of results. Computer programs, which are recommended in great variety in literature (or are being offered commercially) are typically used for the evaluation of pharmacokinetic tests. A definite conclusion about these software is neither possible nor meaningful. Users are confronted with typical problems:

- The description of the programs is insufficient.
- Which mathematical methods were used?
- Under which assumptions are the computer results the solution to the posed problem?
- Can the truth of these assumptions on the data on hand be checked at all?
- Is there a mathematical characterization of the calculation method carried out?
- Is a modification or addition permitted according to the problem?
- Due to technical reasons, the user can't come to a conclusion about the mathematical method presented to him in the form of a computer program.

The most unfavorable situations are:

- The method of parameter calculation is unreasonably chosen.
- The computer program is faulty.

- The user knows neither the calculation principle nor the mode of operation of the program.

The method of least squares (MLS), as well as a weighted MLS are most frequently applied to obtain an approximation. Various numeric procedures can be found in program-technical realizations. A bibliographical reference is not carried out here due to the enormous number of publications that turned up with regard to this topic. Further methods for parameter calculation are used in pharmacokinetics:

- so-called extended MLS estimates (BEAL 1984, YAMAOKA et al. 1986),
- direct numeric integration of defining differential equations (ENGLERT/GOEHRING/WEDEKIND 1984),
- TSCHEBYSCHEW approximation (WERNER 1970, 1982),
- maximum-likelihood estimations (SANDOR/CONROY/HOLLENBERG 1970, LU/MAO, 1993),
- FOURIER transformation (GARDER 1968),
- numeric inversion of the LAPLACE transformation (LUXON 1987),
- moment estimations (ISENBERG/Dyson 1969, YAMAOKA/NAKAGAWA/UNO 1978b, KNOLLE 1984),
- reference to convolution integrals (IGA et al 1986),
- graphic methods, particularly the so-called peeling technology (MEIER/RETTIG/HESS 1981).

The bibliographical references are examples. It is apparent from this obviously incomplete list that various mathematical theories are referred to in the context of parameter calculation.

RUPP (1980) already refers to about 3200 pages of literature in his assessment of the computer programs used in pharmacokinetics. He offers a chronological representation of software development and shortly characterizes the programs. Some of them are currently still being used in up to date versions.

BITTRICH/HABERLAND/ JUST (1979) give a summary of the calculation methods used in the kinetics of chemical reactions.

In BOZLER/VAN ROSSUM (1982, page 211), HARTMANN reports about observed outcomes of the experimental computer comparisons of the often mentioned software packages, PHARMFIT, NONLIN, SAAM and TOPFIT, developed for pharmacokinetics: 'In the discussion, there was agreement on the reliability of these programs when the results are interpreted by

an experienced kineticist'. There was also a discussion about NONLIN between VENG-PEDERSEN (1977, 1978) as a critic and METZLER/ELFRING (1978) as the authors of the software. The comparison of NONLIN and other programs by VALENTINE/HUNTER (1985) should be mentioned. Less common systems should hardly get any critical judgment. MARTIN et al (1984) compare approximations and statistical estimates in that four various methods are applied to the measurements obtained from three patients. A comparative study from HATTON/MASSEY/RUSSEL (1984) therefore seems worth mentioning because the clinical meaningfulness of differences in calculation results is discussed. JOHNSON/MEYERSOHN (1985) compare different calculation methods which regard the parameters of infusion models.
Modern software fulfils more extensive requirements meanwhile. For example the ADAPT software package (D'ARGENIO et. al. 2009) offers a comfortable environment for modeling, parameter estimation and simulation experiments in pharmacokinetics. Weighted least-square estimation, ML estimation, generalized least-square estimation and BAYESIAN estimation are there available to calculate model parameters from a set of individual kinetics. Considerable numerical efforts are involved in these procedures. The specifc pharmacokinetic models, measurement plans and error models are mathematically developed and explained in the user manual of ADAPT.

15.1 Method of least squares parameter calculations

This chapter will comparatively present MLS parameter calculations for individual kinetics. It deals with variants of the method of least squares. The measured drug concentrations at times $t_1, \ldots, t_r$ in the blood $cm_i = cm(t_i)$ should be able to be described by the simplest of the discussed model functions:

$$c(t) = Ae^{at}, \quad a < 0 .$$

The transition to the standardized function $f_{cg}(t)$ (see Chapter 13.2) which is truncated at t_1 and t_r yields $f_{cg}(t) = c(t)/AUC_g$ with $AUC_g = A[\mathrm{e}^{at_1} - \mathrm{e}^{at_r}]/a = h(a)A/a$. The meaning of $h(a)$ is obvious. A is replaced, and

$$z(t) = \frac{AUC_g a e^{at}}{h(a)}$$

has to be fitted to cm_i. AUC_g can be determined from the measurements by different methods. Integrals of interpolating spline functions come into play here. To determine the minimum of the function

$$S\left(h\left(a\right),a\right)=\sum_{i=1}^{r}\left[cm\left(t_i\right)-z\left(t_i\right)\right]^2 \tag{15.1}$$

means calculating the parameter a by MLS. If one takes logarithms of $z(t)$,

$$\ln z\left(t\right)=\ln AUC_g+\ln a+at-\ln h\left(a\right)\ , \tag{15.2}$$

the appropriate MLS-calculation of a is the minimum of

$$S_L\left(h\left(a\right),a\right)=\sum_{i=1}^{r}\left[\ln cm\left(t_i\right)-\ln z\left(t_i\right)\right]^2\ . \tag{15.3}$$

This method leads to a linear system of equations, whose explicit solution formulas are easier to handle than the iterative solution to the original problem. In pharmacokinetics, this is the reason for the wide spread determination of the parameters with regard to the logarithmic model equation. Parameter calculation with regard to the logarithmic model and parameter calculatiion with regard to the original model lead to different values.
WITTSTEIN (1882) already dealt with this and gave a method that corrected this difference.
Briefly, if (x_i, y_i) are observation results and $\hat{y}_i = f(x_i, \alpha_j)$ are function values that are dependent on the parameters α_j of the function f to be fitted, and $\mathcal{T}$ is a differentiable transformation with regard to x, then

$$\sum_{i=1}^{r}\left(y_i-\hat{y}_i\right)^2\approx\sum_{i=1}^{r}W_i\left[\mathcal{T}(y_i)-\mathcal{T}\left(\hat{y}_i\right)\right]^2$$

is true for

$$W_i=\left[\frac{d}{dy}\mathcal{T}\left(y_i\right)\right]^{-2}\ .$$

The weights W_i have to be calculated with regard to the observation results. By minimizing

$$S_W\left(h\left(a\right),a\right)=\sum_{i=1}^{r}\left[\ln cm\left(t_i\right)\right]^2\left[\ln cm\left(t_i\right)-\ln z\left(t_i\right)\right]^2 \tag{15.4}$$

the corrected (according to WITTSTEIN) MLS calculation regarding the logarithmic transformation is obtained. The smaller the distances $|y_i-\hat{y}_i|$, the more exact is the WITTSTEIN correction. However, with the general availability of computers these days, such a method is only historically interesting.
In a computer experiment, which is thought of as an illustrating example, the mentioned variants of the MLS calculation were compared. Corresponding empirical probability density functions were obtained through repeated calculations from simulated measurements. They are dependent on:

- the number r of measurements,
- the location of the measuring times t_i,
- the probability distributions of measurement errors in places t_i,
- the value of the parameter to be calculated,
- the qualities of the numeric method as well as
- the starting values of the iterative procedure.

The last two points have subordinate meaning for the observed one-dimensional estimation problem.
The values $A = 300$, $DOS = 300$, $a = 0.5$ as well as $t_i = i, i = 1, \ldots, 5$ were predefined. The specification of dimensions is avoided since all reference quantities are not actual measurements. The following is true for the 'series of measurements':

$$cm(t_i) = c(t_i) + X_i,\ i = 1, \ldots 5,$$

where X_i is a realization of a random variable (measuring error) with normal distribution $N(0; \{0.05\ c(t_i)\}^2)$ and

$$cml(t_i) = \ln c(t_i) + Y_i,\ i = 1, \ldots 5,$$

where Y_i is a realization of a random variable (measuring error) with normal distribution $N(0; \{0.05\ \ln c(t_i)\}^2)$. With these data one calculated:

1. realizations $\hat{a_1}$ of the MLS-calculation $\underline{\hat{a_1}}$ for a with regard to $cm(t_i)$ according to (15.1),
2. realizations $\hat{a_2}$ of the MLS-calculation $\underline{\hat{a_2}}$ for a with regard to $ln\ cm(t_i) = cml(t_i)$ according to (15.3),
3. realizations $\hat{a_3}$ of the MLS-calculation $\underline{\hat{a_3}}$ for a with regard to $cm(t_i)$ according to (15.4).

A simulation size of 200 seemed to be enough to demonstrate the tendencies. In another simulation experiment of size 235, additionally to the above described realizations $\hat{a}_1$ and $\hat{a}_3$,

4. realizations $\hat{a}_4$ of the MLS-calculation $\underline{\hat{a}}_4$ for a with regard to the transformation $\ln\ cm(t_i) = \ln[c(t_i) + X_i]$ according to (15.3)

were calculated for every 'measuring series' $cm(t_i)$, $i = 1, \ldots, 5$. Moreover, the differences $\hat{a}_1 - \hat{a}_3$ and $\hat{a}_1 - \hat{a}_4$ are investigated.
A composition of the calculated results can be seen in Tables 15.1 and 15.2 as well as in Figures 15.1 and 15.2. The sample statistics show that the calculation methods $\underline{\hat{a}}_1$, $\underline{\hat{a}}_2$ and $\underline{\hat{a}}_4$ lead to different results. Table 15.2

Table 15.1: Sample statistics of the outcomes of three calculation methods (200 runs). Explanations can be found in the text.

Calculation-method	Mean-value	Standard-deviation	Min	Max	Var. coeff.
$\underline{\hat{a}}_1$	0.4996	0.0273	0.4262	0.5832	5.4735
$\underline{\hat{a}}_2$	0.4973	0.0552	0.3198	0.6255	11.0909
$\underline{\hat{a}}_3$	0.4987	0.0270	0.4259	0.5786	5.4077

and Figure 15.2 show how large the difference between parameter values of measuring series from different methods can be. In extensive computer experiments, ELLER/GYOERI (1978) demonstrate that it is a function of the error variance. The agreement of $\underline{\hat{a}}_1$ and $\underline{\hat{a}}_3$ illustrates the method of WITTSTEIN. Which of the calculation methods does the user have to choose for the evaluation of an individual kinetics?

Table 15.2: Sample statistics of the outcomes of three calculation methods (235 runs). Explanations can be found in the text.

Calculation-method	Mean-value	Standard-deviation	Min	Max	Var. coeff.
$\underline{\hat{a}}_1$	0.5020	0.0276	0.4488	0.5640	5.5006
$\underline{\hat{a}}_3$	0.5009	0.0271	0.4490	0.5613	5.4153
$\underline{\hat{a}}_4$	0.4996	0.0156	0.4577	0.5404	3.1271
$\underline{\hat{a}}_1 - \underline{\hat{a}}_4$	0.0023	0.0244	−0.0608	0.0703	1026.2
$\underline{\hat{a}}_1 - \underline{\hat{a}}_3$	0.0011	0.0009	−0.0002	0.0050	82.28

The application of $\underline{\hat{a}}_1$ means that the model $z(t)$ derived from $c(t)$ is considered to be adequate for the situation and that concentrations $cm(t_i)$ are actually measured. Since biochemists indicate the precision of their methods anyway, they have prior knowledge of measurement errors X_i. The calculated parameter value $\hat{a}_1$ can be interpreted as an elimination constant with regard to $z(t)$. Elimination half-life, clearance, and distribution volume can be derived and understood as it is typical in pharmacokinetics. The estimation $\underline{\hat{a}}_2$ of the parameter a means that $\ln z(t)$ is accepted as

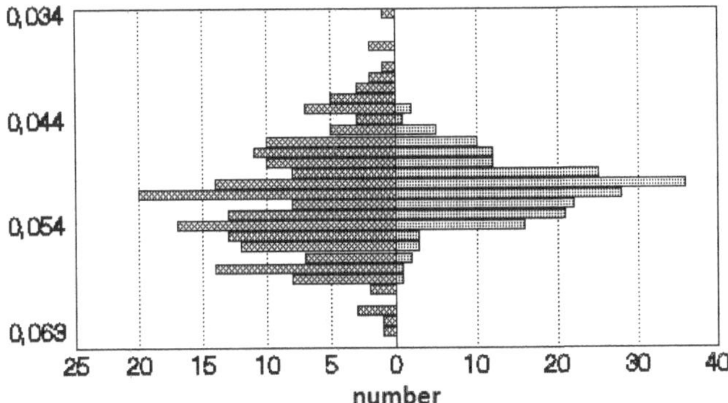

Fig. 15.1: Frequency distributions of 200 MLS calculations $\hat{a}_1$ (right) and $\hat{a}_2$ (left). Explanations can be found in the text.

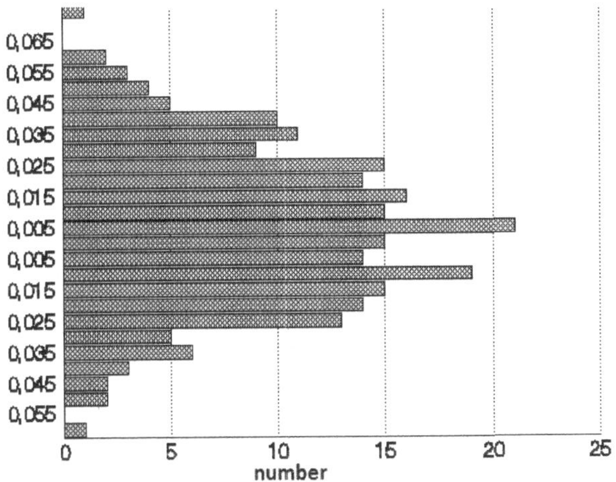

Fig. 15.2: Frequency distribution of 235 differences $\hat{a}_1 - \hat{a}_4$. Explanations can be found in the text.

a pharmacokinetic model and that logarithms of concentrations are measured. Appropriately, Y_i are the errors that arise from measuring logarithms of concentrations. Their probability distributions are determined again by the given precision of methods of measurement. However, terms like elimination constant, clearance, and distribution volume are possible to understand with respect to the linear function $\ln z(t)$. The application of $\underline{\hat{a}}_4$ is inadmissible. If $cm(t_i) = c(t_i) + X_i,\ i = 1, \ldots, 5$ is written in the form

$$cm\,(t_i) = c\,(t_i)\left[1 + \frac{X_i}{c\,(t_i)}\right] \ , i = 1, \ldots, 5,$$

then

$$\ln cm(t_i) = \ln c(t_i) + \ \ln\left[1 + \frac{X_i}{c\,(t_i)}\right] \ , i = 1, \ldots, 5 \ .$$

The error X_i usually has an expected value of zero; the method of analysis used by chemists is "right". From the theory of probability, it is known that the expected value remains unchanged only for exceptions when a random variable is transformed. In addition, it is intuitively clear that by the ln-transformation, the actual error (the precision of the analysis method) is reduced without authorization. Take the standard deviation of $\underline{\hat{a}}_1$ and $\underline{\hat{a}}_4$ in Table 15.2 into account. Precision is feigned here. It arises from the logarithmic transformation. The application of $\underline{\hat{a}}_4$ is nevertheless widespread.

The calculation of parameters of logarithmic values of concentration are given as the only method in some pharmacokinetic monographs (e.g. PFEIFER/BORCHERT 1980, TALLARIDA/MURRAY 1981). Others (e.g. DOST 1968, SHARGEL/YU 1980, MEIER/RETTIG/HESS 1981) introduce more methods. Comparative considerations can be found in DOWD/RIGGS (1965), BOZLER/VAN ROSSUM (1982), MARTIN et al (1984), KOHBERGER (1980), MCINTOSH/MCIINTOSH (1980), BEAL (1984), LAM (1979), MEYER/RICHTER (1980), YAMAOKA/TANAKA/OKUMURA et al. (1986), REITBERG et al. (1985) and BENTZIEN et al (1985). Many authors publish information about computer programs for the parameter calculation of logarithmetised measurements of concentrations (e.g. ROCCHETTI/URSO 1982, VARKONYI/GABANYI/DEUTSCH 1983, HOLMES/WILSON/MCCALL 1986).

15.2 Statistical parameter estimation for an individual kinetics

In literature about pharmacokinetics, parameter calculation for an individual kinetics is also treated as a statistical estimation problem (cf. for example MATIS/WEHRLY/GERALD (1985)). Besides representations of the statistical estimation theory which apparently is thought to further education, the derived propositions are tied to non-verifiable assumptions (cf.

the explanations in Chapter 13.3. about regression models) or can be seen in connection with approximations by means of linearization like e.g. it is explained in the classic book by BARD (1974).

The calculation of the parameters of a pharmacokinetic model is now formulated as a statistical problem of a density parameters estimation. At the beginning of this chapter, it was illustrated in what way the classical pharmacokinetic Two-compartment model for iv bolus administration could be connected to a probability theoretical framework. The random variable X is the residence time of a drug molecule in an organism. The parameters of its probability distribution can be estimated from samples. It is necessary to know its relations to the pharmacokinetic model for the interpretation of results. The residence times of the applied drug molecules shall serve as a concrete sample. To obtain a sample therefore indicates supplying the organism with a number of molecules and registering the residence time of every molecule. Four points of view require discussion with regard to the sampling method:

(1) It must be assumed that the residence times of the single molecules are stochastically independent. This prerequisite is necessary to be able to use basic statements of the statistical estimation theory. On the other hand it is also essential for a theory of distribution processes at a molecular level, see VENG-PEDERSEN (1991) and in a more general setting CLIFFORD and GREEN (1994). One surely will accept the independence at low drug concentrations. Saturation as known for drug transporters or metabolizing enzymes can be an argument against independence. Pharmacologists have to decide the acceptance of the assumption at the end.

(2) A measurement cm_i of concentration at t_i is not realization of the residence time of a molecule. It has a summarizing character. Therefore, standard ML-methods cannot be used for parameter estimations. These require knowledge of the numerical values that the residence time takes on in a concrete sample. Residence times of single molecules are not available from the classical pharmacokinetic experiment.

(3) The molecules however are not actually detected one by one when a concentration is measured. Certain units of mass are counted. The precision of the method of measurement is contemplated here. According to this, the sample size N is considered to be the integral multiple of the unit of administered drug mass. Such a specifcation is also found in kinetics of chemical reactions: The equations formulated for molecules

are only the model of thought for the relation between the amounts of substances that are, for example, measured in Mol.

(4) If the last measurement of concentration yields an observed value considerable above zero, then the lifetime of a part of the elements of the sample would not be recorded by the observation. However, transition to a truncated distribution is possible for the pharmacokinetic problem. This requires a interpretation of results with respect to the truncation.

Methods for the calculation of the mean residence time MRT were already recommended by YAMAOKA/NAKAGAWA/UNO (1978b). The use of the concept MRT in connection with compartment models, however, is not generally possible (see Chapter 13.1).
A summary about non-parametric density estimations is given by WERTZ (1978). Through an instructive example, he covers the differences in nature of parametric and non-parametric , or 'optimal' density estimators (cf. page 21). The book by DEVROYE (1987) which offers a wealth of points of view with regard to the problem of density estimation and justifies the use of corresponding literature absolutely has to be mentioned here.
Aspects of the application of kernel estimators are discussed, for example, by MUELLER (1985). The ability to state local confidence intervals for curve estimates is remarkable. This method is not suitable for use in pharmacokinetics because the measurements are not close enough together. SCHMERLING (1988) shared this information.
The varied minimum-χ^2-method is recommended for the construction of parameter estimates in connection with pharmacokinetic models.
This corresponds with the interpretability of the pharmacokinetic experiment as a sample procedure, makes propositions concerning the qualities of possible estimators, and permits the use of the statistical test theory for the treatment of the problem concerning the selection of a model (see Chapter 9 and Chapter 15.3).

15.2.1 *Varied minimum-χ^2-estimation*

$f_X(t, \alpha_1, \ldots, \alpha_m)$ denotes the density of the continuous random variable X which is dependent on parameters α_j. $I_1, \ldots, I_n$ with $m < n$ is a disjunct partition of the real range of X. The associated probabilities are $p_i = P(X \in I_i)$ with $p_1 + \cdots + p_n = 1$ and $p_i = p_i(\alpha_1, \ldots, \alpha_m)$. B_i is the number of observations in I_i and $E_i = Np_i$ is the number of realizations of

X in I_i expected with regard to f_X in a sample of size N. If

$$\chi^2 = \chi^2(\alpha_1, \ldots, \alpha_m) = \sum_{i=1}^{n} \frac{(B_i - E_i)^2}{E_i}$$

reaches a minimum at $(\hat{\alpha}_1, \ldots, \hat{\alpha}_m)$, then $(\hat{\alpha}_1, \ldots, \hat{\alpha}_m)$ is called the Minimum-χ^2-estimate for $(\alpha_1, \ldots, \alpha_m)$. Suppose p_i is differentiable to α_j, then

$$\frac{\partial}{\partial \alpha_j} \chi^2(\hat{\alpha}_1, \ldots, \hat{\alpha}_m) = 0 \;, \quad j = 1, \ldots, m$$

are necessary so that $(\hat{\alpha}_1, \ldots, \hat{\alpha}_m)$ is an extreme value of χ^2.
The following simpler system of equations can be obtained when the denominator while forming the partial derivatives is viewed as a constant:

$$\sum_{i=1}^{n} \frac{(B_i - E_i)}{p_i} \frac{\partial p_i}{\partial \alpha_j} = 0 \;, \quad j = 1, \ldots, m, m < n \;.$$

Its solution yields the so-called varied minimum-χ^2-estimation of the parameters. A nonlinear system of equations has to be solved. This is only possible by approximation. Gradient procedures work with precision because the needed partial derivatives of the functions to be minimized are not calculated by means of numeric methods. They have to be worked into the program and are calculated as subroutines in their respective places. The estimates of the varied minimum-χ^2-method should be denoted by $(\hat{\alpha}_1, \ldots, \hat{\alpha}_m)$ so that more symbols do not need to be introduced.
The problem of parameter estimation in Two-compartment-iv models is now examined with regard to this method.
The measuring times $t_1, \ldots, t_r$, $r \geq 2$, yield a partition $I_i = (t_i, t_{i+1})$ of (t_1, t_r) in $r-1$ intervals as well as $\mathbb{R}^+$ in $r+1$ intervals. The latter are the ones mentioned above as well as $(0, t_1)$ and (t_r, ∞).
The number B_i observed in I_i drug molecules is proportional to the integral over the interval (t_i, t_{i+1}) of the observed concentration-time function $c_b(t)$. An approximation is required to more accurately denote the observed function $c_b(t)$ from the measurements $cm(t_i)$, $i = 1, \ldots, r$ on hand. Smoothing spline functions are suitable for this. Part I of the book at hand provides material for the development of suitable spline functions.
The application of spline functions for the calculation of the area under concentration time curves was already recommended by YEH/ KWAN (1978) for the use in pharmacokinetics. They compare these procedures with other numeric methods of integration.

$$B_i = \int_{t_i}^{t_{i+1}} Sp(t)\,\mathrm{d}t; \; N = \sum_{i=1}^{r-1} B_i$$

are set. $Sp(t)$ denotes an interpolating, cubic spline function. It is defined such that $Sp(t_i) = cm(t_i)$, $i = 1, \ldots, r$, and that the polynomials of 3rd degree associated with the time intervals agree with its first and second derivatives in t_i. The conditions with regard to the edge points t_1 and t_r influence the qualities of $Sp(t)$. It was determined in extensive tests that the suitable given derivatives in the edge points lead to agreement of the area under the observed exponential function and the area under the interpolating spline function. The mentioned conditions concerning the boundary imply that the spline function outside of (t_1, t_r) is thought to be extended by a linear function with a given slope. It proves to be sufficient, for practical applications, to use the MLS-estimates of the slope of the linear function for the boundary conditions, and to approximate the logarithms of the first or the last measurements, respectively. The number of E_i expected realizations of X in I_i arises from the knowledge of $f_X(t)$ and is calculated as

$$E_i = Np_i = N \int_{t_i}^{t_{i+1}} f_X(t)\, \mathrm{d}t \, .$$

E_∞ denotes the expected value related to the interval (t_r, ∞).
Two problems need to be solved in order to be able to estimate the parameters of the residence time distribution of interest by means of the Minimum-χ^2-method.

Problem 1: $f_X(t) > 0$ for $t \in \mathbb{R}^+$. B_0, or B_∞ that correspond with the intervals $(0, t_1)$, or (t_k, ∞) are zero though.

The results of the parameter estimates are influenced by this. This will be examined in an example: With regard to $c(t) = Ae^{at} = 3e^{-0.5t}$, the parameter a of the residence time distribution $f_X(t) = f_c(t)$ can be estimated via the varied minimum-χ^2-method in the way that it is represented here. The first measurement is at $t_1 = 0$. Ten further measurements are distributed equidistantly on $(0, t_{11})$. As measurements, the calculated $c(t_i)$ were taken so that -0.5 can be expected for a. The last measuring time t_{11} is fixed in such a way that

$$AUCR = \int_{t_{11}}^{\infty} c(t)\, \mathrm{d}t$$

is a predefined percentage of the whole area under the curve AUC. As noted in Table 15.3, the parameter estimate is only satisfactory when the

Table 15.3: Varied minimum-χ^2-estimation for an incomplete observation of the course of $c(t)$. The integrals FC and FSP allow for a comparison between the spline approximation and the given function $c(t)$ (see Part III). Further explanations are found in the text.

AUCR as % of AUC	$\hat{a}$	FSP	FC
1	0.50030	5.93963	5.94000
2	0.50117	5.87981	5.88000
3	0.50263	5.81988	5.82000
4	0.50467	5.75991	5.76000
5	0.50731	5.69994	5.70000
10	0.52877	5.39998	5.40000
20	0.60450	4.80000	4.80000
30	0.71627	4.20000	4.20000
40	0.87014	3.60000	3.60000
50	1.08700	3.00000	3.00000

whole course of $c(t)$ is observed. The given comparison between the areas

$$FC = \int_0^{t_{11}} c(t)\,\mathrm{d}t$$

and

$$FSP = \int_0^{t_{11}} Sp(t)\,\mathrm{d}t$$

demonstrates that the spline approximation is nearly precise in this case. To obtain a better estimate for a, B_∞ has been calculated respectively with the parameters of the preceding iteration.

Table 15.4 illustrates the expected better estimate of a. It also clarifies that the heuristically well-founded procedure cannot offer a satisfactory solution to the problem. The cause is the bad estimation of E_∞ by $B_\infty(\hat{a}_{i-1})$, $\hat{a}_{i-1}$ denotes the estimate for a in the previous iteration of the solution to the minimum problem. Note that the actual integrals, not approximations, are used for B_∞.

Table 15.4: Varied minimum-χ^2-estimation with iterative correction for non-complete observations of the course of $c(t)$. Further explanations can be found in the text.

AUCR as % of AUC	$\hat{a}$	$B_\infty(\hat{a}_{i-1})$	E_∞
1	0.5018	0.05916	0.06000
2	0.5013	0.11902	0.12000
3	0.5010	0.17896	0.18000
4	0.5008	0.23893	0.24000
5	0.5007	0.29892	0.30000
10	0.5004	0.59894	0.60000
20	0.5002	1.19889	1.20000
30	0.5003	1.17983	1.80000
40	0.5013	2.39326	2.40000
50	0.5066	2.95891	3.00000
60	0.5232	3.40278	3.60000
70	0.5612	3.56952	3.90000
80	0.6400	3.31345	4.80000
90	0.8218	2.54149	5.40000

As already mentioned, a way out can be found in the transition to a truncated distribution. t_{11} is the right truncation point. The density

$$f_{X_g}(t) = \frac{1}{1 - e^{at_{11}}} (-a)\, e^{at}$$

of the truncated distribution is the basis for the calculation of p_i and E_i. Truncation implies an increased amount of effort when working out the method of varied minimum-χ^2-estimation because the partial derivatives, with respect to the parameters, have to be taken into account. It is already visible in a simple example through the comparison of

$$\frac{\mathrm{d}}{\mathrm{d}a} p_i = \frac{\mathrm{d}}{\mathrm{d}a}\left(e^{at_i} - e^{at_{i+1}}\right) = t_i e^{at_i} - t_{i+1} e^{at_{i+1}}$$

and

$$\frac{\mathrm{d}}{\mathrm{d}a} p_{ig} = \left[\left(t_{i+1} e^{at_{i+1}} - t_i e^{at_i}\right)\left(1 - e^{at_{11}}\right) - \left(e^{at_i} - e^{at_{i+1}}\right)\left(t_{11} e^{at_{11}}\right)\right] \times \left(1 - e^{at_{11}}\right)^{-2} .$$

Table 15.5: Varied minimum-χ^2-estimation for a non-complete observation of the course of $c(t)$ with truncation. Further explanations can be found in the text.

AUCR as % of AUC	$\hat{a}$	AUCR as % of AUC	$\hat{a}$
1	0.5018	30	0.5001
2	0.5013	40	0.5001
3	0.5010	50	0.5000
4	0.5008	60	0.4999
5	0.5007	70	0.4999
10	0.5004	80	0.4999
20	0.5002	90	0.4999

For the given example, the varied minimum-χ^2-estimations of parameter a by truncation at point t_{11} are listed in Table 15.5. The incompleteness of the observation is denoted as $AUCR$ as a % of AUC. The estimations $\hat{a}$ remain stable.
As mentioned above, another problem has to be discussed:

Problem 2: If $f_X(t)$ represents a linear combination of the densities of two standardized $c(t)$ -functions, then the expected values $E_i = Np_i$ are not only dependent on one of the two functions $c_j(t)$ anymore. In these cases, the observations of only one of the courses $c_1(t)$, or $c_2(t)$, is not enough to indicate the realizations of the random variable. The number of drug molecules observed in a single compartment is not proportional to the number of molecules present in the organism.

The Two-compartment-iv model $(k_{10}, k_{12}, k_{20}, 0)$ is observed as an example. $f_X(t) = Lf_{c_1}(t) + (1 - L)f_{c_2}(t)$ is true for the density with regard to Proposition 13.8. If the observations should analogically be connected through an attempt $Sp(t) = LSp_1(t) + (1 - L)Sp_2(t)$ then the PEARSON-χ^2 for $Sp_j(t) = c_j(t)$ is zero as expected. This combination of the two observed functions is however dependent on L and therefore also dependent on the model parameters to be estimated.

Proposition 15.1. *Let $f_X(t)$ be the density of a residence time distribution with regard to a Two-compartment-iv model corresponding with Propositions*

13.8, 13.9 and 13.10. Suppose the model parameters are known. Let

$$B_i = N \left[L^* \int_{t_i}^{t_{i+1}} f_{c_1}(t)\,\mathrm{d}t + (1 - L^*) \int_{t_i}^{t_{i+1}} f_{c_2}(t)\,\mathrm{d}t \right]$$

with $N = AUC$, $t_1 = 0$ and $t_r = \infty$. L has the same meaning as it does in the mentioned Propositions 13.8, 13.9, and 13.10.
Under these assumptions, the PEARSON-χ^2 *is zero if and only if $L = L^*$ is true.*
Note: $t_r = \infty$ is not practically realizable.

Proof. $L = L^*$ obviously yields $\chi^2 = 0$. Let $\chi^2 = 0$. This is equivalent to $B_i = E_i$ for all $i = 1, \ldots, k-1$. Conversion yields

$$(L - L^*) \int_{t_i}^{t_{i+1}} f_{c_1}(t)\,\mathrm{d}t = (L - L^*) \int_{t_i}^{t_{i+1}} f_{c_2}(t)\,\mathrm{d}t$$

for all i. It can be calculated that the integrals over (t_i, t_{i+1}) for $f_{c_1}(t)$ and $f_{c_2}(t)$ given in Propositions 13.8, 13.9 and 13.10 are respectively different. Therefore $L - L^* = 0$. □

For Two-compartment-iv models with elimination from both compartments, the empirical density of the distribution of the random variable X is obtained by the observation in both compartments only when the model parameters and L are known. With that in mind, the estimation of parameters via the Minimum-χ^2-method is not possible for this pharmacokinetic model.

Definition 15.1. *A pharmacokinetic model with corresponding probability theoretical model is said to be* ***estimable*** *when its parameters are estimable in a statistical sense with regard to the concentration measurements on hand.*

Proposition 15.2. *Of the Two-compartment-iv models (10.3), exactly $(k_{10}, 0, 0, 0)$, $(0, k_{12}, k_{20}, 0)$, $(0, k_{12}, k_{20}, k_{21})$ and $(k_{10}, k_{12}, 0, k_{21})$ are estimatable with regard to the varied Minimum-χ^2-method.*

Proof. This proposition arises from Propositions 13.1 to 13.7 and 15.1, as well as the given explanations. The existence of the respective varied minimum-χ^2-estimator will be studied in this chapter. □

The pharmacokinetic model $(0, k_{12}, k_{20}, 0)$ stands in connection with the model parameters c^0, k_{12} and k_{20}. If the concentration-time function $c_2(t)$ should be fitted to the measurements, then the meanings of the system parameters A, λ_1 and λ_2 are obvious in (see Lemma 10.4 Case 1):

$$c_2(t) = \frac{c^0 k_{12}}{k_{20} - k_{12}} \left(e^{-k_{12}t} - e^{-k_{20}t}\right) = A\left(e^{\lambda_1 t} - e^{\lambda_2 t}\right) .$$

The accompanying stochastic model is described in Proposition 13.4. The following is true for the density:

$$f_X(t) = f_{c_2}(t) = \frac{k_{12}k_{20}}{k_{20} - k_{12}} \left(e^{-k_{12}t} - e^{-k_{20}t}\right) .$$

The model parameters of the stochastic attempt are k_{12} and k_{20}.

Definition 15.2. *An estimable pharmacokinetic model is said to be* ***calculable*** *when the parameters of the pharmacokinetic model can be calculated from the estimated parameters of the associated stochastic model. If an estimable pharmacokinetic model is uniquely associated with a compartment model, then it is called* ***identifiable*** *in the observed class of estimable compartment models.*

Proposition 15.3. *The following propositions are true for the class of Two-compartment-iv models* (10.3) *that are estimable regarding the varied Minimum-χ^2-method:*
$(k_{10}, 0, 0, 0)$ *can be calculated and identified,*
$(0, k_{12}, k_{20}, 0)$ *with* $k_{12} \neq k_{20}$ *can be calculated but not identified,*
$(0, k_{12}, k_{20}, 0)$ *with* $k_{12} = k_{20}$ *can be calculated and identified,*
$(0, k_{12}, k_{20}, k_{21})$ *can't be calculated and cannot be identified and*
$(k_{10}, k_{12}, 0, k_{21})$ *can be calculated and identified.*

Proof. The assertions arise with reference to the residence time distributions affiliated with the models, and to Table 12.1. The model parameter c^0 is calculated with regard to $AUC = B_1 + \cdots + B_{r-1}$ as well as k_{ij}. The residence time densities belonging to $(0, k_{12}, k_{20}, 0)$ with $k_{12} \neq k_{20}$ and $(0, k_{12}, k_{20}, k_{21})$ coincide. □

The uniqueness of the estimated values as well as the quality of the estimates are now examined. The possibility of receiving a more adequate mathematical description by means of truncation was emphasized above.

15.2.2 *Qualities of the varied minimum-χ^2-estimator*

A quality of the Minimum-χ^2-estimator will be given along with the considerations concerning the sample size formulated at the beginning of this chapter. The number B_i in I_i observed drug molecules is proportional to the integral of the observed concentration-time function $c_b(t)$ over this interval. The scale of the concentration measurements is chosen according to the precision of the method used for measuring. Multiplying the scale by L leads to multiplication of the integrals by L. Consequently, B_i, or $N = B_1 + \cdots + B_n$ change into $B_i^* = LB_i$, or $N^* = LN$.

Proposition 15.4. *The given varied minimum-χ^2-estimate of a parameter corresponding with a pharmacokinetic model is invariant under multiplicative changes of units of measurements in which concentrations are given.*

Proof. For $L \neq 0$, the following is true for the system of equations defining the varied minimum-χ^2-estimator:

$$0 = \sum_{i=1}^{n} \frac{B_i - E_i}{p_i} \frac{\partial p_i}{\partial \alpha_j} = L \sum_{i=1}^{n} \frac{B_i - E_i}{p_i} \frac{\partial p_i}{\partial \alpha_j} = \sum_{i=1}^{n} \frac{B_i^* - E_i^*}{p_i} \frac{\partial p_i}{\partial \alpha_j} .$$ □

In general, the weighted MLS calculation of the parameters does not have this property. It is a remarkable advantage which approaches the striven general validity of specified pharmacokinetic parameters.

A fundamental theorem is consulted for the characterization of the statistical qualities of the varied minimum-χ^2-estimator in connection with Two-compartment-iv models (10.3). It dates back to R. A. FISHER, E. A. PEARSON, J. NEYMAN and H. CRAMER. Bibliographical references as well as a proof can be found in CRAMER (1946). It is also shown there (Chapter 33.4) that the propositions about the asymptotic distribution of χ^2 for parameters, which are to be estimated for arbitrary asymptotically normal and asymptotically effective estimators, exist. They are therefore also valid for maximum-likelihood estimators. The problem is also discussed by FISZ (1989, Chapter 12.4). The theorem will now be quoted.

Theorem 15.1. (CRAMER *1946, pages 427-434*) *n real functions $p_1(\alpha_1, \ldots, \alpha_m), \ldots, p_n(\alpha_1, \ldots, \alpha_m)$ of $m < n$ variables $\alpha_1, \ldots, \alpha_m$ are given. Let the following be true for all points $(\alpha_1, \ldots, \alpha_m)$ of an interval $\mathcal{R}$ over $\mathbb{R}^m$ that has not degenerated:*

(1) $\sum_{i=1}^{n} p_i(\alpha_1, \ldots, \alpha_m) = 1$.
(2) $p_i(\alpha_1, \ldots, \alpha_m) > 0$ *for all* $i = 1, \ldots, n$.

(3) *All p_i have continuous partial derivatives*

$$\frac{\partial}{\partial \alpha_j} p_i \quad \text{and} \quad \frac{\partial^2}{\partial \alpha_j \partial \alpha_k} p_i .$$

(4) *The matrix* $J = \left(\frac{\partial p_i}{\partial \alpha_j} \right)$, $i = 1, \ldots, n$ *and* $j = 1, \ldots, m$, *has rank* m.

Let X be a random variable, $I_1, \ldots, I_n$ a disjoint partition of its range, and $p_i^0 = p_i(\alpha_1^0, \ldots, \alpha_m^0) = P(X \in I_i)$, $\alpha^0 = (\alpha_1^0, \ldots, \alpha_m^0)$ an inner point of $\mathcal{R}$. B_i for N realizations of X denotes the number of values in I_i, $B_1 + \cdots + B_n = N$.
Then the system of equalities

$$\sum_{i=1}^{n} \frac{B_i - N p_i}{p_i} \frac{\partial}{\partial \alpha_j} p_i = 0 \, , \quad j = 1, \ldots, m,$$

has a unique solution $\hat{\alpha} = (\hat{\alpha}_1, \ldots, \hat{\alpha}_m)$. It approaches α^0 for $N \to \infty$. The random variable

$$\chi^2 = \sum_{i=1}^{n} \frac{(B_i - N\hat{p}_i)^2}{N\hat{p}_i} \, ,$$

calculated with $p_i = p_i(\hat{\alpha}_1, \ldots, \hat{\alpha}_m)$, is asymptotically χ^2 distributed with $(n - m - 1)$ degrees of freedom.

The varied minimum-χ^2-estimators that are connected with the estimable Two-compartment-iv models (10.3) will now be examined.
$I_1 = [0, t_2), \ldots, I_i = [t_i, t_{i+1}), \ldots, I_{r-1} = [t_{r-1}, \infty)$ define a partition of $\mathbb{R}^+$ given by the measuring times $t_1, \ldots, t_r, r > 2$. The difficulties that arose from B_0, or B_∞ in the practical application of the method were already mentioned (B_0 and B_∞ correspond with the intervals $[0, t_1)$, or $[t_k, \infty)$) . B_i, as well as N are defined in Chapter 15.2.1.
It is assumed that the previous considerations concerning the pharmacokinetical experiment do not lead to the decline of the probability theoretical model presented in Chapter 13.

Proposition 15.5. *A parameter estimation due to the Minimum-χ^2-method exists for the Two-compartment-iv model $(k_{10}, 0, 0, 0)$. It is uniquely determined and converges for $N \to \infty$ towards the parameter k_{10}.*

Proof. The density $f_X(t) = k_{10} e^{-k_{10}t}$ of the residence time distribution in connection with the model yields

$$p_i = \int_{t_i}^{t_{i+1}} f_X(t) \, \mathrm{d}t = e^{-k_{10} t_i} - e^{-k_{10} t_{i+1}}$$

for $i = 2, \ldots, k-2$ and $p_1 = 1 - \mathrm{e}^{-k_{10}t_2}$, $p_{k-1} = \mathrm{e}^{-k_{10}t_{k-1}}$. It is shown that the assumptions of the theorem from CRAMER are fulfilled.

(1) Assumption: It is fulfilled due to the definition of p_i.
(2) Assumption: $p_i > 0$ yields from the qualities of the exponential distribution.
(3) Assumption: It is fulfilled due to the qualities of the exponential distribution.
(4) Assumption: $\frac{\mathrm{d}}{\mathrm{d}k_{10}} p_1 = k_{10}\mathrm{e}^{-k_{10}t_2} \neq 0$ for $k_{10} \neq 0$ and $t_2 \neq 0$.

The proposition then comes from the theorem of CRAMER. □

Proposition 15.6. *A parameter estimation due to the varied Minimum-χ^2-method exists for the Two-compartment-iv model $(0, k_{12}, k_{20}, 0)$ with $k_{12} = k_{20} = k$. It is uniquely determined and converges for $N \to \infty$ towards the parameter k.*

Proof. The density $f_X(t)$ of the residence time distribution in connection with the model in Proposition 13.5 yields

$$p_1 = 1 - (1 + kt_2)\, e^{-kt_2}, \; p_i = (1 + kt_i)\, e^{-kt_i} - (1 + kt_{i+1})\, e^{-kt_{i+1}} \,,$$

$i = 2, \ldots, k-2$, and

$$p_{k-1} = (1 + kt_{k-1})\, e^{-kt_{k-1}} \,.$$

It is shown that the assumptions of the theorem of CRAMER are fulfilled for this model.

(1) Assumption: It is fulfilled because of the definition of p_i.
(2) Assumption: The p_i (the integrals of $f_X(t)$ on the I_i) are truly larger than zero because $f_X(t) > 0$ and $t_1 < t_2 < \cdots < t_k$.
(3) Assumption: The first and second derivatives of p_i for k are continuous again as sums of products of continuous functions.
(4) Assumption: $\frac{\mathrm{d}}{\mathrm{d}k} p_1 = kt_2^2\mathrm{e}^{-kt_2} \neq 0$.

The proposition then results due to CRAMER's theorem. □

Proposition 15.7. *Due to the varied minimum-χ^2-method, a parameter estimation exists for the Two-compartment-iv model $(0, k_{12}, k_{20}, 0)$ where $k_{12} \neq k_{20}$. It is uniquely determined and converges towards the parameter vector (k_{12}, k_{20}) for $N \to \infty$.*

Proof. The density that corresponds with the model in Proposition 13.4:

$$f_X(t) = \frac{k_{12}k_{20}}{k_{20}-k_{12}}\left(e^{-k_{12}t} - e^{-k_{20}t}\right)$$

yields

$$p_1 = \frac{k_{12}k_{20}}{k_{20}-k_{12}}\left[\frac{\left(e^{-k_{20}t_2}-1\right)}{k_{20}} - \frac{\left(e^{-k_{12}t_2}-1\right)}{k_{12}}\right],$$

$$p_i = \frac{k_{12}k_{20}}{k_{20}-k_{12}}\left[\frac{\left(e^{-k_{20}t_{i+1}}-e^{-k_{20}t_i}\right)}{k_{20}} - \frac{\left(e^{-k_{12}t_{i+1}}-e^{-k_{12}t_i}\right)}{k_{12}}\right],$$

$i = 2, \ldots, k-2$, and

$$p_{k-1} = \frac{k_{12}k_{20}}{k_{20}-k_{12}}\left(\frac{e^{-k_{12}t_{k-1}}}{k_{12}} - \frac{e^{-k_{20}t_{k-1}}}{k_{20}}\right).$$

It is shown that the assumptions of the theorem of CRAMER are fulfilled for this model.

(1) Assumption: It is fulfilled due to the definition of the p_i.
(2) Assumption: The p_i are truly greater than zero because $f_X(t) > 0$ and $t_1 < t_2 < \cdots < t_k$.
(3) Assumption: According to the model assumptions, $p_i = p_i(k_{12}, k_{20})$ are defined on the interior of $\mathbb{R}^{2+} = \{(k_{12}, k_{20}) : k_{12} \in \mathbb{R}^+, k_{20} \in \mathbb{R}^+\} \subset \mathbb{R}^2$ and are continuous again as a sum of the products of continuous functions. The partial derivatives of 1st and 2nd order are also products of continuous functions and are therefore continuous functions again. A detailed proof of this proposition can be avoided here. Expressions such as:

$$\begin{aligned}\frac{\partial}{\partial k_{12}}p_i = \frac{k_{20}}{k_{12}-k_{20}}&(t_ik_{12}e^{-k_{12}t_i} - t_ik_{20}e^{-k_{12}t_i} + e^{-k_{12}t_i}\\ &- e^{k_{20}t_i} - k_{12}t_{i+1}e^{-k_{12}t_{i+1}} + k_{20}t_{i+1}e^{-k_{12}t_{i+1}} - e^{-k_{12}t_i}\\ &+ e^{-k_{20}t_{i+1}})\end{aligned}$$

are obtained.
(4) Assumption: It has to be proven that the rank of the functional matrix J is 2. It is sufficient to show that a quadratic submatrix J_u exists where the determinant in every point of the defined range of p_i is different from zero. The following is observed:

$$J_u = \begin{vmatrix} \frac{\partial p_1}{\partial k_{12}} & \frac{\partial p_1}{\partial k_{20}} \\ \frac{\partial p_2}{\partial k_{12}} & \frac{\partial p_2}{\partial k_{20}} \end{vmatrix}.$$

The assumption that $\det K_u = 0$ is true for an internal point of $\mathbb{R}^{2+}$ contradicts the given assumption $0 < t_2$ that comes from the context of the pharmacokinetic experiment. The determinant

$$\begin{aligned}\det J_u = &-[\{(\{k_{12} - k_{20}\}t_2 + 1)e^{-k_{12}t_2} + e^{-k_{20}t_2}\} \\ &\times \{(\{k_{12} - k_{20}\}t_2 - 1)e^{-k_{20}t_2} - (\{k_{12} - k_{20}\}t_3 - 1) \\ &\times e^{-k_{20}t_3} + e^{-k_{12}t_2} - e^{-k_{12}t_3}\} - \{-(\{k_{12} - k_{20}\}t_2 - 1) \\ &\times e^{-k_{20}t_2} - e^{-k_{12}t_2}\}\{(\{k_{12} - k_{20}\}t_2 + 1)e^{-k_{12}t_2} \\ &- (\{k_{12} - k_{20}\}t_3 + 1)e^{-k_{12}t_3} - e^{-k_{20}t_2} + e^{-k_{20}t_3}\}] \\ &\times k_{12}k_{20}[k_{12} - k_{20}]^{(-4)}\end{aligned}$$

is zero if and only if

$$\begin{aligned}&\{(\{k_{12} - k_{20}\}t_2 + 1)\mathrm{e}^{-k_{12}t_2} + \mathrm{e}^{-k_{20}t_2}\} \\ &\quad \times \{(\{k_{12} - k_{20}\}t_2 - 1)\mathrm{e}^{-k_{20}t_2} - (\{k_{12} - k_{20}\}t_3 - 1) \\ &\quad \times \mathrm{e}^{-k_{20}t_3} + \mathrm{e}^{-k_{12}t_2} - \mathrm{e}^{-k_{12}t_3} \\ &\quad = \{-(\{k_{12} - k_{20}\}t_2 - 1)\mathrm{e}^{-k_{20}t_2} - \mathrm{e}^{-k_{12}t_2})\} \\ &\quad \times \{(\{k_{12} - k_{20}\}t_2 + 1)\mathrm{e}^{-k_{12}t_2} - (\{k_{12} - k_{20}\}t_3 + 1) \\ &\quad \times \mathrm{e}^{-k_{12}t_3} - \mathrm{e}^{-k_{20}t_2} + \mathrm{e}^{-k_{20}t_3}\}\end{aligned}$$

is true. The functions obtained in this expression are developed up to the third component of a TAYLOR series at point $(0, 0)$:

$$f_1(k_{12}, k_{20}) = e^{-k_{12}t_i} = 1 - k_{12}t_i + k_{12}^2 t_i^2/2;$$
$$f_2(k_{12}, k_{20}) = e^{-k_{20}t_i} = 1 - k_{20}t_i + k_{20}^2 t_i^2/2;$$
$$f_3(k_{12}, k_{20}) = t_i(k_{12} - k_{20})e^{-k_{12}t_i} = (k_{12} - k_{20})t_i + (k_{12}k_{20} - k_{12}^2)t_i^2;$$
$$f_4(k_{12}, k_{20}) = t_i(k_{12} - k_{20})e^{-k_{20}t_i} = (k_{12} - k_{20})t_i + (-k_{12}k_{20} + k_{20}^2)t_i^2 .$$

These power series absolutely converge for all points of $\mathbb{R}^{2+}$. They can be added term by term and yield the TAYLOR series of the sum of the respective functions. The first three terms are added:

$$\begin{aligned}&\{(-k_{12}k_{20} + k_{12}^2/2 + k_{20}^2/2)t_2^2\} \times \{2 - 2k_{20}t_2 + (3k_{20}^2/2 - k_{12}k_{20} \\ &\quad + k_{12}^2/2)t_2^2 + (k_{12}k_{20} - k_{12}^2/2 - k_{20}^2/2)t_3^2\} \\ &\quad = \{(k_{12}k_{20} - k_{20}^2/2 - k_{12}^2/2)t_2^2\} \times \{2 - 2k_{12}t_3 \\ &\quad + (k_{12}k_{20} - k_{12}^2/2 - k_{20}^2/2)t_2^2 \\ &\quad + (3k_{12}^2/2 + k_{20}^2/2 - k_{12}k_{20})t_3^2\} .\end{aligned}$$

The series should be ordered in the form

$$f_{Taylor}(k_{12}, k_{20}) = L_0 + L_1 k_{12} + L_2 k_{20} + L_3 k_{12}k_{20} + L_4 k_{12}^2 + L_5 k_{20}^2 + \ldots$$

The CAUCHY product series are formed on the left and right side of the equality. They also converge towards the respective product of the represented functions due to the absolute convergence of the function series on $\mathbb{R}^{2+}$. The comparison of coefficients is possible due to the identity theorem for power series. The first three terms of the product series on either side of the equals sign are zero. The following is true for the fourth term:

$$-2t_2^2k_{12}k_{20} = 2t_2^2k_{12}k_{20} \; .$$

A contradition to the assumptions $k_{ij} > 0$ and $t_2 > 0$ arises here. Consequently $\det J_u \neq 0$ for all (k_{12}, k_{20}) from the interior of $\mathbb{R}^{2+}$. The rank of J is 2.

The proposition comes from the theorem of CRAMER. □

Proposition 15.8. *Due to the varied minimum-χ^2-method, a parameter estimation exists for the Two-compartment-iv model $(0, k_{12}, k_{20}, k_{21})$. It is uniquely determined and converges towards the parameter vector (λ_1, λ_2) for $N \to \infty$.*

The proof follows analogously with the one of the previous proposition. λ_1 and λ_2 are explained in Proposition 13.6.
An analogy to the last proposition is not proven for the estimable model $(k_{10}, k_{12}, 0, k_{21})$. Exceptionally confusing expressions were given for the formation of the determinant of a 3×3 function matrix as well as the development of series for functions of three variables. The thoughts behind the proof were already explained. The processing of this formulation possibly can be carried out with the help of a computer algebra system.
The following refers to a truncated residence time distribution.

Proposition 15.9. *An estimation with regard to the varied minimum-χ^2-method exists for the parameter k_{10} of the truncated in the measuring times t_1 and t_r residence time distribution associated with the corresponding One-compartment-iv model $(k_{10}, 0, 0, 0)$. It is uniquely determined and converges, for $N \to \infty$, towards the parameter.*

Proof. The following is obtained from f_{X_g} with regard to Proposition 13.20:

$$p_{ig} = \frac{e^{-k_{10}t_i} - e^{-k_{10}t_{i+1}}}{e^{-k_{10}t_1} - e^{-k_{10}t_r}} \; .$$

The assumption $\frac{\mathrm{d}}{\mathrm{d}k_{10}} p_{ig} = 0$ leads to a contradiction. From TAYLOR expansion and the formation of the CAUCHY product series, a comparison of coefficients yields $(t_r + t_1) = (t_{i+1} + t_i)$ for their third terms, which is not possible for $r > 2$ and $t_i > 0$. The derivatives therefore do not disappear. The other assumptions of the theorem of CRAMER can easily be checked.

□

It is shown above in Chapter 15.1 that transformations of measurements influence the MLS-calculation of parameters. In an example from econometrics, (the example is closely related to the models used here), RAO (1986) leads to the point that the maximum-likelihood estimation remains unchanged under the logarithmic transformation of data. Considerations regarding the invariance of estimators under certain transformations of random variables are found in BOROVKOV (1984, Chapter 2). It is not necessary to examine such questions in further detail with regard to the problems that are dealt with here; the original measurements can be fallen back on for the estimation of parameters.

A practical experience should be given in conclusion: The varied minimum-χ^2-estimate behaves moderately with respect to errors in measurements. With regard to this, it is preferred over the MLS-calculation. This observation will not be discussed in detail.

An example of a parameter estimation by means of the varied minimum-χ^2-method can be found at the end of Chapter 16.

15.3 The varied minimum-χ^2-method applied to population kinetics

The problem of averaging proper nonlinear functions was addressed in Chapter 8. It is of special interest in a population kinetics context. The described varied minimum-χ^2-method particularly offers the possibility of calculating average kinetics. Individual kinetics $c^{(l)}(t)$, $l = 1, \ldots, k$, deliver each $B_i^{(l)}$ observed and $E_i^{(l)}$ expected numbers of realizations of residence times in the respective intervals I_i (see the explanations of the varied minimum-χ^2-estimation above).

The following process can be proposed to determine the varied minimum-χ^2-averaged kinetics $c_{av}(t)$ of the given individual kinetics $c^{(l)}(t)$:

Plan to determine varied minimum-χ^2-averaged kinetics

1. Chose a system of knots t_i, $i = 1, 2, ..., r$, and **fix** it for all measurements!
2. Calculate the parameter $(\hat{\alpha}_1, \ldots, \hat{\alpha}_m)$ of the varied Minimum-χ^2-averaged kinetics $c_{av}(t)$ as the solution to

$$\sum_{l=1}^{k}\sum_{i=1}^{n}\frac{\left(B_i^{(l)}-E_i^{(l)}\right)}{p_i}\frac{\partial p_i}{\partial \alpha_j}=0\ ,\quad j=1,\ldots,m, m<n\ .$$

The requirement of a system of fixed knots t_i, $i = 1, 2, ..., r$, for all measurements is a necessary condition for the proposed averaging method. This way the averaging is similar to the situation one meets at the classical χ^2-test of homogeneity.
As an example, two kinetics $c^{(1)}(t) = A^{(1)}\mathrm{e}^{-k_{10}^{(1)}t}$ and $c^{(2)}(t) = A^{(2)}\mathrm{e}^{-k_{10}^{(2)}t}$ are varied minimum-χ^2-averaged (see Figure 15.3).

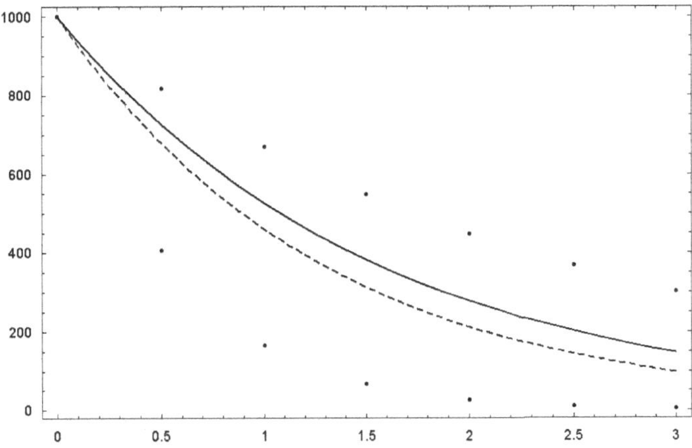

Fig. 15.3: Illustration of the varied minimum-χ^2-averaging (solid line) of two individual kinetics (single dots). The MLS-averaged curve (dotted line) is given for comparison. Explanations can be found in the text.

Chapter 16

Selection of pharmacokinetic models

If a mathematical model is formulated for the timely change of the concentrations of drugs in an organism, then it is to be biochemically, physiologically, etc. verified in a technical context. The a-posteriori selection of a model cannot do without a mathematical characterization of the attempts of the formula and the method for model parameter calculation regarding given data. Well-founded rules for decisions should be formulated. The judgement of the agreement of measurements with a function chosen for the description of them is a generally formulated task. It requires the appropriate choice of respective applications and assesment of distance concepts. For experimental evaluations, statistical methods are preferred in which experimental planning (in the statistical sense), parameter estimation and tests of goodness of fit infer the ideal problem solution. There is an extensive literature to this problem area concerning linear statistical models.

Such an extensive analytic treatment of the discrimination of models is hardly conceivable for nonlinear statistical attempts in relation with pharmacokinetics.

An impression of the selection of models in pharmacokinetically applied methods is given by BARTFAI/MANNERSIK(1972), WAGNER (1975) or MEIER/RETTIG/HESS (1981). BUCKWITZ/HOLZHUETTER and MCINTOSH/MCINTOSH (1980), WAGNER (1993), BOURNE (1995) also give reference to the kinetics of enzymes.

Presently, mathematically founded propositions concerning the quality of the agreement of data and models are still the exception in pharmacokinetics. The following listed rules have the characteristics of agreements for model discrimination:

MEIER/RETTIG/HESS (1981) recommend the sum of squared errors as a reference quantity for the selection of a model. The required reference value is obtained by hand by drawing an approximating curve for the measurements. PFEIFFER/PFLEGEL/BORCHERT (1988) formulate the following regulation: determine the logarithms $\ln[cm(t_i)]$ from the decreasing part of the observed time course of concentration of available measurements $cm(t_i)$, then adapt a linear function of these results due to MLS and determine the value $A_1 := e^S$ from its intersection point $(0, S)$ and with the ordinate. A number A_2 for at least the three last measurements can be calculated in the same way. A one-compartment model is assumed when $(A_1 - A_2)/A_2 < 0.1$ holds. Grounds for this rule are not given. BARTFAI/MANNERSIK (1972) recommend the rejection of the model in question if numeric difficulties appear during a matrix inversion or the applied iterative procedure does not converge for the data set on hand. By suitable variable transformation, the graph of a nonlinear function can become a straight line. The visual distinction of models is supported in this way. For example,

$$v(x) = \frac{a_1 x}{1 + b_1 x} \quad \text{and} \quad \frac{1}{v(x)} = \frac{b_1}{a_1} + \frac{1}{a_1}\frac{1}{x}$$

are equivalent.

The graph of the last function represents a straight line for the coordinates $(1/v(x), 1/x)$. The distinction of

$$w\,(x) = \frac{a_1 x + a_2 x^2}{1 + b_1 x + b_2 x^2}$$

is thus simple. In this case the $(1/w(x), 1/x)$-representation is no longer linear (WALTER 1977).

WAGNER (1975) proceeds: A one-compartment model is assumed if the measurements $cm(t_i)$ can be adequately described by a function $c(t) = Ae^{at}$. A two-compartment model is used if an attempt $c(t) = Ae^{at} + Be^{bt}$ is necessary.

The number of exponential terms $A_i e^{a_i t}$ that are summed up for the describing $c(t)$-function is made concrete by WAGNER (1977) with regard to a specific measure of certainty. The rules for decisions are empiric and are listed in a table. For example, a sum of n exponential terms is adapted to data on hand in accordance with MLS; the given measure of certainty for this is less than 0.75. The extended model is chosen provided that the MLS adaptation of $(n + 1)$ exponential terms produces an improvement in the measure of certainty by more than 10%. BOXENBAUM/RIEGELMANN/ELASHOFF (1974) explain the exact same procedure, though with regard to the sum of weighted quadratic deviations. In the monograph by WAGNER (1993), Chapter 17 is dedicated to questions about model discrimination.

BARDSLEY/MCGINLAY (1987) measure the distance between two functions adapted to the same data in the following way: The square of their differences is integrated over the observation interval. RAO/IYENGA (1984) set

$$y_{obs} = \varrho \left[y^{(1)} - y^{(2)} \right] + \left[y^{(1)} - y^{(2)} \right] /2 \,.$$

y_{obs} are the observed values and $y^{(i)}$ are the calculated values with regard to model (i). ϱ is to be estimated. Model (1) is favored if ϱ is 1/2. Model (2) is favored when ϱ is near $-1/2$.

The obvious empirical characteristic of the given rules for the choice of models immediately allows the user to evaluate the results. However, statistical deciding procedures require a method of criticism on the basis of knowledge and assessment of the assumptions and the mathematical qualities of the test.

Without exceptions, the statistical methods found in literature about choices of models in pharmacokinetics refer to regression models

$$cm_i = c\,(t_i) + \epsilon_i, \quad i = 1, \ldots, r \,.$$

Here $c(t)$ is a model function and the cm_i are the measured drug concentrations. The remainders ϵ_i are understood as random variables. Two methods are differentiable.

1. Nonparametric statistical procedures that make as few assumptions as possible necessary, serve for the selection of models due to the judgement of the remainders. This includes sign and rank tests which are recommended for the application in the pharmacokinetics by MEIER/RETTIG/HESS (1981) and BUCKWITZ/HOLZHUETTER. The possibilities of conditional tests of FISHER and PITMAN based on combinatorical considerations should be noted.

2. The ϵ_i are pressuposed to be stochastically independent. Homoscedasticity is usually accepted. Hypothesis examination is carried out by means of methods which are justified for linear statistical models with regard to the supposition of normal distribution. BARTFAI/MANNERSIK (1972), BOXENBAUM/RIEGELMANN/ELASHOFF (1974) and STEINIJANS (1982) judge the variance of the remainder with the F-test. Mathematically not well-founded, but for its situation however practicable, BARDSLEY/MCGINLEY (1987) assess the application of the F-test for the choice of models. GLADIGAN/VOLLMER (1977) judge the different nature of observed time courses of concentration with the help of an analysis of variance. The sum S of

weighted quadratic remainders, as well as the number m of parameters to be estimated have influence on the criteria of goodness of fit by AKAIKE (1974) ($AIC = S + 2m$) and SCHWARZ (1978) ($SCH = S + m \ln m$). YAMAOKA/NAKAGAWA/UNO (1978a) applied the AKAIKE criterion to pharmacokinetic examples. Normally distributed remainders were assumed. They came to the conclusion that for the observed data sets the AIC and F-Test both lead to the selection of the same model. LUDDEN/BEAL/SHEINER (1994) compare the same criteria, on the basis of Monte-Carlo simulations though.

Once again, attention should be drawn to the peculiarity of the pharmacokinetic experiment: An individual kinetics cannot be repeated under equal conditions and as a rule, includes no more than say 10 measurements at various times. A verification of the assumptions of the distribution with regard to ϵ_i is therefore not possible. In this case, a regression attempt for individual kinetics proves to be unsuitable for the handling of the problem of the selection of a model. This is however specific to the area of application. For example, examinations of the kinetics of enzymes permit the repetition of experiments so that the error model of a regression attempt is verifiable and variance estimates make sense for the situation (BUCKWITZ/HOLZHUETTER). A large group of statistical tests of goodness of fit refer to the comparison of an empirical and a predefined distribution function. Included here are the tests from KOLMOGOROV/SMIRNOV, ANDERSON/DARLING, CRAMER/V. MISES and others. FROSINI (1987) gives an overview. The corresponding test statistics are calculated from the realizations of the random variable in question. As shown in Chapter 15.2, the measurements $cm(t_i)$ cannot be looked at as the realization of the random variable residence time. DEVROYE (1987) deals with distances for densities. He also briefly discusses density estimations and the selection of models in the preface.
From the combination of pharmacokinetic and probability theoretical attempts developed in Chapter 13 and Chapter 15.2 it can be seen that the classic χ^2-test of goodness of fit from K. PEARSON for the selection of models in pharmacokinetics can be recommended when the estimates of parameters yield from the varied minimum-χ^2-method.

Proposition 16.1. *In the case of Two-compartment-iv models* $(k_{10}, 0, 0, 0)$, $(0, k_{12}, k_{20}, 0)$ *and* $(0, k_{12}, k_{20}, k_{21})$, *the parameters of the associated residence time distributions (cf. Chapter* 13 *and Chapter* 15.2 *) are estimat-*

able by means of the varied minimum-χ^2-method. Then the model decision can be performed by means of the χ^2-test of goodness of fit from PEARSON.

Proof. The mentioned models can be estimated with respect to the varied Minimum-χ^2-method (Proposition 15.2). The proofs of Propositions 15.5 to 15.8 and the Theorem by CRAMER yield that the PEARSON test statistic χ^2 is asymptomatically χ^2-distributed. The number of degrees of freedom yields $(r - m - 1)$ where m describes the number of estimated parameters. □

The applicability of the PEARSON-test could not be proven for the estimatable Two-compartment-iv model $(k_{10}, k_{12}, 0, k_{21})$; see the formulated remarks at the end of Proposition 15.8.

Proposition 16.2. *Let the residence time distribution that is truncated at the first and last measurements be assigned to the One-compartment-iv model $(k_{10}, 0, 0, 0)$. It's parameter k_{10} is estimatable via the varied minimum-χ^2-method. The agreement of the measurements and the function $c(t) = c^0 e^{-k_{10}t}$ can then be judged by means of the χ^2-test of goodness of fit. The number of degrees of freedom is $(r - 2)$.*

Proof. One goes back to Proposition 15.9 and argues in the same way as in the preceding proof. □

It is worth mentioning that under truncation, the test decision is given with respect to the observation interval $[t_1, t_r]$! The parameter estimate also has to be interpreted in connection with the probability theoretical model that arose through truncation.
An example will illustrate the test of goodness of fit in connection with the parameter estimation. The model function $c(t) = c^0 e^{-k_{10}t}$ is chosen and truncation is carried out at t_1 and t_r. Next, the 'measured values' are calculated for $c^0 = 250$ and $k_{10} = 0.5$ at measuring times $t_i = i, i = 1, \ldots, 10$. For odd i these are exactly the $c(t_i)$. For even i the $c(t_i)$ are alternatively increased or decreased by JPR percent of their values. With these data, the estimate $\hat{k}_{10}/CHI$ of k_{10} is obtained with regard to the varied minimum-χ^2-method with reference to the residence time distribution with truncation. The corresponding χ^2-value CHI, as well as the accompanying residual sum of squares RSS are calculated. For the model with truncation, the MLS-estimates $\hat{k}_{10}/MLS$, as well as the corresponding CHI and RSS were investigated at the same time. The different ways that the estimation principles work can be seen in Table 16.1. The critical values of the

Table 16.1: Varied minimum-χ^2-estimations $\hat{k}_{10}/CHI$ and MLS-estimations $\hat{k}_{10}/MLS$ together with χ^2-values CHI and the residual sum of squares RSS for the pharmacokinetical model $c(t) = c^0 \mathrm{e}^{-k_{10}t}$ with truncation. As indicated in the text, the data sway around JPR % around the function $c(t) = 250\mathrm{e}^{-0.5t}$. A computer program one can find in Part III.

JPR %	$\hat{k}_{10}/CHI$	CHI	RSS	$\hat{k}_{10}/MLS$	CHI	RSS
0	0.5000	0.00	0.00	0.5000	0.00	0.00
5	0.5058	0.19	42.79	0.4948	0.30	21.02
10	0.5161	0.67	144.25	0.4898	1.18	84.17
20	0.5231	3.03	572.10	0.4801	4.58	337.48
30	0.5343	6.75	1276.78	0.4710	9.98	761.11
40	0.5455	11.91	2252.32	0.4624	17.23	1356.14
50	0.5565	18.49	3493.38	0.4543	26.20	2123.58

tests here are $\chi^2_{0.05,8} = 15.51$ and $\chi^2_{0.01,8} = 20.09$. It is recognizable at which fluctuation of the measurements the test of goodness of fit declines the compatibility with the best fitted model function. Table 16.2 shows the invariance of the estimates as well as that the decision depends on the size of the experiment (cf. Chapter 15.2).

The data, as indicated, cannot be understood as realizations of mathematical samples of the random variable residence time! It also was not intended to investigate power functions of the test. One often uses distribution families like the PEARSON system to make systematizations of the formation of alternative hypotheses. With regard to Proposition 13.17, distributions related to the considered compartment models are only part of the PEARSON system in special cases. For the example that was calculated, the last two intervals would have to be united, and the number of degrees of freedom would have to be reduced by 1 in order to satisfy the known demand that the intervals (t_i, t_{i+1}) should have a corresponding expected value of at least 5 in the χ^2-test.

It is not surprising that the non-rejection of the fitted function does not make a pharmacokinetic model concrete. WESTLAKE (1973) quotes LANCZOS (1956) in connection with pharmacokinetics: The functions

$$f(x) = 0.0951e^{-x} + 0.8607e^{-3x} + 1.557e^{-5x}$$

Table 16.2: Varied minimum-χ^2-estimation for JPR% = 50 dependent from c^0, cf. Table 16.1.

c^0	$\hat{k}_{10}/CHI$	CHI
150	0.5738	9.57
250	0.5738	15.96
350	0.5738	22.34

and

$$g(x) = 2.202e^{-4.45x} + 0.305e^{-1.58x}$$

practically cannot be distinguished on $(0, \infty)$. The value of the difference does not get any larger than 0.0064 here. With respect to stochastic processes, HERRMANN (1978) gives examples for monotonic courses for various models, for which deciding on a model isn't possible solely from the fitted curve.

The pharmacokinetic model cannot be proven to be adequate even when a test of goodness of fit does not reject the compatibility of the data and fitted function. WESTLAKE (1971) explains in an example that parameter calculation and good fit with regard to compartment 1 can go on with extremely large differences between model and observation with regard to compartment 2.

In consideration of the possible questions, the various experimental situations, the variety of the model attempts, and the thus required mathematical treatments, a choice of models does not seem unproblematic when left to a computer. There were already offers regarding this in different software packages (e.g. WAGNER 1975, VEROTTA/RECCHIA/URSO 1986).

Chapter 17

Pharmacokinetics for multiple applications

If the time courses of concentrations of active agents are repeatedly observed for an individual, sequences of individual kinetics are of interest. Such problems appear, for example, in the modeling of urea kinetics in dialysis patients (LOPOT 1990, JÄGER 1989, SARGENT and GOTCH 1980).
The mathematical description of pharmacokinetical processes with multiple applications will be discussed in this chapter. The repeated administration of a drug is the basis for a sure effect for almost all therapeutic strategies. On the other hand, intoxication is possible through the effects of accumulation. In 1924, WIDMARK and TANDBERG already published a mathematical description for the course of concentration of such events with regard to the One-compartment-iv model. Their methods were copied by DOST (1953) with regard to the One-compartment-ev model as well as by GIBALDI/PERRIER(1982) with regard to a many-compartment model. It will be shown that the WIDMARK-TANDBERG sequence of individual kinetics are not associated with such underlaying compartment models anymore. This has the consequence that the validity of the conditions of these attempts cannot be verified. There is hereby a cause to develop a broader method for the modeling of multiple applications. It is only an approach of the description of the events of interest. The fact that both procedures are compared leads to the possibility of obtaining an idea of the range of model error. Its dependence on the dosage interval and invasion constant parameters is derived for the situation given by DOST (1953, 1968). A definition is given first.

Definition 17.1. *After the expiration of a fixed length of time, the dose DOS of a drug is given to an organism in the same way. In the interest of mathematical modeling, it is presumed that all the conditions influencing pharmacokinetic processes are invariant. This is called multiple applications* *from now on. The concentration-time-function that corresponds with the time interval* $[\{n-1\}\tau, n\tau)$ *is denoted by* $c_n(t)$, $n = 2, 3....$

The observation of dosage intervals of equal lengths τ offers advantages in the following formulations but does not present any reductions in principles. One follows WIDMARK/TANDBERG (1924) in all publications known to the authors about multiple dosage. The time courses of drug concentration in blood are obtained by the addition of functions: if $c_1(t)$ describes the course on $[0,\tau)$, then the function $c_2(t) = c_1(t) + c_1(t-\tau)$ is assigned $[\tau, 2\tau)$ and the function

$$\begin{aligned} c_n(t) &= c_1(t) + c_1(t-\tau) + \ldots \\ &\quad + c_1(t - \{n-2\}\tau) + c_1(t - \{n-1\}\tau) \end{aligned} \tag{17.1}$$

is assigned $[\{n-1\}\tau, n\tau)$, $n = 2, ...$
From this, DOST (1953) calculates the courses of concentration $c_n(t)$ for the cases $c_1(t) = Ae^{-k_{el}t}$ as well as $c_1(t) = A[e^{-k_i t} - e^{-k_{el}t}]$ and GIBALDI/PERRIER (1982) calculate the courses of concentration $c_n(t)$ for the case $c_1(t) = A_1 e^{-k_1 t} + \cdots + A_m e^{-k_m t}$. Functions $c_n(t)$ and $c_{n+1}(t)$ are hardly different for sufficiently large t. The function sequence $c_n(t)$ pointwise converges towards a course of concentration $c_\infty(t)$. Is such an attempt to model multiple applications compatible with the given pharmacokinetic model?

Proposition 17.1. $c_1(t)$ *denotes the course of concentration for the one-compartment-iv model. For multiple applications, the functions* (17.1) *satisfy the differential equation* (10.1).

Proof. In accordance with Lemma 10.1, $c_1(t) = c^0 e^{-k_{el}t}$ on $[0,\tau)$ is the solution to the given initial value problem (10.1). With that in mind, the following is true:

$$c_2(t) = c_1(t) + c_1(t)\, e^{k_{el}\tau} = \left(1 + e^{k_{el}\tau}\right) c_1(t) \ .$$

Differentiation yields

$$c_2'(t) = -k_{el}\left(1 + e^{k_{el}\tau}\right) c_1(t) = -k_{el} c_2(t) \ ,$$

so (10.1) is fulfilled.
Generally one has $c_n(t) = [1 + e^{k_{el}\tau} + \cdots + e^{k_{el}\{n-1\}\tau}]c_1(t)$ and therefore $c_n'(t) = -k_{el}c_n(t)$. The inductive proof of the proposition can be omitted here. □

Proposition 17.2. *In accordance with Lemma 10.2, Case 2, the course of concentration for the one-compartment-ev model is* $c_1(t) = A[\mathrm{e}^{-k_i t} - \mathrm{e}^{-k_{el} t}]$. *For multiple application, the functions* (17.1) *do not satisfy the differential equation* (10.2).

Proof. The function $c_2(t)$ is formed from the given function $c_1(t)$ with regard to (17.1). It is assumed that it satisfies the differential equation (10.2). If it is substituted into this differential equation together with its derivative $c_2'(t)$, then the following yields:

$$A\left(k_{el}e^{-k_{el}t} - k_i e^{-k_i t}\right) + A\left(k_{el}e^{-k_{el}\{t-\tau\}} - k_i e^{-k_i\{t-\tau\}}\right)$$
$$= -k_{el}A\left(e^{-k_i t} - e^{-k_{el}t}\right) - k_{el}A\left(e^{-k_i\{t-\tau\}} - e^{-k_{el}\{t-\tau\}}\right) + k_i c_i^0 e^{-k_i t}\,.$$

In simpler form:

$$-Ak_i\left(e^{-k_i t} + e^{-k_i\{t-\tau\}}\right) = -Ak_{el}\left(e^{-k_i t} + e^{-k_i\{t-\tau\}}\right) + k_i c_i^0 e^{-k_i t}\,,$$

so $A(k_{el} - k_i)[1 + \mathrm{e}^{k_i\tau}] = k_i c_i^0$. With the obvious correspondence of A from Lemma 10.2, the last equation reduces itself to $1 + \mathrm{e}^{k_i\tau} = 1$. This relation does not exist for any $k_i > 0$ and any $\tau > 0$. $c_2(t)$ therefore does not satisfy the differential equation (10.2).
Forming the function $c_n(t)$ according to (17.1) and substituting it into the differential equation (10.2) yields $1 + \mathrm{e}^{k_i\tau} + \mathrm{e}^{2k_i\tau} + \cdots + \mathrm{e}^{\{n-1\}k_i\tau} = 1$ and also a contradiction. The complete inductive proof of the proposition does not have to be given here. □

It can also be proven for Case 1 of the One-compartment-ev model, according to Lemma 10.2, that for multiple applications, the courses of concentration due to (17.1) don't satisfy the differential equation (10.2) for the different dosage intervals. If multiple applications are formalized in the given way, then parameters k_i and k_{el} are not interpretable any more in the same way as in the scope of the One-compartment-ev model. Since it confines themselves to the One-compartment-iv model, the modeling of WIDMARK/TANDBERG (1924) of multiple application is compatible with the differential equation model (10.1).

The **requirement** therefore seems meaningful:

> For multiple application, the courses of concentration $c_1, c_2, \ldots, c_n, \ldots$ affiliated with respective dosage intervals have to satisfy the differential equation that defines the pharmacokinetic model.

The schematic generalization of the method formed in (17.1) possibly goes back to DOST (1953). In later literature, the modeling of multiple application, if at all, is always derived in the same way as in his fundamental book. It does not apply when the course of the function according to (17.1) is described in pharmacokinetic literature as a result of superpositions. This concept is well-defined in mathematics and describes methods for determining a solution to a differential equation by the addition or integration of elements of the set of solutions to the same equation. Proposition 17.2 however advises, that already for a simple model the sequence (17.1) does not list solutions to only one differential equation. This will be put in concrete terms for the mentioned example.
The differential equation (10.2)

$$c'(t) + k_{el}c(t) = I_1(t) = k_i c_i^0 e^{-k_i t}, \quad k_i > 0, k_{el} > 0$$

corresponds with the One-compartment-ev model. The initial condition is $c(0) = 0$.
For multiple application, a sequence of equations

$$c_n'(t') + k_{el}c(t') = I_n(t'), t' = t - \{n-1\} \ ,$$

with the initial conditions $c_1(0) = 0$, $c_2(0) = c_1(\tau)$, $c_3(0) = c_2(2\tau)$, ... can be set. The dosage intervals are listed with n. The inhomogeneities are the following:

$$\begin{aligned}
I_1(t') &= k_i c_i^0 e^{-k_i t'}, \\
I_2(t') &= k_i c_i^0 \left(e^{-k_i t'} + e^{-k_i\{t'+\tau\}}\right), \\
I_3(t') &= k_i c_i^0 \left(e^{-k_i t'} + e^{-k_i\{t'+\tau\}} + e^{-k_i\{t'+2\tau\}}\right), \\
&\vdots \ ,
\end{aligned}$$

where, as indicated, t' refers to the dosage interval.
The solution functions as written in (17.1) can be obtained if this sequence of initial value problems is solved. It does not seem necessary to explain these calculations in great detail. The observation that the description of multiple drug application is brought in connection with a sequence of assumptions that does not allow itself to be practically verified is important. One differential equation corresponds exactly with every dosage interval; it is therefore a special pharmacokinetic model. In this sense, the question of how acceptable such a mathematical modeling is, is asked. Satisfactorily indicating the invasion process of a drug knowingly already causes certain experimental problems in simple applications, i.e. observing $I_1(t')$!

However, multiple applications naturally let themselves be put in relation to exactly one differential equation for a compartment model. The initial value problem, as defined by the model of differential equations as well as the respective initial conditions, can be solved for every dosage interval. The concentration available at the beginning of the dosage interval is to be chosen as the initial condition. The result from WIDMARK/TANDBERG (1924) yields for the One-compartment-iv model.

Proposition 17.3. *The One-compartment-ev model* (10.2) *is presupposed with $k_i \neq k_{el}$. Then the* n*th concentration function $c_n(t)$ for $t \geq \{n-1\}\tau$ and defined on $[0, \infty)$ for multiple application $n = 2, 3, \ldots$ with associated dosage interval is given by*

$$\begin{aligned} c_n(t - \{n-1\}\tau) &= c_1(\tau)[e^{k_{el}\tau} + \ldots + e^{(n-1)k_{el}\tau}]e^{-k_{el}\tau} \\ &\quad + c_1(t - \{n-1\}\tau). \end{aligned} \tag{17.2}$$

According to Lemma 10.2, Case 2, *$c_1(.)$ then describes the solution to* (10.2).

Proof. The proof occurs by mathematical induction. Under the given conditions, the general solution to the inhomogeneous linear differential equation (10.2) is $c(t) = Fe^{-k_{el}t} + Ae^{-k_i t}$ with $F \in \mathbb{R}$ and $A = c_i^0/(k_{el} - k_i)$. The initial concentration dependent on the dose DOS of the invasion process is substituted into A. The integration constant F is determined from the initial condition.
Let $n = 1$. The initial condition $c_1(0) = 0$ then yields the relation $F = -A$ so that $c_1(t) = A[e^{-k_i t} - e^{-k_{el}t}]$. $c_1(t)$ is defined on $[0, \infty)$ for $t \geq 0$.
Let $n = 2$. The initial condition is $c_2(t - \tau) = c_1(\tau)$ for $t = \tau$ and yields $F = c_1(\tau) - A$. The function $c_2(t-\tau) = c_1(\tau)e^{k_{el}\tau}e^{-k_{el}t} + c_1(t-\tau)$ defined for $t \geq \tau$ on $[0, \infty)$ yields from the general solution.
Let $n = 3$. The initial condition is $c_3(t - 2\tau) = c_2(t - \tau)$ at $t = 2\tau$. With $F = c_2(\tau) - A$ and $c_2(\tau) = c_2(2\tau - \tau) = c_1(\tau)[e^{-k_{el}\tau} + 1]$ one gets the function $c_3(t - 2\tau) = c_1(\tau)[e^{k_{el}\tau} + e^{2k_{el}\tau}]e^{-k_{el}t} + c_1(t - 2\tau)$ defined by $t \geq 2\tau$ on $[0, \infty)$.
For $c_n(t - \{n-1\}\tau)$, the relation indicated in the proposition exists on the n-th dosage interval. The initial condition $c_{n+1}(t - n\tau) = c_n(t - \{n-1\}\tau)$ should be valid for $c_{n+1}(t - n\tau)$ at point $t = n\tau$. In addition, the function is the solution to the differential equation (10.2) from which $F = c_n(\tau) - A$ yields. One has $c_n(\tau) = c_1(\tau)[e^{-(n-1)k_{el}\tau} + \cdots + e^{-k_{el}\tau} + 1]$. This is substituted into the general solution to the differential equation (10.2), the

function $c_{n+1}(t-n\tau) = c_1(\tau)[e^{k_{el}\tau} + \cdots + e^{nk_{el}\tau}]e^{-k_{el}t} + c_1(t-n\tau)$ defined for $t \geq n\tau$ on $[0,\infty)$ yields. □

Due to their construction, the functions (17.2) satisfy the differential equation (10.2). The parameters k_{el} and k_i are interpretable for multiple application in the same way as they are within the scope of a One-ompartment-ev model. The uniqueness of these functions yields from the famous propositions of the theory of ordinary differential equations.
This method of description of the time courses of concentration for multiple application introduced here is not adequate: The problems with regard to the modeling difficulties are overcome, but the drug amounts that still have not gone from the extravascular space into the compartment at times n, $n = 1, 2, \ldots$, remain unconsidered.
Propositions concerning the convergence of the sequence of functions (17.2) are meaningful for the understating of the courses of concentration for multiple application. Steady-state or balanced kinetics are concepts that are connected with this. The functions are now denoted differently: $t' := t - \{n-1\}\tau$ varies in the boundaries of 0 to ∞ for $\{n-1\}\tau \leq t \leq n$. The functions (17.2) for $n = 2, 3, \ldots$ are written as

$$\begin{aligned} c_n(t') &= c_1(\tau)\left(e^{k_{el}\tau} + \ldots + e^{\{n-1\}k_{el}\tau}\right)e^{-k_{el}(t'+\{n-1\}\tau)} + c_1(t') \\ &= c_1(\tau)\left(e^{-\{n-2\}k_{el}\tau} + \ldots + e^{-k_{el}\tau} + 1\right)e^{-k_{el}t'} + c_1(t')\ . \quad (17.3) \end{aligned}$$

The expression contains the $(n-1)$th partial sum s_{n-1} of the geometrical sequence with the initial term 1 and the quotient $e^{-k_{el}\tau} < 1$. So

$$c_n(t') = c_1(\tau)s_{n-1}e^{-k_{el}t'} + c_1(t'), \qquad n = 2, 3, \ldots$$

with

$$s_{n-1} = \frac{1 - e^{-(n-1)k_{el}\tau}}{1 - e^{-k_{el}\tau}}\ .$$

Proposition 17.4. *$k_{el} \neq k_i$ is assumed for the One-compartment-ev model (10.2) with multiple application. Then the sequence (17.3) of the dosage interval associated concentration functions converges uniformly towards*

$$c_\infty(t') = \frac{c_i^0 k_i}{k_{el} - k_i}\left(e^{-k_i t'} - \frac{1 - e^{-k_i\tau}}{1 - e^{-k_{el}\tau}}e^{-k_{el}t'}\right)\ .$$

As it is typical in pharmacokinetics, this limit function is called a steady-state kinetics.

Proof. The sequence s_{n-1} of partial sums converges towards $[1 - \mathrm{e}^{-k_{el}\tau}]^{-1}$. The sequence of the functions (17.3) converges pointwise. The uniform convergence yields from the boundedness of $c_1(t')$ on $[0, \tau]$ as well as the boundedness and strict monotony of $\mathrm{e}^{-k_{el}t'}$. By the fact that $c_1(.)$ for τ, or t' is calculated in accordance with Proposition 17.3 and is substituted into

$$c_\infty(t') = \frac{c_1(\tau)}{1 - e^{-k_{el}\tau}} e^{-k_{el}t'} + c_1(t') \ ,$$

as indicated in the proposition, this expression obtains its form after some transformations. □

It is well known that in the case of uniform convergence, differentiation and integration can be interchanged with the formation of the limit. For the described situation of a multiple application, the approximation of the bioavailability is perhaps of interest. It is defined with regard to the time integrals of courses of concentration and can be obtained for the steady-state kinetics by the explained passage to the limit.
The function $c_\infty(t')$ particularily satisfies the differential equation (10.2). The corresponding courses of concentration over time $y_n(t')$ and $y_\infty(t')$ derived by Dost (1953, pages 253 ff.) are quoted here for the purpose of the comparison with functions $c_n(t')$, or $c_\infty(t')$ given in (17.3), or Proposition 17.4:

$$y_n(t') = \frac{ak_1}{k_1 - k_2}\left(\frac{1 - e^{-k_2\tau\{n+1\}}}{1 - e^{k_2\tau}} e^{-k_2t'} - \frac{1 - e^{-k_1\tau\{n+1\}}}{1 - e^{-k_1\tau}} e^{-k_1t'}\right)$$

and

$$y_\infty(t') = \frac{ak_1}{k_1 - k_2}\left(\frac{e^{-k_2t'}}{1 - e^{-k_2\tau}} - \frac{e^{-k_1t'}}{1 - e^{-k_1\tau}}\right) .$$

The corresponding parameter names are $a = c_i^0$, $k_1 = k_i$ and $k_2 = k_{el}$.

Proposition 17.5. *The following relation exists between the steady-state kinetics* $c_\infty(t')$ *of the One-compartment-ev model by multiple application mentioned in Proposition* 17.4 *and the corresponding function* $y_\infty(t')$ *given by* Dost *(1953):*

$$c_\infty(t') = \left(1 - e^{-k_i\tau}\right) y_\infty(t') \ .$$

Proof. The relation is obviously correct. □

Only finite $\tau > 0$ and $k_i > 0$ are of interest so that $c_\infty(t') < y_\infty(t')$ is always true. The difference between the two courses is easily calculated. Its relevance must be judged in the respective application case. Both functions correspond to fairly contrary attempts. $c_\infty(t')$ is derived with the knowledge that the actual invasion process is incompletely recorded. In connection with $y_\infty(t')$, it was presupposed and accepted that the invasion process changes itself qualitatively (but not observably) with time.
Two situations have to be distinguished: If the pharmacokinetic parameters are known, then $y_\infty(t')$ lies above the steady-state kinetics $c_\infty(t')$ which corresponds with the compartment model. If parameters are calculated from courses with regard to $y_\infty(t')$, then c_i^0 becomes overestimated. The calculation of the initial dose DOS^* also has to be corrected. With regard to $y_\infty(t')$ one gets

$$DOS^* = DOS \cdot R_y = \frac{DOS}{(1 - e^{-k_{el}\tau})\,(1 - e^{k_i\tau})}$$

and

$$DOS^* = DOS \cdot R_c = \frac{DOS}{1 - e^{-k_{el}\tau}}$$

can be executed with regard to $c_\infty(t')$. This does not have to be proven here.

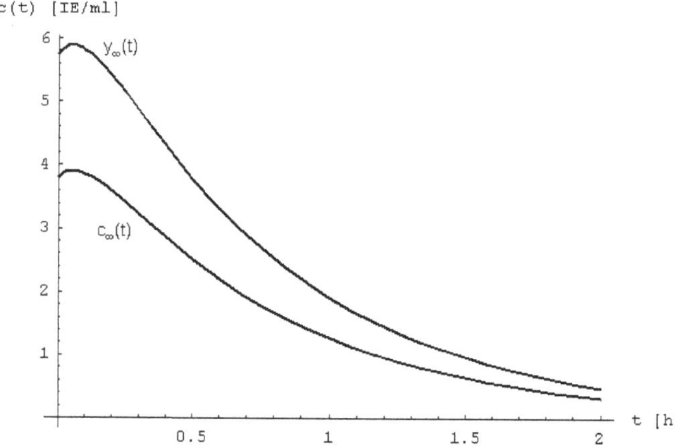

Fig. 17.1: Comparison of steady-state kinetics for multiple application: $y_\infty(t)$ indicates the course from DOST (1968, S.179). $c_\infty(t)$ is the course for the same parameter but in accordance with Proposition 17.4.

Sections 60 and 62 from DOST (1968) are used as an example. Penicillin is applied intramuscularly in a dose that results in the initial concentration of 1 I.E./ml. The invasion constant is $k_i = 8.7h^{-1}$, and the elimination constant is $k_{el} = 1.37h^{-1}$. For $\tau = 0.5h$, $c_\infty(t')$ lies below $y_\infty(t')$ by 1.3%. For $\tau = 0.125h$, $c_\infty(t')$ lies below $y_\infty(t')$ by 33.7%. Figure 17.1 illustrates this proposition. The initial doses are different in the following way: $R_y = 1.01R_c$ is true in the first case, and $R_y = 1.50R_c$ is true in the second case. The thoughts about the modelling of multiple applications developed in this section justify a more extensive look at the topic as well as examinations of the clinical relevance of the differences that are shown.

PART 3

Mathematica® programs for selected problems

Program list

```
Off[Remove::rmnsm]; Remove["Global`*"];
(*-----------------------------------------------------*)
(* This procedure is constructing the interpolating    *)
(* natural spline of degree n = 2 k-1.                 *)
(*-----------------------------------------------------*)
IntnatSplDegree[m_, x_, y_, k_] := Module[{H, R},
   H = Table[0, {i, 1, m + k}, {j, 1, m + k}];
   R = Table[0, {i, 1, m + k}]; Do[R[[i]] = y[[i]], {i, 1, m}];
   Do[{If[j == 1, H[[m+j,k+i]] = 1, H[[m+j,k+i]] = x[[i]]^(j-1)];
     If[j == 1, H[[i,j]] = 1, H[[i,j]] = x[[i]]^(j-1)]}, {i, 1, m}, {j, 1, k}];
   Do[H[[i,k+j]] = (x[[i]] - x[[j]])^(2 k-1), {i, 1, m}, {j, 1, i}];
   c = N[LinearSolve[H, R]];
  ]; (* IntnatSplDegree *)
(*-----------------------------------------------------*)

(* The data : Example 2.1 *)
m = 13; k = 4;
x = {-1, -0.5, 0, 0.5, 1, 1.5, 2, 2.5, 3, 3.5, 4, 4.5, 5};
y = {15, 8.0625, 5, 3.5625, 3, 4.0625, 9, 21.5625, 47, 92.0625,
   165, 275.5625, 435};

IntnatSplDegree[m, x, y, k];

Sf[t_, i_] := Sum[c[[j]] t^(j-1), {j, 1, k}] + Sum[c[[k+j]] (t - x[[j]])^(2 k-1), {j, 1, i}];
SBild = Table[Plot[Sf[q, i], {q, x[[i]], x[[i+1]]},
            DisplayFunction -> Identity], {i, 1, m - 1}];
PP = Table[Point[{x[[i]], y[[i]]}], {i, 1, m}];
Punkte = Graphics[{PointSize[0.007], PP}, PlotRange -> All];
Show[{SBild, Punkte}, DisplayFunction -> $DisplayFunction,
      PlotRange -> All, Frame -> True, AspectRatio -> 0.75];
Do[{Abl = D[Sf[q, i], {q, 3}], SBild[[i]] = Plot[Abl, {q, x[[i]], x[[i+1]]},
     DisplayFunction -> Identity]}, {i, 1, m - 1}];
Show[SBild, DisplayFunction -> $DisplayFunction,
      PlotRange -> All, Frame -> True, AspectRatio -> 0.75];
```

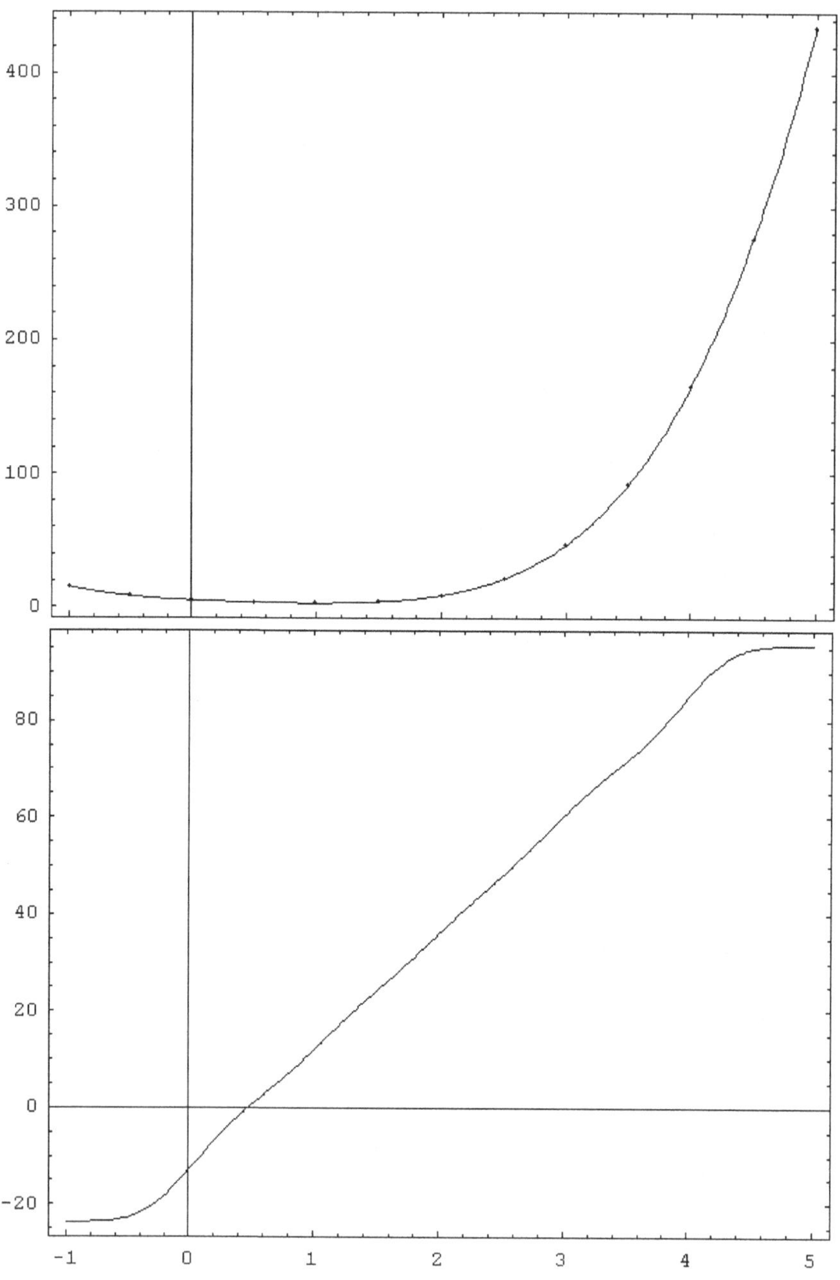
400
300
200
100
0
80
60
40
20
0
-20
-1
0
1
2
3
4
5

```
(* This program solves the interpolation problem (T)   *)
(* in C^2[x1, xm] with GK + μ^2 AQ = min! (see Chapter 3.1) *)
(* GK:=∫_{x1}^{xm} f''(x)^2 dx  and  AQ:=∫_{x1}^{xm} f'(x)^2 dx.   *)
(* The solution is a spline s(x) under tension          *)
(* s(x)=a_i e^(-μ(x-x_i))+b_i e^(μ(x-x_i))+c_i (x-x_i)+d_i, x∈[x_i, x_{i+1}].   *)
(*---------------------------------------------------------*)
(* The data: Example 3.4 *)
m = 10; μ = 10;
x = {9.3, 10.2, 11.4, 11.8, 13.1, 13.4, 14, 14.2, 15.3, 17.2};
y = {15.9, 20.7, 18.6, 20.4, 19, 18, 22.5, 21.5, 23.9, 26.2};
dx = Table[(x[[i+1]] - x[[i]]), {i, 1, m - 1}] ;
dy = Table[(y[[i+1]] - y[[i]]), {i, 1, m - 1}];
MA = Table[0, {i, 1, m - 2}, {j, 1, m - 2}];
R = Table[0, {i, 1, m - 2}];
Do[R[[i]] = dy[[i+1]]/dx[[i+1]] - dy[[i]]/dx[[i]], {i, 1, m - 2}]
MA[[1,1]] = -1/(dx[[1]] μ^2) - 1/(dx[[2]] μ^2) + Coth[dx[[1]] μ]/μ + Coth[dx[[2]] μ]/μ;
MA[[1,2]] = 1/(dx[[2]] μ^2) - (2 e^(dx[[2]] μ))/((-1 + e^(2 dx[[2]] μ)) μ);
Do[{MA[[i,i-1]] = 1/(dx[[i]] μ^2) - (2 e^(dx[[i]] μ))/((-1 + e^(2 dx[[i]] μ)) μ);
    MA[[i,i]] = -1/(dx[[i]] μ^2) - 1/(dx[[i+1]] μ^2) + Coth[dx[[i]] μ]/μ +
        Coth[dx[[i+1]] μ]/μ;
    MA[[i,i+1]] = 1/(dx[[i+1]] μ^2) - (2 e^(dx[[i+1]] μ))/((-1 + e^(2 dx[[i+1]] μ)) μ)}, {i, 2, m - 3}];
MA[[m-2,m-2]] = -1/(dx[[m-2]] μ^2) - 1/(dx[[m-1]] μ^2) + Coth[dx[[m-2]] μ]/μ +
    Coth[dx[[m-1]] μ]/μ;
```

$$MA_{[[m-2,m-3]]} = \frac{1}{dx_{[[m-2]]}\,\mu^2} - \frac{2\,e^{dx_{[[m-2]]}\,\mu}}{\left(-1 + e^{2\,dx_{[[m-2]]}\,\mu}\right)\mu};$$

```
Q = LinearSolve[MA, R]; (* Q=s ''(xi)          *)
Q = Prepend[Q, 0];       (* Q[[1]]=0 und       *)
Q = Append[Q, 0];        (* Q[[m]]=0 anfügen *)
```

$$a = \text{Table}\left[\frac{\left(e^{2\,\mu\,dx_{[[i]]}}\,Q_{[[i]]} - e^{\mu\,dx_{[[i]]}}\,Q_{[[i+1]]}\right)}{\left(e^{2\,\mu\,dx_{[[i]]}} - 1\right)\mu^2},\ \{i, 1, m-1\}\right];$$

$$b = \text{Table}\left[\frac{e^{\mu\,dx_{[[i]]}}\,Q_{[[i+1]]} - Q_{[[i]]}}{\left(e^{2\,\mu\,dx_{[[i]]}} - 1\right)\mu^2},\ \{i, 1, m-1\}\right];$$

$$c = \text{Table}\left[\frac{dy_{[[i]]}}{dx_{[[i]]}} + \frac{(Q_{[[i]]} - Q_{[[i+1]]})}{\mu^2\,dx_{[[i]]}},\ \{i, 1, m-1\}\right];$$

$$d = \text{Table}\left[y_{[[i]]} - \frac{Q_{[[i]]}}{\mu^2},\ \{i, 1, m-1\}\right];$$

$$Sf[t\_, i\_] := a_{[[i]]}\,e^{-\mu\,(t - x_{[[i]]})} + b_{[[i]]}\,e^{\mu\,(t - x_{[[i]]})} + c_{[[i]]}\,(t - x_{[[i]]}) + d_{[[i]]};$$

```
SBild = Table[Plot[Sf[q, i], {q, x[[i]], x[[i+1]]},
            DisplayFunction → Identity], {i, 1, m - 1}];
PP = Table[Point[{x[[i]], y[[i]]}], {i, 1, m}];
Punkte = Graphics[{PointSize[0.01], PP}, PlotRange → All];
Show[{SBild, Punkte}, DisplayFunction → $DisplayFunction,
  Prolog → Thickness[0.002], PlotRange → All, Frame → True,
  AspectRatio → 0.75];
Do[{Abl = D[Sf[q, i], q], SBild[[i]] = Plot[Abl, {q, x[[i]], x[[i+1]]},
         DisplayFunction → Identity]}, {i, 1, m - 1}];
Show[{SBild}, DisplayFunction → $DisplayFunction,
       Prolog → Thickness[0.002], Frame → True,
       AspectRatio → 0.75];
```

$$GK = \sum_{i=1}^{m-1} \int_{x_{[[i]]}}^{x_{[[i+1]]}} D[D[Sf[q, i], q], q]^2\, dq;$$

$$AQ = \sum_{i=1}^{m-1} \int_{x_{[[i]]}}^{x_{[[i+1]]}} D[Sf[q, i], q]^2\, dq;$$

```
Print["GK = ", GK, "  , AQ = ", AQ];
```

```
(* This procedure is constructing the interpolating        *)
(* cubic spline s (x) with s' (x1)=a1l and s' (xm)=a1r.     *)
(* Input data are the number m of points (xi,yi),a1l,a1r,*)
(* the list x of xi and the list y of yi, repectively.     *)
(*--------------------------------------------------------*)
IntCubSpline1Abl[m_, x_, y_, a1l_, a1r_] := Module[{dx, dy, Z, R},
   a = Table[0, {i, 1, m}]; b = a; c = a; d = a;
   dx = Table[x[[i + 1]] - x[[i]], {i, 1, m - 1}];
   dy = Table[y[[i + 1]] - y[[i]], {i, 1, m - 1}];
   Z = Table[0, {i, 1, m}, {j, 1, m}];
   Do[Z[[i + 1, i + 1]] = 2 * (dx[[i]] + dx[[i + 1]]), {i, 1, m - 2}];
   Do[Z[[i + 1, i]] = dx[[i]], {i, 1, m - 1}];
   Do[Z[[i, i + 1]] = dx[[i]], {i, 1, m - 1}];
   Z[[1, 1]] = 2 * dx[[1]]; Z[[m, m]] = 2 * dx[[m - 1]];
   R = Table[3 * (dy[[i]] / dx[[i]] - dy[[i - 1]] / dx[[i - 1]]),
     {i, 2, m - 1}];
   R = Prepend[R, 3 * (dy[[1]] / dx[[1]] - a1l)];
   R = Append[R, 3 * (a1r - dy[[m - 1]] / dx[[m - 1]])];
   b = LinearSolve[Z, R]; d = y;
   a = Table[(b[[i + 1]] - b[[i]]) / (3 * dx[[i]]), {i, 1, m - 1}];
   c = Table[dy[[i]] / dx[[i]] - dx[[i]] / 3 * (2 *b[[i]] + b[[i + 1]]),
     {i, 1, m - 1}];
  ]; (* IntCubSpline1Abl *)

(* The data: Example 3.2 *)
m = 8; a1l = 10; a1r = 10;
x = { 2, 10, 20, 30, 40, 50, 60,  90};
y = {30, 25,  2, 22, 40, 68, 76, 125};
IntCubSpline1Abl[m, x, y, a1l, a1r];

Sf[t_, i_] := a[[i]] (t - x[[i]]) ^3 + b[[i]] (t - x[[i]]) ^2
             + c[[i]] (t - x[[i]]) + d[[i]];
SBild = Table[Plot[Sf[q, i], {q, x[[i]], x[[i + 1]]},
              DisplayFunction → Identity], {i, 1, m - 1}];
PP = Table[Point[{x[[i]], y[[i]]}], {i, 1, m}];
Punkte = Graphics[{PointSize[0.007], PP}, PlotRange → All];
Show[{SBild, Punkte}, DisplayFunction → $DisplayFunction,
      PlotRange → All, Frame → True, AspectRatio → 0.75];
```

```
(* This procedure is constructing the interpolating      *)
(* cubic spline s (x) with s "(x1)=a2l and s" (xm)=a2r.    *)
(* Input data are the number m of points (xi,yi),a2l,a2r,*)
(* the list x of xi and the list y of yi, repectively.   *)
(*-------------------------------------------------------*)
IntCubSpline2Abl[m_, x_, y_, a2l_, a2r_] := Module[{dx, dy, Z, R},
   a = Table[0, {i, 1, m}]; b = a; c = a; d = a;
   dx = Table[x[[i + 1]] - x[[i]], {i, 1, m - 1}];
   dy = Table[y[[i + 1]] - y[[i]], {i, 1, m - 1}];
   Z = Table[0, {i, 1, m - 2}, {j, 1, m - 2}];
   Do[Z[[i, i]] = 2 * (dx[[i]] + dx[[i + 1]]), {i, 1, m - 2}];
   Do[Z[[i + 1, i]] = dx[[i + 1]], {i, 1, m - 3}];
   Do[Z[[i, i + 1]] = dx[[i + 1]], {i, 1, m - 3}];
   R = Table[3 * (dy[[i]] / dx[[i]] - dy[[i - 1]] / dx[[i - 1]]),
     {i, 2, m - 1}];
   R[[1]] = R[[1]] - dx[[1]] * a2l;
   R[[m - 2]] = R[[m - 2]] - dx[[m - 1]] * a2r; d = y;
   b = LinearSolve[Z, R]; b = Prepend[b, a2l]; b = Append[b, a2r];
   a = Table[(b[[i + 1]] - b[[i]]) / (3 * dx[[i]]), {i, 1, m - 1}];
   c = Table[dy[[i]] / dx[[i]] - dx[[i]] / 3 * (2 * b[[i]] + b[[i + 1]]),
     {i, 1, m - 1}];
  ]; (* IntCubSpline2Abl *)

(* The data: Example 3.1 *)
m = 9; a2l = -18; a2r = 30;
x = { -3, -2, -1, 0, 1, 2,  3,  4,   5};
y = {-27, -8, -1, 0, 1, 8, 27, 64, 125};

IntCubSpline2Abl[m, x, y, a2l, a2r]

Sf[t_, i_] := a[[i]] (t - x[[i]]) ^3 + b[[i]] (t - x[[i]]) ^2
            + c[[i]] (t - x[[i]]) + d[[i]];
SBild = Table[Plot[Sf[q, i], {q, x[[i]], x[[i + 1]]},
            DisplayFunction → Identity], {i, 1, m - 1}];
PP = Table[Point[{x[[i]], y[[i]]}], {i, 1, m}];
Punkte = Graphics[{PointSize[0.007], PP}, PlotRange → All];
Show[{SBild, Punkte}, DisplayFunction → $DisplayFunction,
      PlotRange → All, Frame → True, AspectRatio → 0.75];
```

```
Off[Remove::rmnsm]; Remove["Global`*"];
(*-----------------------------------------------------*)
(* This procedure is constructing the interpolating   *)
(* cubic spline s (x) witch AQ=∫x1^xm s' (x)^2 dx = min!.  *)
(* Input data are the number m of (xi,yi), the list   *)
(* x of xi and the list y of yi, repectively.         *)
(*-----------------------------------------------------*)
S3minAQ[m_, x_, y_] := Module[{Z, ZI, ZITr, H, EE, β, R, K, st},
   dx = Table[(x[[i+1]] - x[[i]]), {i, 1, m - 1}];
   dy = Table[(y[[i+1]] - y[[i]]), {i, 1, m - 1}];
   Z = Table[0, {i, 1, m}, {j, 1, m}]; Z[[1,1]] = 1; Z[[m,m]] = 1;
   Do[{Z[[i,i]] = 2 (dx[[i]] + dx[[i-1]]), Z[[i,i-1]] = dx[[i-1]],
       Z[[i,i+1]] = dx[[i]]}, {i, 2, m - 1}];
   ZI = Inverse[Z]; ZITr = Transpose[ZI];
   H = Table[0, {i, 1, m}, {j, 1, m}];
   H[[1,1]] = 4 dx[[1]]^3; H[[1,2]] = 7 / 2 dx[[1]]^3;
   Do[{H[[i,i-1]] = 7 / 2 dx[[i-1]]^3, H[[i,i]] = 4 (dx[[i-1]]^3 + dx[[i]]^3),
      H[[i,i+1]] = 7 / 2 dx[[i]]^3}, {i, 2, m - 1}];
   H[[m,m-1]] = 7 / 2 dx[[m-1]]^3; H[[m,m]] = 4 dx[[m-1]]^3;
   EE = ZITr.H.ZI; β = Table[0, {i, 1, m}];
   Do[β[[k]] = 3 (dy[[k]]/dx[[k]] - dy[[k-1]]/dx[[k-1]]), {k, 2, m - 1}];
   R = {-Sum[EE[[1,k]] β[[k]], {k, 2, m-1}], -Sum[EE[[m,k]] β[[k]], {k, 2, m-1}]};
   K = {{EE[[1,1]], EE[[1,m]]}, {EE[[1,m]], EE[[m,m]]}};
   st = LinearSolve[K, R]; (* solve equation (3.22) *)
   β[[1]] = st[[1]]; β[[m]] = st[[2]]; b = LinearSolve[Z, β];
   d = y; a = Table[(b[[i+1]] - b[[i]])/(3 dx[[i]]), {i, 1, m - 1}];
   c = Table[dy[[i]]/dx[[i]] - dx[[i]]/3 (2 b[[i]] + b[[i+1]]), {i, 1, m - 1}];
  ]; (* S3minAQ *)
(*-----------------------------------------------------*)
```

```
(* The data: Example 3.3 *)
m = 6;
x = {-1, 0.0, 1, 2.00,  4, 5.0};
y = {-1, 3.5, 1, 0.26, -1, 1.5};

S3minAQ[m, x, y];
```

$$GK = \frac{4}{3} \sum_{i=1}^{m-1} dx_{[[i]]} \left(b_{[[i]]}^2 + b_{[[i]]} b_{[[i+1]]} + b_{[[i+1]]}^2\right);$$

$$AQ = \frac{1}{45} \sum_{i=1}^{m-1} dx_{[[i]]}^3 \left(4\, b_{[[i]]}^2 + 7\, b_{[[i]]} * b_{[[i+1]]} + 4\, b_{[[i+1]]}^2\right) + \sum_{i=1}^{m-1} \frac{dy_{[[i]]}^2}{dx_{[[i]]}};$$

$$\text{Print}\left[\text{"GK=}\int_{x1}^{xm} s''(x)^2 dx \ = \text{", GK, "\textbackslash nAQ=}\int_{x1}^{xm} s'(x)^2 dx \ = \text{", AQ}\right];$$

$$Sf[t\_, i\_] := a_{[[i]]} (t - x_{[[i]]})^3 + b_{[[i]]} (t - x_{[[i]]})^2 + c_{[[i]]} (t - x_{[[i]]}) + d_{[[i]]};$$

```
SBild = Table[Plot[Sf[q, i], {q, x[[i]], x[[i+1]]},
            DisplayFunction → Identity], {i, 1, m - 1}];
PP = Table[Point[{x[[i]], y[[i]]}], {i, 1, m}];
Punkte = Graphics[{PointSize[0.01], PP}, PlotRange → All];
Show[{SBild, Punkte}, DisplayFunction → $DisplayFunction,
     PlotRange → All, Prolog → Thickness[0.002], Frame → True,
     AspectRatio → 0.75];
Do[{Abl = D[Sf[q, i], q],
    SBild[[i]] = Plot[Abl, {q, x[[i]], x[[i+1]]},
              DisplayFunction → Identity]}, {i, 1, m - 1}];
Show[{SBild}, DisplayFunction → $DisplayFunction,
  Prolog → Thickness[0.002], Frame → True, AspectRatio → 0.75];
```

$$GK=\int_{x1}^{xm} s''(x)^2 dx = 129.005$$

$$AQ=\int_{x1}^{xm} s'(x)^2 dx = 39.0302$$

```
Off[Remove::rmnsm]; Remove["Global`*"];
(*-----------------------------------------------------*)
(* This procedure solves the problem (OP3):             *)
(* μ*∫_{x1}^{xm} s''(x)^2 dx + ∑_{i=1}^{m} (yi - s(xi))^2/ki^2 =min!   *)
(* Input data are the smoothing parameter μ, m, the     *)
(* list x of xi,the list y of yiand the ki, repectively. *)
(*-----------------------------------------------------*)
CubSplineSmooth[m_, x_, y_, k_, μ_] := Module[{Lx, Ly, u, v, w, z1},
   a = Table[0, {i, 1, m}]; b = a; c = a; d = a;
   Lx = Table[0, {i, 1, m}]; Ly = Lx; u = 1 / (x[[2]] - x[[1]]);
   v = 1 / (x[[3]] - x[[2]]); z1 = (y[[2]] - y[[1]]) * u;
   Do[{ a[[i]] = 2 * (1 / u + 1 / v) / 3 + 2 μ ((k[[i]] * u) ^2 +
          (k[[i + 1]] * (u + v)) ^ 2 + (k[[i + 2]] * v) ^2);
     If[ i ≤ m - 3, { w = 1 / (x[[i + 3]] - x[[i + 2]]);
        d[[i]] = 1 / v / 3 - 2 * μ * v * ((u + v) k[[i + 1]] ^2 +
            (v + w) k[[i + 2]] ^2)}];
     z2 = (y[[i + 2]] - y[[i + 1]]) * v; b[[i + 1]] = z2 - z1;
     c[[i]] = 2 * μ * v * w * k[[i + 2]] ^ 2; u = v; v = w; z1 = z2},
    {i, 1, m - 2}];
   Lx[[1]] = a[[1]]; Ly[[1]] = d[[1]];
   Ly[[2]] = d[[2]] - Ly[[1]] c[[1]] / Lx[[1]];
   Lx[[2]] = a[[2]] - Ly[[1]] ^2 / Lx[[1]];
   Do[{ Ly[[i]] = d[[i]] - Ly[[i - 1]] c[[i - 1]] / Lx[[i - 1]];
     Lx[[i]] =
      a[[i]] - Ly[[i - 1]] ^2 / Lx[[i - 1]] - c[[i - 2]] ^2 / Lx[[i - 2]]},
    {i, 3, m - 2}];
   b[[2]] = b[[2]] / Lx[[1]]; b[[1]] = 0; b[[m]] = 0; c[[m - 3]] = 0;
   Do[b[[i + 1]] =
     (b[[i + 1]] - c[[i - 2]] b[[i - 1]] - Ly[[i - 1]] b[[i]]) / Lx[[i]],
    {i, 2, m - 2}];
   Do[b[[i + 1]] = b[[i + 1]] - (Ly[[i]] b[[i + 2]] +
              c[[i]] b[[i + 3]]) / Lx[[i]], {i, m - 3, 1, -1}];
   v = 1 / (x[[2]] - x[[1]]); d[[1]] = y[[1]] - 2 μ b[[2]] * v * k[[1]] ^2;
   Do[{ u = v; v = 1 / (x[[i + 1]] - x[[i]]); d[[i]] = y[[i]] -
       2 μ * k[[i]] ^2 (b[[i - 1]] u - b[[i]] (u + v) + b[[i + 1]] v)},
    {i, 2, m - 1}]; d[[m]] = y[[m]] - 2 μ * k[[m]] ^2 * b[[m - 1]] * v;
   Do[{ u = x[[i + 1]] - x[[i]]; a[[i]] = (b[[i + 1]] - b[[i]]) / (3 * u);
      c[[i]] = (d[[i + 1]] - d[[i]]) / u - u / 3 * (2 * b[[i]] + b[[i + 1]])},
     {i, 1, m - 1}];
  ]; (* CubSplineSmooth *)
```

```
(* The data: Example 4.1 *)
m = 8;
x = { 2, 10, 20, 30, 40, 50, 60,  90};
y = {30, 25,  2, 22, 40, 68, 76, 125};
k = Table[1, {i, 1, m}];

μ = 19; CubSplineSmooth[m, x, y, k, μ];

Sf[t_, i_] := a[[i]] (t - x[[i]]) ^3 + b[[i]] (t - x[[i]]) ^2 +
             c[[i]] (t - x[[i]]) + d[[i]];
SBild = Table[Plot[Sf[q, i], {q, x[[i]], x[[i + 1]]},
             DisplayFunction → Identity], {i, 1, m - 1}];
PP = Table[Point[{x[[i]], y[[i]]}], {i, 1, m}];
Punkte = Graphics[{PointSize[0.007], PP}, PlotRange → All];
Show[{SBild, Punkte}, DisplayFunction → $DisplayFunction,
     PlotRange → All, Frame → True, AspectRatio → 0.75];
```

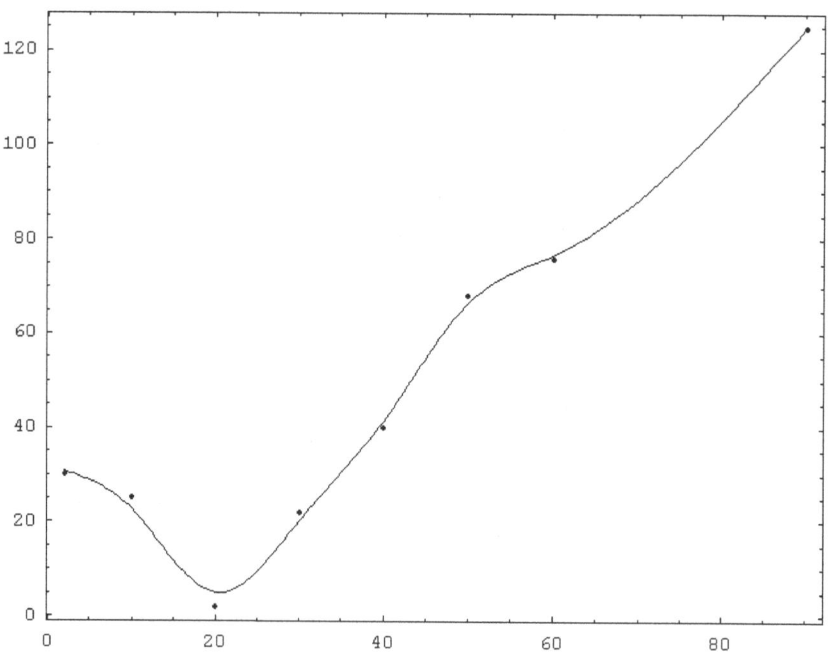

```
Off[Remove::rmnsm]; Remove["Global`*"];
(* -------------------------------------------------- *)
(* This procedure solves the problem (R3):              *)
(* ∫_{x1}^{xm} s''(x)^2 dx = min! where  Σ_{i=1}^{m}(yi - s(xi))^2≤S  *)
(* Input data are S, m, the list x of xi and the list *)
(* y of yi, repectively.                                *)
(* -------------------------------------------------- *)
CubSplineSmooth[m_, x_, y_, μ_] := Module[{Lx, Ly, u, v, w, z1},
   Lx = Table[0, {i, 1, m}]; Ly = Lx; u = 1 / (x[[2]] - x[[1]]);
   v = 1 / (x[[3]] - x[[2]]); z1 = (y[[2]] - y[[1]]) * u;
   Do[{ a[[i]] = 2 * (1 / u + 1 / v) / 3 + 4 * μ * (u * u + u * v + v * v);
     If[ i ≤ m - 3, { w = 1 / (x[[i + 3]] - x[[i + 2]]);
                 d[[i]] = 1 / v / 3 - 2 * μ * v * (u + 2 * v + w)}];
     z2 = (y[[i + 2]] - y[[i + 1]]) * v; b[[i + 1]] = z2 - z1;
     c[[i]] = 2 * μ * v * w; u = v; v = w; z1 = z2},
    {i, 1, m - 2}];
   Lx[[1]] = a[[1]]; Ly[[1]] = d[[1]];
   Ly[[2]] = d[[2]] - Ly[[1]] c[[1]] / Lx[[1]];
   Lx[[2]] = a[[2]] - Ly[[1]] ^2 / Lx[[1]];
   Do[{ Ly[[i]] = d[[i]] - Ly[[i - 1]] c[[i - 1]] / Lx[[i - 1]];
     Lx[[i]] =
      a[[i]] - Ly[[i - 1]] ^2 / Lx[[i - 1]] - c[[i - 2]] ^2 / Lx[[i - 2]]},
    {i, 3, m - 2}];
   b[[2]] = b[[2]] / Lx[[1]]; b[[1]] = 0; b[[m]] = 0; c[[m - 3]] = 0;
   Do[b[[i + 1]] =
     (b[[i + 1]] - c[[i - 2]] b[[i - 1]] - Ly[[i - 1]] b[[i]]) / Lx[[i]],
    {i, 2, m - 2}];
   Do[b[[i + 1]] = b[[i + 1]] - (Ly[[i]] b[[i + 2]] +
               c[[i]] b[[i + 3]]) / Lx[[i]], {i, m - 3, 1, -1}];
   v = 1 / (x[[2]] - x[[1]]); d[[1]] = y[[1]] - 2 μ b[[2]] * v;
   Do[{ u = v; v = 1 / (x[[i + 1]] - x[[i]]);
     d[[i]] = y[[i]] - 2 μ (b[[i - 1]] u - b[[i]] (u + v) + b[[i + 1]] v)},
    {i, 2, m - 1}]; d[[m]] = y[[m]] - 2 μ b[[m - 1]] * v;
  ]; (* CubSplineSmooth *)

ZIELF := Module[{aa}, CubSplineSmooth[m, x, y, μ];
   SSE = Sum[(y[[i]] - d[[i]]) ^2, {i, 1, m}] - S;]; (* ZIELF *)
(*-----------------------------------------------------*)
```

```
(* The data: Example 4.2 *)
m = 81; S = 506.25;
x = {415.76, 417.92, 420.1, 422.28, 424.46, 426.64, 428.82, 431,
   433.18, 435.36, 437.54, 439.72, 441.9, 444.08, 446.26, 448.44,
   450.62, 452.8, 454.98, 457.16, 459.34, 461.52, 463.7, 465.88,
   468.06, 470.24, 472.42, 474.6, 476.78, 478.96, 481.14, 483.32,
   485.5, 487.68, 489.86, 492.04, 494.22, 496.4, 498.58, 500.76,
   502.94, 505.12, 507.3, 509.48, 511.66, 513.84, 516.02, 518.2,
   520.38, 522.56, 524.74, 526.9, 529.1, 531.28, 533.46`, 535.64,
   537.82, 540., 542.18, 544.36, 546.54, 548.72, 550.9, 553.08,
   555.26, 557.44, 559.62, 561.8, 563.98, 566.16, 568.34, 570.52,
   572.7, 574.88, 577.06, 579.24, 581.42, 583.6, 585.78, 587.96,
   590.14};
y = {36, 39, 42, 43, 45, 42, 43, 43, 46, 47, 45, 49, 55, 56, 58, 62, 62,
   58, 58, 54, 66, 66, 70, 76, 72, 75, 78, 78, 75, 79, 87, 89, 90, 85,
   89, 86, 101, 98, 95, 92, 96, 103, 100, 99, 111, 113, 111, 114, 117,
   112, 108, 104, 105, 97, 97, 96, 97, 103, 88, 88, 90, 84, 66, 59, 58,
   55, 50, 50, 52, 51, 50, 53, 50, 47, 47, 53, 51, 57, 57, 65, 57};
a = Table[0, {i, 1, m}]; b = a; c = a; d = a;
μ = 10^10; ZIELF; Smax = SSE + S; If[SSE < 0,
 {Print["S must not be greater then ", Smax], Interrupt[]}];
μ = 10^-10; SSEold = SSE; (* seeking for an initial value *)
(Do[{ZIELF; If[SSE < 0, {SSEold = SSE, μ = 10 μ}, Goto[go]]},
   {i, 1, 20}]; Label[go]);
μ1 = μ / 10; μ = Abs[SSEold (μ - μ1) / (SSE - SSEold) + μ1];
Print["initial value is μ = ", N[μ]]; IT = 0;
(* Newton iteration *)
(Label[WEI]; If[IT == 0, SSEold = 0, SSEold = SSE];
  IT++; z = μ; dif = Abs[μ / 10^6]; μ = μ + dif;
  ZIELF; f = SSE; μ = μ - 2 dif; ZIELF; f = f - SSE; Abl = f / 2 / dif;
  If[Or[IT > 20, Abl == 0], Goto[SP]]; μ = z; ZIELF; μ = Abs[μ - SSE / Abl];
  If[And[μ < 10^20, Abs[1 - (SSE + S) / (SSEold + S)] > 10^-6], Goto[WEI]];
  Label[SP]) ;
If[Abs[1 - (SSE + S) / (SSEold + S)] > 10^-6,
  Print["No convergence ! SSE =", SSE + S, " , Abl = ", Abl]];
Do[{ u = x[[i + 1]] - x[[i]]; a[[i]] = (b[[i + 1]] - b[[i]]) / (3 u);
   c[[i]] = (d[[i + 1]] - d[[i]]) / u - u / 3 (2 b[[i]] + b[[i + 1]])},
  {i, 1, m - 1}];
```

```
SSE = Sum[(y[[i]] - d[[i]])^2, {i, 1, m}];
Print["Lagrange multiplier p = 1/μ = ", 1/μ, "\nSSE = ", SSE];
Sf[t_, i_] := a[[i]] (t - x[[i]])^3 + b[[i]] (t - x[[i]])^2 +
            c[[i]] (t - x[[i]]) + d[[i]];
SBild = Table[Plot[Sf[q, i], {q, x[[i]], x[[i + 1]]},
            DisplayFunction → Identity], {i, 1, m - 1}];
PP = Table[Point[{x[[i]], y[[i]]}], {i, 1, m}];
Punkte = Graphics[{PointSize[0.007], PP}, PlotRange → All];
Show[{SBild, Punkte}, DisplayFunction → $DisplayFunction,
      PlotRange → All, Frame → True, AspectRatio → 0.75];
```

```
initial value is μ = 0.659108

Lagrange multiplier p = 1/μ = 0.0739027
SSE = 506.25
```

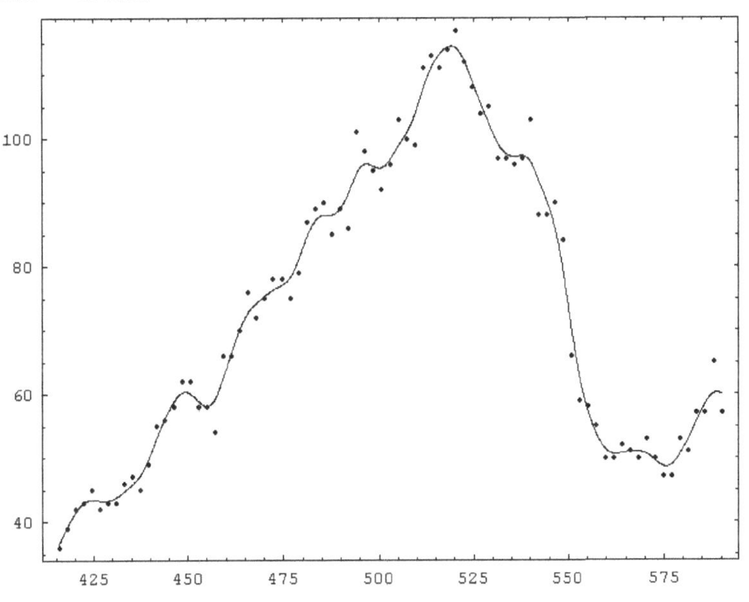

```
Off[Remove::rmnsm]; Remove["Global`*"];
(* ------------------------------------------------ *)
(* This procedure solves the problem (GK3):          *)
(* ∑_{i=1}^m (y_i - s (x_i))^2 = min! where ∫_{x_1}^{x_m} s'' (x)^2 dx ≤ T   *)
(* Input data are T, m, the list x of x_i and the list *)
(* y of y_i, repectively.                             *)
(* ------------------------------------------------ *)
CubSplineSmooth[m_, x_, y_, μ_] := Module[{Lx, Ly, u, v, w, z1},
   Lx = Table[0, {i, 1, m}]; Ly = Lx; u = 1 / (x[[2]] - x[[1]]);
   v = 1 / (x[[3]] - x[[2]]); z1 = (y[[2]] - y[[1]]) * u;
   Do[{ a[[i]] = 2 * (1 / u + 1 / v) / 3 + 4 * μ * (u * u + u * v + v * v);
     If[ i ≤ m - 3, { w = 1 / (x[[i + 3]] - x[[i + 2]]);
                  d[[i]] = 1 / v / 3 - 2 * μ * v * (u + 2 * v + w) }];
     z2 = (y[[i + 2]] - y[[i + 1]]) * v; b[[i + 1]] = z2 - z1;
     c[[i]] = 2 * μ * v * w; u = v; v = w; z1 = z2},
    {i, 1, m - 2}];
   Lx[[1]] = a[[1]]; Ly[[1]] = d[[1]];
   Ly[[2]] = d[[2]] - Ly[[1]] c[[1]] / Lx[[1]];
   Lx[[2]] = a[[2]] - Ly[[1]] ^2 / Lx[[1]];
   Do[{ Ly[[i]] = d[[i]] - Ly[[i - 1]] c[[i - 1]] / Lx[[i - 1]];
     Lx[[i]] =
      a[[i]] - Ly[[i - 1]] ^2 / Lx[[i - 1]] - c[[i - 2]] ^2 / Lx[[i - 2]]},
    {i, 3, m - 2}];
   b[[2]] = b[[2]] / Lx[[1]]; b[[1]] = 0; b[[m]] = 0; c[[m - 3]] = 0;
   Do[b[[i + 1]] =
     (b[[i + 1]] - c[[i - 2]] b[[i - 1]] - Ly[[i - 1]] b[[i]]) / Lx[[i]],
    {i, 2, m - 2}];
   Do[b[[i + 1]] = b[[i + 1]] - (Ly[[i]] b[[i + 2]] +
                 c[[i]] b[[i + 3]]) / Lx[[i]], {i, m - 3, 1, -1}];
  ]; (* CubSplineSmooth *)

ZIELF := Module[{aa}, CubSplineSmooth[m, x, y, μ];
   GK = 4 / 3 Sum[(x[[i + 1]] - x[[i]]) (b[[i]] ^2 +
          b[[i]] b[[i + 1]] + b[[i + 1]] ^2), {i, 1, m - 1}] - T;
  ]; (* ZIELF *)
(*-------------------------------------------------------*)
```

```
m = 8; T = 1; (* The data: Example 4.4 *)
x = {1, 2, 3, 4, 5, 6, 8, 11.0};
y = {0, 1, 4, 3, 1, 2, 2,  3.5};
a = Table[0, {i, 1, m}]; b = a; c = a; d = a;
μ = 0; ZIELF; GKmax = GK + T; If[GK < 0,
 {Print["S must not be greater then ", GKmax], Interrupt[]}];
μ = 10^-10; GKold = GKmax - T; (* seeking vor an initial value *)
(Do[{ZIELF; If[GK > 0, {GKold = GK, μ = 10 μ}, Goto[go]]},
   {i, 1, 20}]; Label[go]);
μ1 = μ / 10; μ = Abs[GKold (μ - μ1) / (GK - GKold) + μ1];
IT = 0; Print["Startwert ist μ = ", N[μ]];
(* Newton iteration *)
(Label[WEI]; If[IT == 0, GKold = 0, GKold = GK];
  IT++; z = μ; dif = Abs[μ / 10^6]; μ = μ + dif;
  ZIELF; f = GK; μ = μ - 2 dif; ZIELF; f = f - GK; Abl = f / 2 / dif;
  If[Or[IT > 20, Abl == 0], Goto[SP]]; μ = z; ZIELF; μ = Abs[μ - GK / Abl];
  If[And[μ < 10^20, Abs[1 - (GK + T) / (GKold + T)] > 10^-6], Goto[WEI]];
  Label[SP]) ;
If[Abs[1 - (GK + T) / (GKold + T)] > 10^-6,
  Print["No convergence ! GK =", GK + T, "  , Abl = ", Abl]];
v = 1 / (x[[2]] - x[[1]]); d[[1]] = y[[1]] - 2 μ b[[2]] * v;
Do[{ u = v; v = 1 / (x[[i + 1]] - x[[i]]);
  d[[i]] = y[[i]] - 2 μ (b[[i - 1]] u - b[[i]] (u + v) + b[[i + 1]] v)},
 {i, 2, m - 1}]; d[[m]] = y[[m]] - 2 μ * b[[m - 1]] * v;
Do[{ u = x[[i + 1]] - x[[i]]; a[[i]] = (b[[i + 1]] - b[[i]]) / (3 u);
   c[[i]] = (d[[i + 1]] - d[[i]]) / u - u / 3 (2 * b[[i]] + b[[i + 1]])},
  {i, 1, m - 1}];
GK = Sum[4 / 3 (x[[i + 1]] - x[[i]]) (b[[i]] ^ 2 + b[[i]] b[[i + 1]] +
        b[[i + 1]] ^ 2), {i, 1, m - 1}];
Print["Lagrange parameter p = μ = ", μ, "\nGK = ", GK];
Sf[t_, i_] := a[[i]] (t - x[[i]]) ^ 3 + b[[i]] (t - x[[i]]) ^ 2 +
             c[[i]] (t - x[[i]]) + d[[i]];
SBild = Table[Plot[Sf[q, i], {q, x[[i]], x[[i + 1]]},
             DisplayFunction → Identity], {i, 1, m - 1}];
PP = Table[Point[{x[[i]], y[[i]]}], {i, 1, m}];
Punkte = Graphics[{PointSize[0.007], PP}, PlotRange → All];
Show[{SBild, Punkte}, DisplayFunction → $DisplayFunction,
     PlotRange → All, Frame → True, AspectRatio → 0.75];
```

```
Off[Remove::rmnsm]; Remove["Global`*"];
(*-------------------------------------------------------*)
(* This procedure solves the problem (OP3) with ki=1,    *)
(* i=1,2,...,m.                                          *)
(* Input data are the number m of points (xi,yi) and     *)
(* the list x of xi and the list y of yi, repectively.   *)
(* The smoothing parameter μ is the minimum of CV(μ),    *)
(* Theorem 4.6 is used.                                  *)
(*-------------------- procedures ---------------------*)
ZIELF := Module[{HH},
   HH = Inverse[IdentityMatrix[m] + μ*H];
   d = HH.y; CV = Sum[((y[[i]] - d[[i]])/(1 - HH[[i,i]]))^2, {i, 1, m}];]; (* ZIELF *)
PARTABL := Module[{ di, f},
   TestZIELF := Module[{dj},
     μ = x1; dj = Abs[x1/10^5] + 10^-20; μ = μ + dj;
     ZIELF; f = CV; μ = μ - 2*dj; ZIELF; f = (f - CV)/2/dj; μ = x1;
    ]; (* TestZIELF *)
   x1 = x9; TestZIELF; f9 = f; di = Abs[x1/10^5] + 10^-20; x1 = x1 + di;
   TestZIELF; AA = f; x1 = x9; di = Abs[x1/10^5] + 10^-20; x1 = x1 - di;
   TestZIELF; AA = (AA - f)/2/di; x1 = x9;]; (* PARTABL *)
OPTIFAK := Module[{fm, df, sm, i, j},
   TEST := Module[{gg}, μ = x9 - f1*h; ZIELF[μ];]; (* TEST *)
   f1 = 0.6; fm = 0; df = 0.5; sm = 10^20; TEST; i = 0;
   While[And[CV < sm, i < 20],
    {sm = CV; fm = f1; i = i + 1; f1 = f1 + df; TEST}];
   f1 = fm - df; TEST; i = 0;
   While[And[CV < sm, i < 20],
    {sm = CV; fm = f1; i = i + 1; f1 = f1 - df; TEST}];
   Do[{(df = df/2; f1 = fm + df; TEST;
      If[CV < sm, { sm = CV; fm = f1; Goto[W]}];
        f1 = fm - df; TEST;
      If[CV < sm , {sm = CV; fm = f1}];
      Label[W];)},
    {j, 1, 7}];
   f1 = fm;]; (* OPTIFAK *)
(*-------------------------------------------------------*)
```

```
x = {1, 2, 3, 4, 5, 6, 7}; y = {1, 2, 4, 2, 2, 5, 3};
dx = Table[x[[i+1]] - x[[i]], {i, 1, m - 1}];
Q = Table[0, {i, 1, m - 2}, {j, 1, m}];
Do[Q[[i,i]]  = 1 / dx[[i]], {i, 1, m - 2}];
Do[{Q[[i,i+1]] = - (1 / dx[[i]] + 1 / dx[[i+1]]), Q[[i,i+2]] = 1 / dx[[i+1]]},
  {i, 1, m - 2}];
QT = Transpose[Q]; Z = Table[0, {i, 1, m - 2}, {j, 1, m - 2}];
Do[Z[[i,i]] = 2 * (dx[[i]] + dx[[i+1]]), {i, 1, m - 2}];
Do[{Z[[i+1,i]] = dx[[i+1]], Z[[i,i+1]] = dx[[i+1]]}, {i, 1, m - 3}];
ZI = Inverse[1 / 3 * Z]; Z = ZI.Q; H = 2 * QT.Z;
μ = 0.001; x9 = μ; x1 = μ; it = 0;
(Label[WEI]; it = it + 1; PARTABL;
  If[AA == 0, Goto[SP]]; μ = x9; h = f9 / AA; OPTIFAK;
  x9 = x9 - f1 * h; μ = Abs[x9]; ZIELF;
  If[CV == 0 , Goto[SP]]; If[it == 1, CValt = 2 * CV];
  Print[" Iteration  ", it ,
   "\n Smoothing parameter   μ = ", μ,
   "\n Objectiv function CV(μ)  = ", CV, "\n "];
  If[Abs[CV / CValt] > 0.9999999, Goto[SP]];
  If[it > 8, Goto[SP]]; If[CV > CValt, Goto[SP]];
  CValt = CV; Goto[WEI];
  Label[SP];) ;
b = Z.d; b = Prepend[b, 0]; b = Append[b, 0]; (* b1=bm=0 *)
a = Table[(b[[i+1]] - b[[i]]) / (3 * dx[[i]]), {i, 1, m - 1}];
c = Table[(d[[i+1]] - d[[i]]) / dx[[i]] - dx[[i]] / 3 * (2 * b[[i]] + b[[i+1]]), {i, 1, m - 1}];
Sf[t_, i_] := a[[i]] (t - x[[i]]) ^3 + b[[i]] (t - x[[i]]) ^ 2 +
          c[[i]] (t - x[[i]]) + d[[i]];
SBild = Table[Plot[Sf[q, i], {q, x[[i]], x[[i +1]]},
          DisplayFunction → Identity], {i, 1, m - 1}];
PP = Table[Point[{x[[i]], y[[i]]}], {i, 1, m}];
Punkte = Graphics[{PointSize[0.007], PP}, PlotRange → All];
Show[{SBild, Punkte}, DisplayFunction → $DisplayFunction,
    PlotRange → All, Frame → True, AspectRatio → 0.75];
```

```
Iteration  1
Smothing parameter   μ = 0.0150006
Objectiv function   SS = 46.2159

Iteration  2
Smothing parameter   μ = 0.156827
Objectiv function   SS = 32.9574

Iteration  3
Smothing parameter   μ = 0.156549
Objectiv function   SS = 32.9707
```

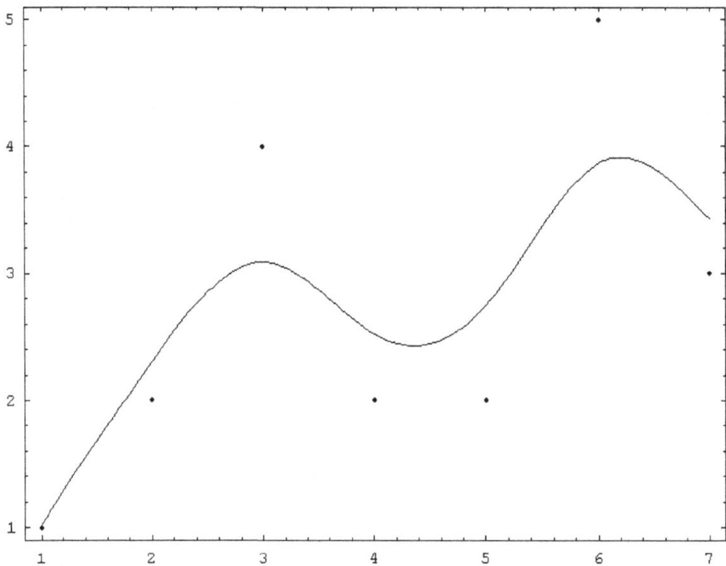

```
Off[Remove::rmnsm]; Remove["Global`*"];
(* This procedure solves the problem (OP2):                    *)
(* μ*∫_{x1}^{xm} s''(x)^2 dx + Σ_{i=1}^m (yi - s(xi))^2 = min!  *)
(* with ki=1,i=1,2,...,m.                                       *)
(* Input data  are the smoothing parameter μ, m  and           *)
(* the list x of xi and the list y of yi, repectively.          *)
(*-------------------------------------------------------------*)
SmoothS2GK[m_, x_, y_, μ_] := Module[{DD, w, invF, Z, K, H, Sd,
    s1, s2}, DD = Table[0, {i, 1, m}, {j, 1, m}]; invF = DD;
   Z = DD; dx = Table[x[[i+1]] - x[[i]], {i, 1, m - 1}];
   w = Table[1/dx[[i]], {i, 1, m - 1}]; Sd = Sum[w[[i]], {i, 1, m-1}]; s1 = 2 Sd - 2 w[[1]];
   DD[[1,1]] = -(w[[1]] + s1)/(dx[[1]] Sd); Z[[1,1]] = w[[1]]; Z[[1,2]] = -w[[1]];
   Do[{Z[[k,k-1]] = -w[[k-1]], Z[[k,k+1]] = -w[[k]], Z[[k,k]] = -Z[[k,k-1]] - Z[[k,k+1]],
     DD[[k,k-1]] = -2 / dx[[k-1]], DD[[k,k]] = 2 / dx[[k-1]],  zw = (w[[k-1]] + s1)/dx[[k-1]],
     s1 = s1 - 2*w[[k]], DD[[1,k]] = (-1)^k/Sd (zw + (w[[k]] + s1)/dx[[k]])},
    {k, 2, m - 1}];
   DD[[1,m]] = (-1)^m w[[m-1]]/(dx[[m-1]] Sd); DD[[m,m-1]] = -2/dx[[m-1]]; DD[[m,m]] = 2/dx[[m-1]];
   Z[[m,m-1]] = -w[[m-1]]; Z[[m,m]] = w[[m-1]];
   Do[Do[invF[[k,i]] = (-1)^(i+k), {k, i, m}], {i, 1, m}]; K = invF.DD;
   K = Transpose[K].Z.K; H = IdentityMatrix[m] + μ K;
   c = LinearSolve[H, y]; s1 = 2*Sd; s2 = 0;
   Do[{s1 = s1 - 2*w[[i]], s2 = s2 + (-1)^(i+1) (w[[i]] + s1) (c[[i+1]] - c[[i]])/dx[[i]]},
    {i, 1, m - 1}]; a = Table[0, {i, 1, m}]; b = a; b[[1]] = s2/Sd;
   Do[{b[[i+1]] = 2 (c[[i+1]] - c[[i]])/dx[[i]] - b[[i]], a[[i]] = (b[[i+1]] - b[[i]])/(2 dx[[i]])},
    {i, 1, m - 1}] ]; (* SmoothS2GK *)
```

```
(* The data: Example 7.1 *)
m = 10;
x = {1, 2, 2.5,   3,   4, 4.2, 4.5,   5,   6, 7};
y = {1, 2, 2.0, 2.3, 1.8, 1.5, 1.1, 1.3, 1.5, 1};

μ = 1; SmoothS2GK[m, x, y, μ];
```

$$SSE = N\left[\sum_{i=1}^{m} (y_{[[i]]} - c_{[[i]]})^2\right];$$

$$GK = \sum_{i=1}^{m-1} (b_{[[i]]}^2 - 2 * b_{[[i]]} * b_{[[i+1]]} + b_{[[i+1]]}^2) / dx_{[[i]]};$$

```
Print["μ = ", μ, "\nGK  = ", GK, "\nSSE = ", SSE,
      "\nμ*GK+SSE = ", μ GK + SSE];

Sf[t_, i_] := a[[i]] * (t - x[[i]]) ^2 + b[[i]] * (t - x[[i]]) + c[[i]];
SBild = Table[Plot[Sf[q, i], {q, x[[i]], x[[i+1]]},
        DisplayFunction → Identity], {i, 1, m - 1}];
PP = Table[Point[{x[[i]], y[[i]]}], {i, 1, m}];
Punkte = Graphics[{PointSize[0.01], PP}, PlotRange → All];
Show[{SBild, Punkte}, DisplayFunction → $DisplayFunction,
      Prolog → Thickness[0.002], Frame → True,
      AspectRatio → 0.75, PlotRange → All];
Do[{Abl = D[Sf[q, i], q],
     SBild[[i]] = Plot[Abl, {q, x[[i]], x[[i+1]]},
     DisplayFunction → Identity]}, {i, 1, m - 1}];
Show[{SBild}, DisplayFunction → $DisplayFunction,
     Prolog → Thickness[0.002], Frame → True,
     AspectRatio → 0.75];
```

```
μ = 1
GK  = 0.271234
SSE = 0.668596
μ*GK+SSE = 0.939829
```

```
(* This procedure solves the problem (R2) (ki=1):          *)
(* ∫_{x1}^{xm} s2''(x)^2 dx = min! where Σ_{i=1}^{m}(yi - s2(xi))^2 ≤ S  *)
(* Input data are S, m, the list x of xi and the list      *)
(* y of yi, repectively.                                   *)
(*---------------------------------------------------------*)
S2givenS[m_, x_, y_, T_] := Module[{DD, w, F, Z, K, H, zw, Sd, s1},
  dx = Table[x[[i+1]] - x[[i]], {i, 1, m - 1}];
  DD = Table[0, {i, 1, m}, {j, 1, m}]; F = DD; Z = DD;
  dx = Table[x[[i+1]] - x[[i]], {i, 1, m - 1}];
  w = Table[1/dx[[i]], {i, 1, m - 1}]; Sd = Sum[w[[i]], {i, 1, m-1}]; s1 = 2*Sd - 2*w[[1]];
  DD[[1,1]] = -(w[[1]] + s1)/(dx[[1]]*Sd); F[[1,1]] = 1; Z[[1,1]] = 1/dx[[1]]; Z[[1,2]] = -1/dx[[1]];
  Do[{F[[k,k-1]] = 1; F[[k,k]] = 1; Z[[k,k-1]] = -1/dx[[k-1]]; Z[[k,k+1]] = -1/dx[[k]];
    Z[[k,k]] = -Z[[k,k-1]] - Z[[k,k+1]]; DD[[k,k-1]] = -2/dx[[k-1]]; DD[[k,k]] = 2/dx[[k-1]];
    zw = (w[[k-1]] + s1)/dx[[k-1]]; s1 = s1 - 2*w[[k]];
    DD[[1,k]] = (-1)^k/Sd * (zw + (w[[k]] + s1)/dx[[k]])}, {k, 2, m - 1}];
  DD[[1,m]] = ((-1)^m * w[[m-1]])/(dx[[m-1]]*Sd); DD[[m,m-1]] = -2/dx[[m-1]]; DD[[m,m]] = 2/dx[[m-1]];
  Z[[m,m-1]] = -1/dx[[m-1]]; Z[[m,m]] = 1/dx[[m-1]];
  F[[m,m-1]] = 1; F[[m,m]] = 1; F = Inverse[F];
  K = Transpose[DD].Transpose[F].Z.F.DD;
  H = IdentityMatrix[m] + 1/p*K; Hinv = Inverse[H]; c = Hinv.y;
  FQS = Sum[(y[[i]] - c[[i]])^2, {i, 1, m}]; Loes = NSolve[FQS == S, p];
  (Label[wei];
   pz = Last[Loes]; pz = pz[[1]][[2]];
   Loes = Drop[Loes, {Length[Loes], Length[Loes]}];
   If[Head[pz] == Complex, Goto[wei]];
   If [pz < 0, Goto[wei]]; );
```

```
  p = pz; c = N[c]; GK = N[c.K.c]; Ziel = p*GK + FQS;
  s1 = 2*Sd; s2 = 0; b = Table[0, {i, 1, m}];
  Do[{s1 = s1 - 2*w[[i]], s2 = s2 + (-1)^(i+1) * (w[[i]] + s1) * (c[[i+1]] - c[[i]])/dx[[i]]},
   {i, 1, m - 1}];
  b[[1]] = s2/Sd; Do[b[[i+1]] = 2 * (c[[i+1]] - c[[i]])/dx[[i]] - b[[i]], {i, 1, m - 1}];
  a = Table[(b[[i+1]] - b[[i]])/(2*dx[[i]]), {i, 1, m - 1}]; ]; (* S2givenS *)
(*---------------------------------------------------------------*)
(* The data: Example 7.3 *)
m = 6; x = {1, 5, 8, 10, 15, 18}; y = {5, 1, 4,  8, 10,  9};
S = 10; S2givenS[m, x, y, S];
Sf[t_, i_] := a[[i]] * (t - x[[i]]) ^2 + b[[i]] * (t - x[[i]]) + c[[i]];
SBild = Table[Plot[Sf[q, i], {q, x[[i]], x[[i +1]]},
            DisplayFunction → Identity], {i, 1, m - 1}];
PP = Table[Point[{x[[i]], y[[i]]}], {i, 1, m}];
Punkte = Graphics[{PointSize[0.007], PP}, PlotRange → All];
Show[{SBild, Punkte}, DisplayFunction → $DisplayFunction,
     PlotRange → All, Frame → True, AspectRatio → 0.75];
```

```
(*-----------------------------------------------------*)
(* This procedure solves the problem (OP2'):          *)
(* μ*∫_{x1}^{xm} s'(x)^2 dx + Σ_{i=1}^m (yi - s(xi))^2 = min!   *)
(* with ki=1,i=1,2,...,m.                              *)
(* Input data  are the smoothing parameter μ, m  and   *)
(* the list x of xi and the list y of yi, repectively. *)
(*-----------------------------------------------------*)
Smooths2AQ[m_, x_, y_, μ_] := Module[{DD, w, invF, Z, K, H, zw,
   Sd, s1, s2}, DD = Table[0, {i, 1, m}, {j, 1, m}];
  invF = DD; Z = DD; dx = Table[x[[i+1]] - x[[i]], {i, 1, m - 1}];
  w = Table[dx[[i]], {i, 1, m - 1}]; Sd = Sum[w[[i]], {i, 1, m-1}]; s1 = 2 Sd - 2 w[[1]];
  DD[[1,1]] = -(w[[1]] + s1)/(dx[[1]] Sd); Z[[1,1]] = w[[1]]; Z[[1,2]] = w[[1]] / 2;
  Do[{Z[[k,k-1]] = w[[k-1]] / 2, Z[[k,k+1]] = w[[k]] / 2, Z[[k,k]] = w[[k-1]] + w[[k]],
    DD[[k,k-1]] = -2 / dx[[k-1]], DD[[k,k]] = 2 / dx[[k-1]], zw = (w[[k-1]] + s1)/dx[[k-1]],
    s1 = s1 - 2 * w[[k]], DD[[1,k]] = (-1)^k/Sd (zw + (w[[k]] + s1)/dx[[k]])},
   {k, 2, m - 1}];
  DD[[1,m]] = ((-1)^m w[[m-1]])/(dx[[m-1]] Sd); DD[[m,m-1]] = -2/dx[[m-1]]; DD[[m,m]] = 2/dx[[m-1]];
  Z[[m,m-1]] = w[[m-1]] / 2; Z[[m,m]] = w[[m-1]]; Z = Z / 3;
  Do[Do[invF[[k,i]] = (-1)^(i+k), {k, i, m}], {i, 1, m}];
  K = invF.DD; K = Transpose[K].Z.K;
  H = IdentityMatrix[m] + μ K; c = LinearSolve[H, y];
  s1 = 2 Sd; s2 = 0;
  Do[{s1 = s1 - 2 w[[i]], s2 = s2 + (-1)^(i+1) (w[[i]] + s1) (c[[i+1]] - c[[i]])/dx[[i]]},
   {i, 1, m - 1}]; a = Table[0, {i, 1, m}]; b = a; b[[1]] = s2/Sd;
  Do[{b[[i+1]] = 2 (c[[i+1]] - c[[i]])/dx[[i]] - b[[i]], a[[i]] = (b[[i+1]] - b[[i]])/(2 dx[[i]])},
   {i, 1, m - 1}]]; (* Smooths2AQ *)
```

```
m = 10; (* The data *)
x = {1, 2, 2.5,   3,   4, 4.2, 4.5,   5,   6, 7};
y = {1, 2, 2.0, 2.3, 1.8, 1.5, 1.1, 1.3, 1.5, 1};
μ = 1; Smooths2AQ[m, x, y, μ];
```

$$SSE = N\left[\sum_{i=1}^{m} (y_{[[i]]} - c_{[[i]]})^2\right];$$

$$AQ = \frac{1}{3} \sum_{i=1}^{m-1} dx_{[[i]]}\ (b_{[[i]]}{}^2 + b_{[[i]]}\ b_{[[i+1]]} + b_{[[i+1]]}{}^2);$$

```
Print["μ = ", μ, "\nAQ  = ", AQ, "\nSSE = ", SSE,
    "\nμ*AQ+SSE = ", μ AQ + SSE];
```

$$Sf[t\_, i\_] := a_{[[i]]}\ (t - x_{[[i]]})^2 + b_{[[i]]}\ (t - x_{[[i]]}) + c_{[[i]]};$$

```
SBild = Table[Plot[Sf[q, i], {q, x[[i]], x[[i+1]]},
     DisplayFunction → Identity], {i, 1, m - 1}];
PP = Table[Point[{x[[i]], y[[i]]}], {i, 1, m}];
Punkte = Graphics[{PointSize[0.01], PP}, PlotRange → All];
Show[{SBild, Punkte}, DisplayFunction → $DisplayFunction,
    Prolog → Thickness[0.002], Frame → True,
    AspectRatio → 0.75, PlotRange → All];
```

```
μ = 1
AQ  = 0.381979
SSE = 0.576868
μ*AQ+SSE = 0.958847
```

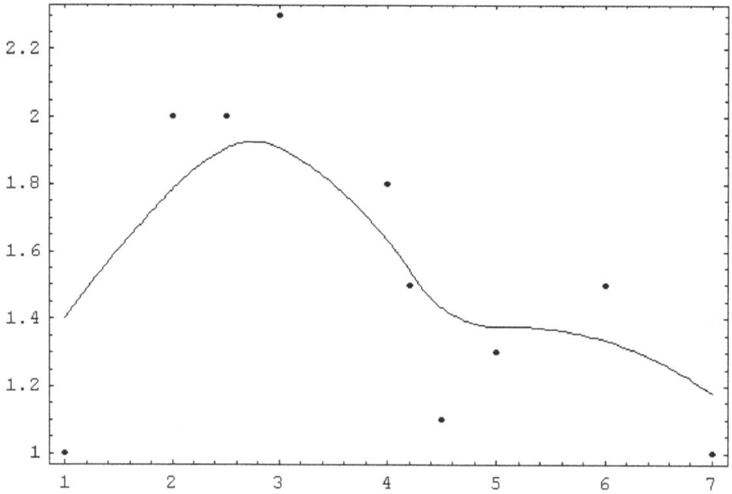

```
Off[Remove::rmnsm]; Remove["Global`*"];
(*-----------------------------------------------------------*)
(* This procedure solves the problem (S1):                    *)
(* μ*∫_{x1}^{xm} s''(x)^2 dx + Σ_{i=1}^{m} k_i (d_i - s'(x_i))^2 = min!    *)
(* Input data  are the smoothing parameter μ,m,k_i>0,         *)
(* the list x of x_i and the list b of b_i, repectively.     *)
(*-----------------------------------------------------------*)
S2smoothder1[m_, x_, d_, k_, μ_] := Module[{dx, H, R},
   H = Table[0, {i, 1, m}, {j, 1, m}]; R = Table[0, {i, 1, m}];
   dx = Table[x[[i+1]] - x[[i]], {i, 1, m - 1}];
   H[[1,1]] = μ/dx[[1]] + k[[1]]; H[[1,2]] = -μ/dx[[1]]; R[[1]] = k[[1]] d[[1]];
   Do[H[[i,i-1]] = -μ/dx[[i-1]]; H[[i,i+1]] = -μ/dx[[i]]; R[[i]] = k[[i]] d[[i]];
    H[[i,i]] = μ/dx[[i-1]] + μ/dx[[i]] + k[[i]], {i, 2, m - 1}];
   H[[m,m]] = μ/dx[[m-1]] + k[[m]]; H[[m,m-1]] = -μ/dx[[m-1]]; R[[m]] = k[[m]] d[[m]];
   b = LinearSolve[H, R];
   a = Table[(b[[i+1]] - b[[i]])/dx[[i]], {i, 1, m - 1}];
  ]; (* S2smoothder1 *)
(*-----------------------------------------------------------*)

(* The data: See figure 7.10 *)
m = 20;
x = {0, 1, 2, 3, 4, 5, 6, 7, 8, 9, 10, 11, 12, 13, 14, 15, 16,
   17, 18, 20};
d = {-0.45, 1.14, 1.02, 0.39, -0.55, -0.99, 0.4, 1.72, 2.47, 0.58,
   0.78, 0.98, 1.08, 3.28, 2.18, 1.12, 0.58, 1.03, 2.22, -0.45};
k = Table[1, {i, 1, m}];

μ = 1; S2smoothder1[m, x, d, k, μ];
```

```
Sf[t_, i_] := a[[i]] * (t - x[[i]]) + b[[i]];
SBild = Table[Plot[Sf[q, i], {q, x[[i]], x[[i+1]]},
        DisplayFunction → Identity], {i, 1, m - 1}];
PP = Table[Point[{x[[i]], d[[i]]}], {i, 1, m}];
Punkte = Graphics[{PointSize[0.01], PP}, PlotRange → All];
Show[{SBild, Punkte}, DisplayFunction → $DisplayFunction,
      Prolog → Thickness[0.002], Frame → True,
      AspectRatio → 0.75, PlotRange → All];
```

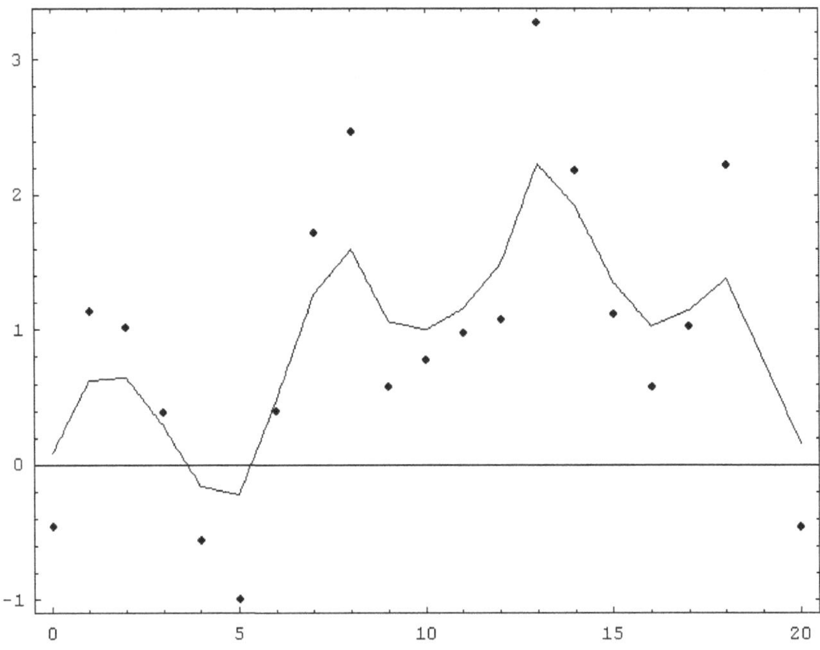

```
Off[Remove::rmnsm]; Remove["Global`*"];
(*-----------------------------------------------*)
(* With this computer  program one can calculate *)
(* Table 15.3 and in addition the Minimum-χ²-es- *)
(* timation of a.                                *)
(*-----------------------------------------------*)
(* The data: c (t)=c0e^-at *)
r = 11;  AUCR = 10;  a = 0.5;  c0 = 3;
(*-----------------------------------------------*)
SPLINEAUC[m_, x_, y_, al_, ar_] := Module[{dx, dy, b, Z, R},
   Z = Table[0, {i, 1, m}, {j, 1, m}];
   dx = Table[x[[i + 1]] - x[[i]], {i, 1, m - 1}];
   dy = Table[y[[i + 1]] - y[[i]], {i, 1, m - 1}];
   Do[Z[[i + 1, i + 1]] = 2 (dx[[i]] + dx[[i + 1]]), {i, 1, m - 2}];
   Do[Z[[i + 1, i]] = dx[[i]], {i, 1, m - 1}];
   Do[Z[[i, i + 1]] = dx[[i]], {i, 1, m - 1}];
   Z[[1, 1]] = 2 dx[[1]]; Z[[m, m]] = 2 dx[[m - 1]];
   R = Table[3 (dy[[i]] / dx[[i]] - dy[[i - 1]] / dx[[i - 1]]),
     {i, 2, m - 1}];
   R = Prepend[R, 3 (dy[[1]] / dx[[1]] - al)];
   R = Append[R, 3 (ar - dy[[m - 1]] / dx[[m - 1]])];
   b = LinearSolve[Z, R];
   (* Int. cubic spline s with s' (t1)=al and s' (tm)=ar *)
   Beob = Table[0, {i, 1, m - 1}]; AUC = 0;
   Do[(* Beob[[i]]=∫_{xi}^{xi+1} s (x) dx *)
    {Beob[[i]] = 0.5 (x[[i + 1]] - x[[i]]) (y[[i]] + y[[i + 1]]) -
       1 / 12 (x[[i + 1]] - x[[i]]) ^ 3 (b[[i]] + b[[i + 1]]),
     AUC = AUC + Beob[[i]]}, {i, 1, m - 1}];];
(* SPLINEAUC *)
```

```
AUCR = AUCR / 100; tend = -Log[AUCR] / a; step = tend / (r - 1);
t = Table[(i - 1) * step, {i, 1, r}]; k = a;
c = Table[c0 * Exp[-a * t[[i]]], {i, 1, r}];
al = -a * c0 * Exp[-a t[[1]]]; ar = -a * c0 * Exp[-a t[[r]]];
a =.; Funk[t_] := c0 * Exp[-a t];

SPLINEAUC[r, t, c, al, ar];

Erw = Table[0, {i, 1, r - 1}];
```

$$\text{Do}\left[\text{Erw}[\![i]\!] = \text{AUC}\,\frac{(\text{Funk}[t[\![i]\!]] - \text{Funk}[t[\![i+1]\!]])}{\text{c0}},\ \{i, 1, r-1\}\right];$$

$$\text{a =.; varChi} = \sum_{i=1}^{r-1} \frac{\text{D}[(\text{Erw}[\![i]\!] - \text{Beob}[\![i]\!])^2, a]}{\text{Erw}[\![i]\!]};$$

$$\text{FC} = \int_0^{\text{tend}} \text{c0} * \text{Exp}[-k * s]\, ds;$$

```
loes = FindRoot[varChi == 0, {a, 0.1}]; a = loes[[1, 2]];
Print["AUCR=", 100 AUCR, " , FSP=", AUC, " , FC=", FC];
Print["Varied Minimum-χ^2-estimated a = ", a];
```

$$\text{a =.; Chi} = \sum_{i=1}^{r-1} \frac{(\text{Erw}[\![i]\!] - \text{Beob}[\![i]\!])^2}{\text{Erw}[\![i]\!]};$$

```
loes = FindMinimum[Chi, {a, 0.1}]; a = loes[[2, 1, 2]];
Print["Minimum-χ^2-estimated a = ", a];
```

```
AUCR=10 , FSP=5.39998 , FC=5.4

Varied Minimum-χ^2-estimated a = 0.528771

Minimum-χ^2-estimated a = 0.52759
```

```
Off[Remove::rmnsm]; Remove["Global`*"];
(*--------------------------------------------------*)
(*   With this program one can calculate Table 16.1   *)
(*--------------------------------------------------*)
(* The data *)
r = 10; JPR = 50; c0 = 250; k10 = 0.5;
(*--------------------------------------------------*)
SPLINEAUC[m_, x_, y_, al_, ar_] := Module[{dx, dy, b, Z, R},
   Z = Table[0, {i, 1, m}, {j, 1, m}];
   dx = Table[x[[i + 1]] - x[[i]], {i, 1, m - 1}];
   dy = Table[y[[i + 1]] - y[[i]], {i, 1, m - 1}];
   Do[Z[[i + 1, i + 1]] = 2 (dx[[i]] + dx[[i + 1]]), {i, 1, m - 2}];
   Do[Z[[i + 1, i]] = dx[[i]], {i, 1, m - 1}];
   Do[Z[[i, i + 1]] = dx[[i]], {i, 1, m - 1}];
   Z[[1, 1]] = 2 dx[[1]]; Z[[m, m]] = 2 dx[[m - 1]];
   R = Table[3 (dy[[i]] / dx[[i]] - dy[[i - 1]] / dx[[i - 1]]),
     {i, 2, m - 1}];
   R = Prepend[R, 3 (dy[[1]] / dx[[1]] - al)];
   R = Append[R, 3 (ar - dy[[m - 1]] / dx[[m - 1]])];
   b = LinearSolve[Z, R];
   (* Int. cubic spline s with s' (x1)=al and s' (xm)=ar *)
   Beob = Table[0, {i, 1, m - 1}]; AUC = 0;
   Do[(* Beob[[i]]=∫_{xi}^{xi+1} s (x) dx *)
    {Beob[[i]] = 0.5 (x[[i + 1]] - x[[i]]) (y[[i]] + y[[i + 1]]) -
       1 / 12 (x[[i + 1]] - x[[i]]) ^ 3 (b[[i]] + b[[i + 1]]),
     AUC = AUC + Beob[[i]]}, {i, 1, m - 1}];]; (* SPLINEAUC *)
```

```
JPR = N[JPR / 100]; a = k10; t = Table[i, {i, 1, r}];
c = Table[c0 * Exp[-a * t[[i]]], {i, 1, r}];
Do[c[[2 i]] = c[[2 i]] - (-1)^i * JPR * c[[2 i]], {i, 1, r / 2}];
al = -a * c0 * Exp[-a t[[1]]]; ar = -a * c0 * Exp[-a t[[r]]];

SPLINEAUC[r, t, c, al, ar];

Erw = Beob; k10 =.; Funk[t_] := c0 * Exp[-k10 * t];
```

$$\text{Do}\left[\text{Erw}[\![i]\!] = \text{AUC}\ \frac{(\text{Funk}[t[\![i]\!]] - \text{Funk}[t[\![i+1]\!]])}{(\text{Funk}[t[\![1]\!]] - \text{Funk}[t[\![r]\!]])},\ \{i, 1, r-1\}\right];$$

$$V = \sum_{i=1}^{r-1} \frac{D[(\text{Erw}[\![i]\!] - \text{Beob}[\![i]\!])^2, k10]}{\text{Erw}[\![i]\!]};$$

$$\text{CHI} = \sum_{i=1}^{r-1} \frac{(\text{Erw}[\![i]\!] - \text{Beob}[\![i]\!])^2}{\text{Erw}[\![i]\!]};\ \text{RSS} = \sum_{i=1}^{r} (c_{[\![i]\!]} - \text{Funk}[t[[i]]])^2;$$

```
loes = FindRoot[V == 0, {k10, 0.4}]; a = loes[[1, 2]];
Print["JPR=", 100 JPR, " , k10/CHI=", a, " , CHI=",
      N[CHI /. k10 -> a], " , RSS=", N[RSS /. k10 -> a]];
(*-------------------------------------------------------*)
loes = FindMinimum[RSS, {k10, 0.4}]; a = loes[[2, 1, 2]];
Print["JPR=", 100 JPR, " , k10/MLS=", a, " , CHI=",
      N[CHI /. k10 -> a], " , RSS=", N[RSS /. k10 -> a]];
```

```
JPR=50. , k10/CHI=0.556508 , CHI=18.4891 , RSS=3493.38

JPR=50. , k10/MLS=0.454307 , CHI=26.203 , RSS=2123.58
```

```
Off[Remove::rmnsm]; Remove["Global`*"];
(*-----------------------------------------------*)
(* With this computer  program one can calculate  *)
(* an χ^2-averaged k of two cinetics c1 (t)=A1e^-k1 t *)
(* and c2 (t)=A2e^-k2 t.                          *)
(*-----------------------------------------------*)
A1 = 1000; A2 = 1000; k1 = 1; k2 = 2;
r = 6;
t = Table[(i - 1), {i, 1, r}];
c1 = Table[A1 e^(-k1 t[[i]]), {i, 1, r}]; (* data of c1 (t) *)
c2 = Table[A2 e^(-k2 t[[i]]), {i, 1, r}]; (* data of c2 (t) *)
(*-----------------------------------------------*)
SPLINEAUC[m_, x_, y_, al_, ar_] := Module[{dx, dy, b, Z, R},
  Z = Table[0, {i, 1, m}, {j, 1, m}];
  dx = Table[x[[i + 1]] - x[[i]], {i, 1, m - 1}];
  dy = Table[y[[i + 1]] - y[[i]], {i, 1, m - 1}];
  Do[Z[[i + 1, i + 1]] = 2 (dx[[i]] + dx[[i + 1]]), {i, 1, m - 2}];
  Do[Z[[i + 1, i]] = dx[[i]], {i, 1, m - 1}];
  Do[Z[[i, i + 1]] = dx[[i]], {i, 1, m - 1}];
  Z[[1, 1]] = 2 dx[[1]]; Z[[m, m]] = 2 dx[[m - 1]];
  R = Table[3 (dy[[i]] / dx[[i]] - dy[[i - 1]] / dx[[i - 1]]),
    {i, 2, m - 1}];
  R = Prepend[R, 3 (dy[[1]] / dx[[1]] - al)];
  R = Append[R, 3 (ar - dy[[m - 1]] / dx[[m - 1]])];
  b = LinearSolve[Z, R];
  (* Int. cubic spline s with s' (t1)=al and s' (tm)=ar *)
  Beob = Table[0, {i, 1, m - 1}]; AUC = 0;
```

```
Do[(* Beob[[i]]=∫_{x_i}^{x_{i+1}} s (x) dx *)
    {Beob[[i]] = 0.5 (x[[i + 1]] - x[[i]]) (y[[i]] + y[[i + 1]]) -
      1 / 12 (x[[i + 1]] - x[[i]]) ^ 3 (b[[i]] + b[[i + 1]]),
     AUC = AUC + Beob[[i]]}, {i, 1, m - 1}];]; (* SPLINEAUC *)

al = -k1 * A1 * Exp[-k1 t[[1]]]; ar = -k1 * A1 * Exp[-k1 t[[r]]];
SPLINEAUC[r, t, c1, al, ar];
AUC12 = AUC; Beob12 = Beob;
al = -k2 * A2 * Exp[-k2 t[[1]]]; ar = -k2 * A2 * Exp[-k2 t[[r]]];
SPLINEAUC[r, t, c2, al, ar];
AUC12 = AUC12 + AUC; Beob12 = Beob12 + Beob;
Erw12 = Table[0, {i, 1, r - 1}];
Funk[t_] := (c01 + c02) / 2 e^(-k t);
Do[Erw12[[i]] = AUC12 (Funk[t[[i]]] - Funk[t[[i + 1]]]) / (Funk[t[[1]]] - Funk[t[[r]]]), {i, 1, r - 1}];
Chi = Sum_{i=1}^{r-1} (Beob12[[i]] - Erw12[[i]])^2 / Erw12[[i]];
k = FindMinimum[Chi, {k, 0.5}][[2, 1, 2]];
Print["χ^2-averaged k = ", k]; k =.;
varChi = Sum_{i=1}^{r-1} D[(Beob12[[i]] - Erw12[[i]])^2, k] / Erw12[[i]];
k = FindRoot[varChi == 0, {k, 0.5}][[1, 2]];
Print["varied χ^2-averaged k = ", k];
χ^2-averaged k = 1.19225
varied χ^2-averaged k = 1.19553
```

Bibliography

Adomian, G. (1983). *Stochastic Systems* (New York).

Agrafiotis, G. (1985). On the stochastic theory of compartments: the two compartment and the mammilary compartment system, *Biometr. J.* **27**, pp. 823–830.

Akaike, H. (1974). A new look at the statistical model identification, *IEEE Trans. Autom. Control. AC* **19**, pp. 716–723.

Allen, D. (1974). The relationship between variable selektion and data augmentation and a method for prediction, *Technometrics* **16**, pp. 125–127.

Anderson, D. (1983). *Compartmental Modeling and Tracer Kinetics* (Springer Verlag, Berlin/Heidelberg).

Ansalone, P. and Laurent, P. (1968). A general method for the construction of interpolating or smoothing spline functions, *Numer. Math.* **12**, pp. 66–82.

Atkins, G. (1969). *Multicomponent Models for Biological Systems* (London).

Bard, Y. (1974). *Nonlinear Parameter Estimation* (New York).

Bardsley, W. and McGinlay, P. (1987). The use of non-linear regression analysis and the F-test for model discrimination with dose-response curves and ligand binding data, *J. Theoret. Biol.* **126**, pp. 183–201.

Bartfai, T. and Mannersik, B. (1972). A procedure based on statistical criteria for discrimination between steady state kinetic models, *FEBS Lett.* **26**, pp. 252–256.

Beal, S. (1984). Population pharmacokinetic data and parameter estimation based on their first two statistical moments, *Drug Metab. Rev.* **15**, pp. 173–193.

Bellman, R. (1983). *Mathematical Methods in Medicine* (Singapore).

Benet, L. (1972). General treatment of linear mammilary models with elimination from any compartment as used in pharmacokinetics, *J. Pharm. Sci.* **61**, pp. 536–541.

Benet, L. (1985). Mean residence time in the central compartment, *J. Pharmacokin. Biopharm.* **13**, pp. 555–558.

Bentzien, G., Kaufmann, B., Schneider, B. and Ritschel, W. (1985). Simulation studies of errors of parameter estimates in pharmacokinetics, *Arzneimittelforsch./Drug Res.* **35**, pp. 7–14.

Bertalanffy, L. v. (1941). Stoffwechseltypen und Wachstumstypen, *Biol. Zent.Bl.*

61, pp. 510–532.

Biebler, K.-E. (1989). *Contributions to the Pharmacokinetics (in German)*, Thesis, Universität Greifswald, Greifswald.

Biebler, K.-E. (1999). Mathematical analysis of compartment models (in German), *Greifswalder Seminarberichte Heft 8* , pp. 0–165.

Bittrich, H., Haberland, D. and Just, G. (1979). *Methoden Chemisch-Kinetischer Berechnungen* (Leipzig).

Bodenstein, M. (1913). Eine Theorie der photochemischen Reaktionsgeschwindigkeiten, *Z. Phys. Chem.* **85**, p. 329.

Böhmer, K. (1974). *Spline-Funktionen* (B.G.Teubner, Stuttgart).

Boldt, G. (1987). *Mathematische Probleme des Wirkstofftransports in der Pharmazie*, Diplomarbeit, Ernst-Moritz-Arndt-Universität/Sektion Mathematik, Greifswald.

Borovkov, A. (1984). *Mathematical Statistics (russian).* (Nauka, Moscow).

Bourne, D. (1995). *Mathematical Modeling of Pharmacokinetic Data* (Technomic Publ. Comp., Lancaster).

Boxenbaum, H. (1982). Literature growth in pharmacokinetics, *J. Pharmacokin. Biopharm.* **10**, pp. 335–348.

Boxenbaum, H., Riegelmann, S. and Elashoff, R. (1974). Statistical estimation in pharmacokinetics, *J. Pharmacokin. Biopharm.* **3**, pp. 123–148.

Bozler, G. and van Rossum, J. (1982). *Pharmacokinetics during Drug Development: Data Analysis and Evaluation Techniques* (Stuttgart).

Bronstein, I., Semendjajew, K., Musiol, G. and Mühlig, H. (1997). *Taschenbuch der Mathematik* (Harri Deutsch Verlag, Thun/Frankfurt/M.), 3. überarbeitete und erweiterte Auflage der Neubearbeitung.

Brown, A. (1902). *J. Chem. Soc.* **81**, p. 373.

Brown, R. (1980). Compartmental system analysis: state of art, *IEEE Trans. BME* **27**, pp. 1–11.

Buckwitz, D. and Holzhuetter, H. (1987). *A New Method to Discriminate between Enzyme-Kinetic Models* (personal communication).

Campello, L. and Cobelli, C. (1978). Parameter estimation of biological stochastic compartmental models - an application, *IEEE Transactions of Biomedical Engineering* **25**, pp. 139–146.

Chanter, D. (1985). The determination of mean residence time using statistical moments: Is it correct? *J. Pharmacokin. Biopharm.* **13**, pp. 93–100.

Chau, N. (1977). Linear pharmacokinetic models: Geometric construction to determine transfer and elimination constants, *J. Pharmacokin. Biopharm.* **5**, pp. 147–159.

Cheng, H., Gillespie, W. and Jusko, W. (1994). Mean residence time concepts for nonlinear pharmacokinetic systems (review article), *Biopharmaceutics and Drug Disposition* **15**, pp. 627–641.

Chernoff, H. and Lehmann, E. (1954). The use of maximum likelihood estimates in χ^2-tests for goodness of fit, *The Annals of Mathematical Statistics* **25**, pp. 579–586.

Cherrault, Y. and Guillez, A. (1981). Parameters identification in biologial systems, *Kybernetes* **10**, pp. 265–269.

Clifford, P. and Green, N. (eds.) (1994). Diffusion kinetics in microscopic nonhomogeneous systems, *Contemporary Problems in Statistical Physics* (ed.: GH Weiss, Philadelphia, PA).

Cramer, H. (1946). *Mathematical Methods of Statistics* (Princeton).

Cutler, D. (1978a). Linear analysis in pharmacokinetics, *J. Pharmacokin. Biopharm.* **6**, pp. 265–282.

Cutler, D. (1978b). Theory of the mean absorption time, an adjunct to conventional bioavailability studies, *J. Pharm. Pharmacol.* **30**, pp. 476–478.

Cutler, D. (1979). A linear recirculation model, *J. Pharmacokin. Biopharm.* **7**, pp. 101–116.

Cutler, D. (1981). Properties of the recirculation model: matrix description and conditions for a monotonic decreasing single pass response, *J. Pharmacokin. Biopharm.* **9**, pp. 217–223.

D'Argenio, D., Schumitzky, A. and Wang, X. (2009). *ADAPT 5 Users Guide: Pharmacokinetic/Pharmacodynamic Systems Analysis Software* (Biomedical Simulations Resource, Los Angeles).

de Boor, C. (2001). *A Practical Guide to Splines* (Springer-Verlag, New York).

Delforge, J. (1981). Necessary and sufficient structural condition for local identifiability of a system with linear compartments, *Math. Biosci.* **54**, pp. 159–180.

Dettli, L., Gladtke, E. and Heimann, G. (1980). 25 years pharmacokinetics, *Pharmacokinetics* , pp. 1–6.

DeVore, R. and Lorentz, G. (1993). *Construktive Approximation - Polynomials and Splines Approximation* (Springer-Verlag, Berlin).

Devroye, L. (1987). *A Course in Density Estimation* (Boston/Basel/Stuttgart).

Ditlevsen, S. and De Gaetano, A. (2005). Mixed effects in stochastic differential equation models, *REVSTAT* **3**, pp. 137–153.

Dost, F. (1953). *Der Blutspiegel* (Leipzig).

Dost, F. (1968). *Grundlagen der Pharmakokinetik* (Stuttgart).

Dowd, J. and Riggs, D. (1965). A comparison of estimates of Michaelis-Menten kinetic constants from various linear transformations, *J. Biol. Chem.* **240**, pp. 863–869.

Eisenfeld, J. (1979). Relationship between stochastic and differential models of compartmental systems, *Math. Biosci.* **43**, pp. 289–305.

Eisenfeld, J. (1981). On mean residence times in compartments, *Math. Biosci.* **57**, pp. 265–278.

Eisenfeld, J. and DeLisi, C. (1985). *Mathematics and Computers in Biomedical Applications* (Amsterdam-New York).

Elderton, W. and Johnson, N. (1969). *Systems of Frequency Curves* (Cambridge).

Eller, J. and Gyoeri, I. (eds.) (1978). *Investigating the goodness of parameter estimates obtained via linearizing transformation (hungarian)*, Proc 9th Colloquium v. Neumann (Soc. Comp. Sci., Szeged).

Englert, R., Goehring, R. and Wedekind, J. (1984). KOMPART - ein interaktives Simulationssystem für pharmakokinetische Kompartimentsysteme, *EDV Med. Biol.* **15**, pp. 1–4.

Eubank, R. (1988). *Spline Smoothing and Nonparametric Regression* (Dekker,

New York).

Faber, G. (1914). Über die interpolatorische Darstellung stetiger Funktionen, *Jahresbericht der DMV* **23**, pp. 192–210.

Faddy, M. (1985). Nonlinear stochastic compartmental models, *IMA J. Math. Appl. Med. Biol.* **2**, pp. 287–297.

Feddersen, C. (1994). *Splinefunktionen in Hilberträumen und im Reellen*, Master thesis, Universität Greifswald und FHS Heilbronn, Greifswald und Heilbronn.

Feldmann, U. and Schneider, B. (1976). A general approach to multicompartment analysis and models of the pharmacodynamics, *Mathematical Models in Medicine* , pp. 243–281Eds J. Berger, W.Böhler. Berlin-Heidelberg-New York.

Fisher, R. (1936). Has mendel's work been rediscovered? *Ann. Sci.* **1**, pp. 115–137.

Fisz, M. (1963). *Probability Theory and Mathematical Statistics* (Wiley, New York).

Fisz, M. (1989). *Wahrscheinlichkeitsrechnung und Mathematische Statistik*, 11th edn. (Berlin).

Frosini, B. (ed.) (1987). *On the Distribution and Power of a Goodness-of-Fit Statistic with Parametric and Nonparametric Applications*, Goodness-of-fit (ed.: P Revesz, K Sarkadi, PK Sen., Amsterdam/Oxford/New York).

Garder, D. (1968). Resolution of multicomponent exponential decay curves using fourier transforms, *Ann. New York Acad. Sci.* **108**, pp. 195–203.

Gibaldi, M. and Perrier, D. (1982). *Pharmacokinetics* (New York/Basel).

Gibson, D., Taylor, M. and Colburn, W. (1987). Curve fitting and unique parameter identification, *J. Pharm. Sci.* **76**, pp. 658–659.

Gillesie, W. and Veng-Pedersen, P. (1985). The discrimination of mean residence time using statistical moments: it is correct, *J. Pharmacokin. Biopharm.* **13**, pp. 549–554.

Gladigan, V. and Vollmer, K.-O. (1977). Beschreibung des pharmakokinetischen Verhaltens von Etozelin und dessen Hauptmetaboliten, *Arzneimittelforsch./Drug Res.* **27**, p. 1786.

Gladtke, E. and Hattingberg, H. (1977). *Pharmakokinetik* (Berlin/Heidelberg/New York).

Golomb, M. (1962). *Lectures on Theory of Approximation* (Argonne National Laboratory, Applied Math. Div., Argonne, IL).

Grass, P. and Habermehl, A. (1980). Digitale Simulation eines mathematischen Modells zum Jodstoffwechsel, *Modelle in der Medizin* (ed.: HJ Jesdinsky, V Weidtmann, Berlin/Heidelberg/New York 1980), pp. 413–424.

Greville, T. (1969). *Theory and Applications of Spline Functions* (Academic Press, New York/San Francisco/London).

Gyoeri, I. and Eller, J. (1981). Compartmental systems with pipes, *Math. Biosci.* **53**, pp. 223–247.

Hagemann, R. (1984). Gibt es Zweifel an Mendels Forschungsergebnissen? *Wiss. Fortschr.* **34**, pp. 69–71.

Hänler, A. (1993). An error estimate for quadratic splines, in *Math. Kolloq.*,

Vol. 46 (Rostock), pp. 60–64.
Hastie, T. and Tibshirani, R. (1990). *Generalized Additive Models* (Chapman and Hall, London etc.).
Hatton, R., Massey, K. and Russel, W. (1984). Comparison of the predictions of one- and two-compartment microcomputer programs for long time tobramycin therapy, *Ther. Drug Monit.* **6**, pp. 432–437.
Hearon, J. and London, W. (1972). Path length and initial derivatives in arbitrary and Hessenberg compartmental systems, *Math. Biosci.* **14**, pp. 121–134.
Henri, V. (1903). *Lois generales de l'action des diastases* (Paris).
Herrmann, H. (1978). Ein pharmakologischer Prozess und seine stochastische Beschreibung, *Biometr. J.* **20**, pp. 795–819.
Himmelstein, K. and Lutz, R. (1979). A review of the applications of physiologically based pharmacokinetic modeling, *J. Pharmacokin. Biopharm.* **7**, pp. 127–145.
Holladay, J. C. (1957). A smoothest curve approximation, *Math. Tables Aids Comput.* **11**, pp. 233–243.
Holmes, R., Wilson, R. and McCall, J. (1986). Electronic spredsheet program for estimating two-compartment intravenous pharmacokinetic parameters by least squares linear regression analysis, *Int. J. Biomed. Comput.* **18**, pp. 203–212.
Humak, K. (1983). *Statistische Methoden der Modellbildung II.* (Berlin).
Iga, K., Okagawa, Y., Yashiki, T. and Shimamoto, T. (1986). Estimation of drug absorption rates using a deconvolution method with nonequal sampling times, *J. Pharmacokin. Biopharm.* **14**, pp. 213–225.
Iosifescu, M. and Tautu, P. (1973). *Stochastic Processes and Applications in Biology and Medicine, II.* (Berlin/Heidelberg/New York).
Isaacson, E. and Keller, H. (1966). *Analysis of Numerical Methods* (Wiley, New York).
Isenberg, I. and Dyson, R. (1969). The analysis of flourescence decay by a method of moments, *Biophys. J.* **9**, pp. 1337–1350.
Jacquez, J. (1972). *Compartmental Analysis in Biology and Medicine* (Elsevier), 1st edition.
Jacquez, J. (1988). *Compartmental Analysis in Biology and Medicine* (Ann Arbor, University of Michigan Press), 2nd edition.
Jäger, B. (1989). *Optimized partition of the weekly dialysis time (in German)*, Thesis, Universität Greifswald, Greifswald.
Jarre, F. and Stoer, J. (2003). *Optimierung* (Springer, Berlin).
Jelliffe, R., Schumitzky, A. and Van Guilder, M. (2000). Population pharmacokinetics/pharma-codynamics modeling: parametric and nonparametric methods, *Therapeutic Drug Monitoring* **22**, pp. 354–365.
Jeromin, B. (1972). *Glättung mittels Spline-Funktionen*, Thesis, TU Dresden, Dresden.
Johnson, S. and Mayersohn, M. (1985). Comparison of fitting methods for the analysis of plasma concentration - time data resulting from constant rate intravenous infusion, *Biopharm. Drug Disp.* **6/3**, pp. 313–323.
Kac, M. (1979). A mathematician looks at medicine, *Am. J. Med.* **66**, pp. 725–

726.

Kaizu, H., Hirata, H. and Kurata, T. (1983). Counter examples of a sufficient condition for local identifiability in a linear compartment system, *Math. Biosci.* **64**, pp. 193–201.

Kajiya, F., Kodama, S. and Abe, H. (1985). *Compartmental Analysis* (Basel/Muenchen/Paris/ London/New York/Tokyo/Sydney).

Kanyar, B., Eller, J. and Gyoeri, I. (1980). Parameter estimation of the radio-cardiogramm using compartmental models with pipes, *Mathematical and computational methods in physiology* (ed.: L. Fedina, B. Kanyar, M. Kollai, Budapest), pp. 229 - 238.

Kehlen, H., Kuschel, F. and Sackmann, H. (1977). *Grundlagen der Chemischen Kinetik*, 2nd edn. (Berlin).

Keilson, J., Kester, A. and Waterhouse, C. (1978). A circulatory model for human metabolism, *J. Theor. Biol.* **74**, pp. 535–547.

Keller, G., Kesting, G. and Roesler, U. (1988). On the asymptotic behaviour of first passage times for diffusions, *Probab. Theor. Rel. Fields* **77**, pp. 379–395.

Kernevez, J. (1980). *Enzyme Mathematics* (Amsterdam/New York).

Knoke, M. and Thierbach, U. (1997). Anonymisierte Messdaten des H_2-Atemtests bei alkoholindizierten Leberzirrhotikern, *Personal Communication*, Ernst-Moritz-Arndt-Universität Greifswald, Zentrum für Innere Medizin.

Knolle, H. (1984). Eine neue Methode der Parameterschätzung in der Pharmakokinetik, *EDV Biol. Med.* **15**, pp. 97–99.

Kobza, J. (1990). Some properties of interpolating quadratic spline. *Acta Univ. Palacki. Olomuc., Fac. Rerum Nat. 97, Math.* **29**, pp. 45–64.

Kobza, J. (1993). Quadratic splines smoothing the first derivatives, *Applications of Mathematics* **37**, 2, pp. 149–156.

Kobza, J. and Kucera, R. (1993). Fundamental quadratic splines and applications. *Acta Univ. Palacki. Olomuc., Fac. Rerum Nat., Math.* **110(32)**, pp. 81–98.

Kohberger, R. (1980). Statistical estimation of the direct linear plot method for estimation of enzyme kinetic parameters, *Anal. Biochem.* **101**, pp. 1–6.

Krisztin, T. (1984). Convergence of solutions of a nonlinear integrodifferential equation arising in compartmental systems, *Acta. Sci. Math.* **47**, pp. 471–485.

Kuhn, H. and Tucker, A. (1951). Nonlinear programming, *Proc. Berkeley Sympos. Math. Statist. Probability* , pp. 481–492.

Kümmerle, H. (1978). *Methoden der Klinischen Pharmakologie* (Muenchen/Wien/ Baltimore).

Künzi, H. P. and Krelle, W. (1962). *Nichtlineare Programmierung* (Springer-Verlag, Berlin/Göttingen/ Heidelberg).

Kusuoka, H., Maeda, H. and Kodama, S. (eds.) (1985). Structural identifiability of linear compartmental systems, *Compartmental Analysis* (ed.: Kajiya F, Kodama S, Abe H., Basel/Muenchen/Paris/London/New York/Tokyo/Sydney).

Lahres, H. (1964). *Einführung in die Diskreten Markov-Prozesse und ihre Anwendungen* (Teubner, Leipzig).

Lam, C. (1979). Comparative study of parameter estimation procedures in enzymic kinetics, *Comput. Biol. Med.* **9**, pp. 145–153.

Lanczos, C. (1956). *Applied Analysis* (Englewood Cliffs (N.J.)).

Landaw, E. and Katz, D. (1985). Comments on mean residence time determination, *J. Pharmacokin. Biopharm.* **13**, pp. 543–547.

Lange, K. (2000). *Numerical Analysis for Statisticians* (Springer-Verlag, New York Berlin Heidelberg).

Lansky, P. (1996). A stochastic model for circulatory transport in pharmacokinetics, *Math. Biosci.* **132**, pp. 141–167.

Lasch, J. (1987). *Enzymkinetik* (Jena).

Lineweaver, H. and Burk, D. (1934). The determination of enzyme dissociation constants, *J. Am. Chem. Soc.* **56**, pp. 658–666.

Lopot, F. (1990). *Urea Kinetic Modeling* (Verlinde, Ruddervoorde).

Lu, D. and Mao, F. (1993). An interactive program for pharmacokinetic modeling, *J. Pharm. Sci.* **82**, pp. 537–542.

Ludden, T., Beal, S. and Sheiner, L. (1994). Comparison of the Akaike Information Criterion, the Schwarz criterion and the F-test as guides to model selections, *J. Pharmacokinetic. Biopharm.* **22**, pp. 431–445.

Luxon, B. (1987). Solution of large compartmental models using numerical transform inversion, *Bull. Math. Biol.* **49**, pp. 395–402.

Maeß, B. and Maeß, G. (1984). Interpolating quadratic splines with norm-minimal curvature, *Rostock, Math. Kolloq.* **26**, pp. 83–88.

Maeß, G. and Frischmuth, K. (1991). Parametric quadratic splines with minimal curvature, *Z. Anal. Anwend.* **10**, 2, pp. 255–262.

Maesz, G. (1988). *Vorlesungen über Numerische Mathematik II.* (Berlin).

Martin, E., Moll, W., Schmid, P. and Dettli, L. (1984). Problems and pitfalls in estimating average pharmacokinetic parameters, *Eur. J. Clin. Pharmacol.* **26**, pp. 595–602.

Martschei, S. (2001). Anonymisierte Daten aus dem Greifswalder Gesundheitsamt über das Wachstum von Kindern, personal communication, Ernst-Moritz-Arndt-Universität Greifswald, Institut für Biometrie und Medizinische Informatik.

Matis, J. and Carter, M. (1972). Multi-compartimental analysis in steady state as a stochastic process, *Acta Biotheoret.* **21**, pp. 2–2.

Matis, J. and Patten, B. (1979). *Compartmental Analysis of Ecosystem Models* (Intl Cooperative Pub House).

Matis, J. and Wehrly, T. (1979). Stochastic models of compartmental systems, *Biometrics* **35**, pp. 199–207.

Matis, J. and Wehrly, T. (1981). The one-compartment models with clustering, *Bull. Math. Biol.* **43**, pp. 651–664.

Matis, J. and Wehrly, T. (1998). A general approach to non-markovian compartmental models, *J. Pharmacokinet. Biopharm.* **26**, pp. 437–456.

Matis, J., Wehrly, T. and Gerald, K. (1985). Use of residence time moments in compartimental analysis, *Am. J. Physiol.* **249**, pp. E409–E415.

Matis, J., Wehrly, T. and Metzler, C. (1983). On some stochastic formulations and related statistical moments of pharmacokinetic models, *J. Pharmacokin.*

Biopharm. **11**, pp. 77–92.
McIntosh, J. and McIntosh, R. (1980). *Mathematical Modeling and Computers in Endocrinology* (Berlin/Heidelberg/New York).
Meier, J., Rettig, H. and Hess, H. (1981). *Biopharmazie: Theorie und Praxis der Pharmakokinetik* (Berlin/Heidelberg/New York).
Meir, A. (1979). Approximation by quadratic splines, *Acta Math. Acad. Sci. Hungarica* **33**, pp. 155–157.
Mendel, G. (1866). Versuche über Pflanzenhybriden, *Verh Naturforsch. Ver. Brünn* **4**, pp. 3–47.
Metzler, C. and Elfring, G. (1978). Curve fitting and modeling in pharmacokinetics: A response, *J. Pharmacokin. Biopharm.* **6**, pp. 443–446.
Meyer, D. and Richter, O. (1980). Numerische Aspekte der Modellanpassung an experimentelle Daten am Beispiel der Streptokinasebehandlung, *Modelle in der Medizin: Theorie und Praxis* (ed.: HJ Jesdinsky, V Weidtmann, Berlin/Heidelberg/New York), pp. 606–627.
Michaelis, L. and Menten, M. (1913). Die Kinetik der Invertierung, *Biochem. Z.* **49**, p. 33.
Mueller, H. (1985). Nichtparametrische Regression für die Analyse von Verlaufskurven, Neuere Verfahren der nichtparametrischen Statistik, *Data Mining und Statistik in Hochschule und Wirtschaft* (Ed.: G.Ch.Pflug, Universität Potsdam).
Myschkis, A. (1955). *Lineare Differentialgleichungen mit Nacheilendem Argument* (Berlin).
Natanson, I. (1955). *Konstruktive Funktionentheorie* (Akademie-Verlag, Berlin).
Parr, W. (1981). Minimum distance estimation: a bibliography, *Communications in Statistics. Theory and Methods* **10**, pp. 1205–1224.
Parthasarathy, P. and Mayilswami, P. (1981). Stochastic compartmental model with branching particles, *Bull. Math. Biol.* **43**, pp. 347–360.
Pearson, K. (1894). Contributions to the mathematical theory of evolutions, II.skew variation in homogenous material. *Philos. Trans. Royal Soc. London. Ser A.* **185**, pp. 71–110.
Peschel, M. and Mende, W. (1986). *The Predator-Prey Model: Do We Live in a Volterra World?* (Berlin).
Pfeifer, S. and Borchert, H. (1980). *Pharmakokinetik und Biotransformation* (Berlin).
Piotrovskii, V. (1987). Pharmacokinetik stochastic model with Weibull-distributed residence times of drug molecules in the body, *Eur. J. Clin. Pharmacol.* **32**, pp. 515–523.
Rana, S. S. (1989). Quadratic spline interpolation, *J. Approximation Theory* **57**, 3, pp. 300–305.
Rana, S. S. (1990). Discrete quadratic splines, *Int. J. Math. Sci.* **13**, 2, pp. 343–348.
Rao, M. and Iyenga, S. (1984). Application techniques in modeling of complex biological system, *Computer Modeling of Complex Biological System* (ed.: S Iyenga, Cleveland), pp. 29 – 54.
Rao, U. (1986). A note on the bias and variance of the maximum likelihood

estimator of the growth rate parameter, *Biometr. J.* **28**, pp. 763–765.
Rasch, D. (1978). *Einführung in die Mathematische Statistik* (Berlin).
Reinsch, C. (1967). Smoothing by spline functions, *Numer. Math.* **10**, pp. 177–183.
Reitberg, D., Smith, I. and Loeve, S. (1985). A rapid, universal TI-59 model-independent pharmacokinetic analysis program based on statistical moment theory, *Drug Intell. Clin. Pharm.* **19**, pp. 125–134.
Rice, J. A. and Silverman, B. (1991). Estimating the mean and covariance structure nonparametrically when the data are curves, *J. R. Statist. Soc. B* **53**, pp. 233–243.
Rivlin, J. (1969). *An Introduction to the Approximation of Functions* (Blaidell Publishing in Waltham, Massachusetts).
Rocchetti, M. and Urso, R. (1982). A microcomputer program for estimation of first oder absorption rate in a one-compartment model by the Wagner-Nelson method, *Comput. Programs Biomed.* **14**, pp. 3–6.
Rossum, J. (1977). *Kinetics of Drug Action* (Berlin/Heidelberg/New York).
Rossum, J. and Ginneken, C. (1980). Pharmacokinetic systems dynamics, Pharmacokinetics (ed.: E Gladtke, G Heimann, Stuttgart), pp. 53–73.
Rupp, D. (1980). Survey of calculation methods in pharmacokinetics, *Pharmacokinetics* (ed.: E Gladtke, G Heimann, Stuttgart), pp. 151–164.
Sandor, T., Conroy, M. and Hollenberg, N. (1970). The application of the method of maximum likelihood to the analysis of tracer kinetic data, *Math. Biosci.* **9**, pp. 149–159.
Sargent, J. and Gotch, F. (1980). Mathematic modeling of dialysis therapy, *Kidney Int.* **10**, pp. 2–10.
Saxena, A. (1989). Interpolation by quadratic splines. *Ganita* **38**, 1/2, pp. 76–90.
Schaback, W. and Werner, H. (1970). *Praktische Mathematik II* (Springer).
Schall, R. and Luus, H. (1992). Comparison of absorption rates in bioequaivalence studies of immidiate release drug, *Int. J. Clin. Pharmacol. Ther. Toxicol.* **30**, pp. 153–159.
Scharf, J. (1981). Möglichkeiten der mathematischen Formulierung von Wachstumsprozessen, *Gegenbaurs Morph. Jahrb.* **127**, pp. 706–740.
Scheler, W. (1980). *Grundlagen der Allgemeinen Pharmakologie* (Jena).
Schmerling, S. (1988). Kernschätz-Methoden, Martin-Luther-Universität/Sektion Mathematik (Halle), lecture.
Schönberg, I. (1946). Contributions to the problem of approximation of equidistant data by analytic functions. *Quart. Appl. Math.* **4**, pp. 45–99, 112–141.
Schreiber, K. (2006). *Untersuchungen zu Enzymkinetischen Modellen: Theorie und Mathematische Lösungsverfahren*, Master thesis, Institut für Mathematik und Informatik, Universität Greifswald.
Schuette, I. (1979). über einige Wachstumsmodelle, die Nachwirkungen berücksichtigen, *Math-Nat. R.* **XXVIII**, pp. 587–596.
Schwarz, G. (1978). Estimating the dimension of a model, *Ann statist* : 461-464.
Shargel, L. and Yu, A. (1980). *Applied Biopharmaceutics and Pharmacokinetics* (New York).
Sharma, A. and Tzimbalario, J. (1977). Quadratic splines, *J. Approximation The-*

ory **19**, pp. 186–193.

Sheiner, L. (1984). Analysis of pharmacokinetic data using parametric models I, *J. Pharmacokin. Biopharm.* **12**, pp. 93–117.

Sheiner, L. (1985). Analysis of pharmacokinetic data using parametric models II, *J. Pharmacokin. Biopharm.* **13**, pp. 514–540.

Sheiner, L. (1986). Analysis of pharmacokinetic data using parametric models III, *J. Pharmacokin. Biopharm.* **14**, pp. 539–555.

Sheiner, L. and Beal, S. (1981). Evoluation of methods for estimating population pharmacokinetic parameters, *J. Pharmacokin. Biopharm.* **9**, pp. 635–665.

Smith, C., Lansky, P. and Lung, T. (1997). Cycle-time and residence-time density approximations in a stochastic model for circulary transport, *Bull. Math. Biol.* **59**, pp. 1–27.

Späth, H. (1995). *One Dimensional Spline Interpolation Algorithms* (A.K. Peters, Wellesley, MA).

Spelucci, P. (1993). *Numerische Verfahren der Nichtlinearen Optimierung* (Birkhäuser Verlag, Basel).

Steinijans, V. (1982). Double exponential concentration-time curves: a mathematical approach to certain time-dependencies in repetitive-dose studies, *Pharmacokinetics during drug development: Data Analysis and Evaluation Techniques* (ed.: J Bozler, Stuttgart), pp. 191 – 203.

Tallarida, R. and Murray, R. (1981). *Manual of Pharmacologic Calculations with Computer Programs* (New York/Heidelberg/Berlin).

Teorell, T. (1937). Kinetics of distribution of substances administered to the body, *Arch. Int. Pharmacodyn.* **57**, pp. 205–240.

Thakur, L. S. (1990). A direct algorithm for optimal quadratic splines, *Numer. Math.* **57**, 4, pp. 313–332.

Tod, M. and Roccisani, J. (1996). Implementation of OSOP, an algorithm for the estimation of optimal sampling times in pharmacokinetics by the ED, EID and API criteria, *Comput. Methods Programs Biomed.* **50**, pp. 13–22.

Vajda, S. (1981). Structural equivalence of linear systems and compartmental models, *Math. Biosci.* **55**, pp. 39–64.

Valentine, J. and Hunter, S. (1985). INTRAV and ORAL:BASIC interactive computer programs for estimating pharmacokinetic parameters, *J. Pharmaceut. Sci.* **74**, pp. 113–119.

Varkonyi, P., Gabanyi, Z. and Deutsch, T. (1983). A calculator package for pharmacokinetic application, *Comput. Programs Biomed.* **17**, pp. 277–283.

Varon, R. (1995). General linear compartment model with zero output III, *Biosystems* **36**, pp. 145–156.

Veng-Pedersen, P. (1977). Curve fitting and modeling in pharmacikinetics and some practical experiences with NONLIN and a new program FUNFIT, *J. Pharmacokin. Biopharm.* **5**, pp. 513–531.

Veng-Pedersen, P. (1978). Curve fitting and modeling in pharmacokinetics: Reply from the author, *J. Pharmacokin. Biopharm.* **6**, pp. 447–449.

Veng-Pedersen, P. (1989a). Mean time parameters in pharmacokinetics: Definition, computation and clinical implications, part I, *Clin. Pharmacokin.* **17**, pp. 345–366.

Veng-Pedersen, P. (1989b). Mean time parameters in pharmacokinetics: Definition, computation and clinical implications, part II, *Clin. Pharmacokin.* **17**, pp. 424–440.

Veng-Pedersen, P. (1991). Stochastic interpretation of linear pharmacokinetics: A linear systemanalysis approach, *J. Pharmaceut. Sci.* **80**, pp. 621–631.

Veng-Pedersen, P. and Gillespie, W. (1986). A note on appropriate constraints on the initial input response when applying deconvolution, *J. Pharmacokin Biopharm.* **14**, pp. 441–447.

Verhulst, P. (1838). Notice sur la loi que la population suit dans son accroisement, *Corr. Math. Phy.* **10**, pp. 113–121.

Verotta, D., Reccia, M. and Urso, R. (1986). MODDIS: A microcomputer program for model discrimination, *Comp. Methods Programs Biomed.* **22**, pp. 209–218.

Wagner, J. (1974). A modern view of pharmacokinetics, Pharmacology and pharmacokinetics (ed.: Teorell, RL Dedrick, PG Condliffe, New York/London), pp. 27–68.

Wagner, J. (1975). Do you need a pharmacokinetic model, and if so, which one? *J. Pharmacokin. Biopharm.* **3**, pp. 457–480.

Wagner, J. (1977). Pharmacokinetic parameters estimated from intravenous data by uniform methods and some of their uses, *J. Pharmacokin. Biopharm.* **5**, pp. 161–182.

Wagner, J. (1981). History of pharmacokinetics, *Pharmacol. Ther.* **12**, pp. 537–562.

Wagner, J. (1993). Pharmacokinetics for the academical scientist, *Technomic Publ. Comp.* .

Wagner, J., Szpunar, G. and Ferry, J. (1985). Michaelis-Menten elimination kinetics: Areas under curves, steady state concentrations and clearances for compartment models with different types of input, *Biopharm. Drug Dispos.* **6**, pp. 177–200.

Wahba, G. and Wold, S. (1975). A completely automatic french curve, *Commun. Statist.* **4**, pp. 1–17.

Walter, C. (1977). Contributions of enzyme models, Mathematical models in biological discovery (ed.: DL Salomon, C Walter, Berlin/Heidelberg/New York), pp. 31–95.

Weiss, M. (1983). Use of gamma distributed residence times in pharacokinetics, *Eur. J. Clin. Pharmacol.* **25**, pp. 695–702.

Weiss, M. (1984). Model independent assessment of accumulation kinetics based on moments of drug disposition curves, *Eur. J. Clin. Pharmacol.* **27**, pp. 355–35.

Weiss, M. (1990). *Theoretische Pharmakokinetik* (Berlin).

Weiss, M. and Foerster, W. (1979). Pharmacokinetic model based on circualtory transport, *Eur. J. Clin. Pharmacol.* **16**, pp. 287–293.

Werner, H. (1970). Tschebyscheff-approximation with sums of exponentials, Approximation theory (ed.: A. Talbot, London/New York), pp. 106–136.

Werner, H. (1982). Beispiele mathematischer Modelle in der Medizin und ihre numerische Behandlung, Dtsch Akad der Naturforscher Leopoldina (Symp

'Numerische Mathematik und ihre Anwendungen', Halle).

Wertz, W. (1978). *Statistical Density Estimation: A Survey* (Goettingen).

Westlake, W. (1971). Problems associates with analysis of pharmacokinetic models, *J. Pharmaceut. Sci.* **60**, pp. 882–885.

Westlake, W. (1973). Use of statistical methods in evaluation of in vivo performance of dosage forms, *J. Pharmaceut. Sci.* **62**, pp. 1579–1589.

Widmark, E. and Tandberg, J. (1924). Ueber die Bedingungen für die Akkumulation indifferenter Narkotika. Theoretische Berechnungen, *Biochem. Z.* **147**, p. 358.

Wiggin, A. (1960). On a multicompartment migration model with chronic feeding, *Biometrics* **16**, pp. 642–658.

Wittstein, T. (1882). Ein Zusatz zur Methode der kleinsten Quadrate, *Z. Math. Phys.* **XXVII**, pp. 315–317.

Wodny, M. (1998). Selected fitting problems (in German), *Greifswalder Seminarberichte Heft 5* (GinkgoPark Mediengesellschaft).

Wodny, M., Jäger, B. and Biebler, K.-E. (2003). Ausgleichende natürliche kubische Splines und die Schätzung des Glättungsparameters in SAS, *Proceedings der 7. Konferenz der SAS-Anwender in Forschung und Entwicklung, Data Mining und Statistik in Hochschule und Wirtschaft* (Shaker Verlag, Aachen).

Wolf, M., Heinzel, G., Koss, F. and Bozler, G. (1977). Modellentwicklung in der Pharmakokinetik, II. Teil: Verallgemeinerte theoretische Darstellung der vollständigen Integration linearer Kompartmentmodelle beliebiger Struktur, *Arzneimittelforsch./Drug Res.* **27**, pp. 900–903.

Yamaoka, K., Nakagawa, T. and Uno, T. (1978a). Application of Akaike's information criterion AIC in the evaltation of linear pharmacokinetic equations, *J. Pharmacokin. Biopharm.* **6**, pp. 165–176.

Yamaoka, K., Nakagawa, T. and Uno, T. (1978b). Statistical moments in pharmacokinetics, *J. Pharmacokin. Biopharm.* **6**, pp. 547–557.

Yamaoka, K., Tanaka, H., Okumura, K., Yasuhara, M. and Hori, R. (1986). An analysis program MULTI(ELS) based on extended nonlinear least squares method for microcomputers, *J. Pharmacobiodyn.* **9**, pp. 161–173.

Yeh, K. and Kwan, K. (1978). A comparison of numerical integrating algorithms by trapezoidal, lagrange and spline approximation, *J. Pharmacokin. Biopharm.* **6**, pp. 79–92.

Yu, J. and Wehrly, T. (2004). An approach to the residence time distribution for stochastic multicompartment models, *Math. Biosci.* **191**, pp. 185–205.

Yuh, L. and Beal, S. (1994). Population Pharmacokinetic/Pharmacodynamic Methodology and Applications: A Bibliography, *Biometrics* **50**, pp. 566–575.

Index